Applied Biology and Biochemistry in Animal Science

Applied Biology and Biochemistry in Animal Science

Edited by **Mia Steers**

SYRAWOOD
PUBLISHING HOUSE

New York

Published by Syrawood Publishing House,
750 Third Avenue, 9th Floor,
New York, NY 10017, USA
www.syrawoodpublishinghouse.com

Applied Biology and Biochemistry in Animal Science
Edited by Mia Steers

International Standard Book Number: 978-1-68286-048-9 (Hardback)

Printed in the United States of America.

Contents

Preface   IX

Chapter 1   **Impact of Age, Activity and Diet on the Conditioning Performance in the Ant *Myrmica ruginodis* Used as a Biological Model (Hymenoptera, Formicidae)**   1
Marie-Claire Cammaerts and Geoffrey Gosset

Chapter 2   **Comparison of the Internal Transcribed Spacer (ITS) Regions Between Hybrid and Their Parents in Scallop**   12
Biao Wu, Aiguo Yang, Jiakun Yan, Wandong Xu, Tao Yu, Jiteng Tian, Zhihong Liu, Liqing Zhou and Xiujun Sun

Chapter 3   **Bio-Active Compounds Composition in Edible Stinkbugs Consumed in South-Eastern Districts of Zimbabwe**   19
Robert Musundire, Johnson C. Zvidzai and Cathrine Chidewe

Chapter 4   **Major Metabolic Diseases Affecting Cows in Transition Period**   29
Leilson R. Bezerra, Cezario B. de Oliveira Neto, Marcos J. de Araújo, Ricardo L. Edvan, Wagner D. C. de Oliveira and Fabrício B. Pereira

Chapter 5   **Effects of Delayed First Feeding on Larval Growth and Survival of Yesso Scallop (*Patinopecten yessoensis*)**   39
Zhongqiang Cai, Xiujun Sun and Aiguo Yang

Chapter 6   **Efficacy of Two Commercial Systems for Identification of Clinical and Environmental *Escherichia coli***   47
Hussein H. Abulreesh

Chapter 7   **Why Cyanobacteria Produce Toxins? Evolutionary Game Theory Suggests the Key**   57
Beatriz Baselga-Cervera, Camino García-Balboa, Eduardo Costas and Victoria López-Rodas

Chapter 8   ***Psoroptes sp.* Infestation in Sulawesi Bear Cuscus (*Ailurops ursinus*) in Indonesia**   71
Purwanta, Mihrani, Sartika Juwita, Ahmad Nadif, Ali Ma'shum and Muh Arby Hamire

Chapter 9   **Decline of *Diporeia* in Lake Michigan: Was Disease Associated With Invasive Species the Primary Factor?**   75
Courtney S. Cave and Kevin B. Strychar

Chapter 10   **The Effects of Anabasine and the Alkaloid Extract of *Nicotiana glauca* on Lepidopterous Larvae**   82
Michelle Zammit, Claire Shoemake, Everaldo Attard
and Lilian M. Azzopardi

Chapter 11   **Breeding Biology of Blackheaded Wagtail *Motacilla feldegg* Michahhelles, 1830 (Passeriformes, Motacillidae, Motacillinae) in South of Russia**   90
E. A. Artemieva and I. V. Muraviev

Chapter 12   **Evaluating the Viability of Lactic Acid Bacteria and Nutritional Quality of *Hibiscus Sabdariffa* Stored Under Natural Condition**   99
Eguono Esther Anomohanran

Chapter 13   **Polymorphism and Damage of Aphids (*Homoptera: Aphidoidea*)**   105
Alla Vereschagina and Elena Gandrabur

Chapter 14   **Occurrence of *Trichograma* Parasitoids in Eggs of Soybean Lepidopteran Pests in Mato Grosso, Brazil**   120
Angélica Massaroli, Alessandra Regina Butnariu and
Augusta Karkow Doetzer

Chapter 15   **In Vitro Culture of Fibroblast-Like Cells From Sheep Ear Skin Stored at 25-26°C for 10 Days After Animal Death**   127
Mahipal Singh and Xiaoling Ma

Chapter 16   **Studying the Effect of the *Ziziphora tenuior* L. Plant on Some Biochemical Factors of Serum in Rats**   134
Hamed Soleyman Dehkordi, Hamid Iranpour Mobarakeh,
Mohsen Jafarian Dehkordi and Faham Khamesipour

Chapter 17   **Green Roof Performance Towards Good Habitat for Butterflies in the Compact City**   139
Lee-Hsueh Lee and Jun-Cheng Lin

Chapter 18   **Microbiological Assay of Folic Acid Content in Some Selected Bangladeshi Food Stuffs**   149
Tanjina Rahman, Mohammed Mehadi Hassan Chowdhury,
Md. Tazul Islam and M. Akhtaruzzaman

Chapter 19   **The Potential of Galangal (*Alpinia galanga* Linn.) Extract against the Pathogens that Cause White Feces Syndrome and Acute Hepatopancreatic Necrosis Disease (AHPND) in Pacific White Shrimp (*Litopenaeus vannamei*)**   158
Tidaporn Chaweepack, Boonyee Muenthaisong, Surachart Chaweepack
and Kaeko Kamei

Chapter 20   *Pseudomonas aeruginosa* – **Pathogenesis and Pathogenic Mechanisms**   **168**
Alaa Alhazmi

Permissions

List of Contributors

Preface

The world is advancing at a fast pace like never before. Therefore, the need is to keep up with the latest developments. This book was an idea that came to fruition when the specialists in the area realized the need to coordinate together and document essential themes in the subject. That's when I was requested to be the editor. Editing this book has been an honour as it brings together diverse authors researching on different streams of the field. The book collates essential materials contributed by veterans in the area which can be utilized by students and researchers alike.

Primarily dealing with animal breeding, behavior and welfare, animal science is an extremely important branch of livestock management and zoology. The book will provide practical knowledge to the readers about various applications of animal science through topics like physiology, evolution, pathology, genetics, etc. which have been lucidly covered in this book. It is a collective contribution of an internationally renowned panel of experts. This text is a complete source of knowledge on this field.

Each chapter is a sole-standing publication that reflects each author's interpretation. Thus, the book displays a multi-facetted picture of our current understanding of applications and diverse aspects of the field. I would like to thank the contributors of this book and my family for their endless support.

Editor

Impact of Age, Activity and Diet on the Conditioning Performance in the Ant *Myrmica ruginodis* Used as a Biological Model (Hymenoptera, Formicidae)

Marie-Claire Cammaerts[1] & Geoffrey Gosset[1]

[1] Faculté des Sciences, Université Libre de Bruxelles, Bruxelles, Belgium

Correspondence: Marie-Claire Cammaerts, Faculté des Sciences, DBO, CP 160/12, Université Libre de Bruxelles, 50, Av. F. D. Roosevelt, 1050 Bruxelles, Belgium. E-mail: mtricot@ulb.ac.be

Abstract

Myrmica ruginodis Nylander 1846 workers either of different ages, or having their food at different distances from the nest, or receiving different diet were submitted to identical operant conditioning. Very young ants (callow ants) could not be visually conditioned. Ants one to two years old rapidly acquired visual conditioning, presenting a score of 90% after 70 hrs and going on presenting a score of 80 %, 369 hrs after training ended. Old ants acquired visual conditioning with difficulty, reaching only a score of 65% and retaining nothing of their conditioning. Ants having their food at 5 cm from their nest acquired 85% of conditioning in 2 days and memorized 75% of it; those having their food at 10 cm from their nest acquired 80% of conditioning in 3 days and retained 65% of it; the ants having their food at 15 cm from their nest (e.g. in the vicinity of their cemeteries) acquired 75% of conditioning in 3½ days and remembered 60% of it. Conditioning is thus more efficiently acquired when individuals' activity is limited. Ants having received no sugar could not acquire conditioning; those having eaten sugar could be easily conditioned. The quantity of sugar provided slightly impacted the ants' conditioning: the ants consumed only what they needed. The influence of the three factors examined - age, activity, diet - allows us to understand some trends of the ants' life and is valuable in the context of other animal species.

Keywords: callow ants, carbohydrate, memory, learning, training

1. Introduction

Habituation and imprinting, then conditioning and learning are animals' essential abilities which, with several other abilities such as kin recognition and navigating, often allow their survival (Pearce, 1997). Classical and operant conditioning has largely been studied in many vertebrate and invertebrate species. Famous authors can be cited: Vertebrates: Pavlov (1927), Watson (1925, 1932), Skinner (1938, 1966); Invertebrates: Matsumoto, Menzel, Sandoz and Giurfa (2012), Laloi et al. (1999).

The conditioning method impacts individuals' performances. For instance, harnessed ants are generally very quickly conditioned (Guerrieri & d'Ettorre, 2010) while free moving ones acquire similar conditioning more slowly (Schwarz & Cheng, 2010). The same occurs for bees (Giurfa, 2003). Other factors may also influence the animals' conditioning score and memory, i.e. age, activity, diet.

It is known that age and experience influence ants' kin recognition (Gronenberg, Heeren, & Hölldobler, 1996) and that ants' spatio-temporal learning depends on the individuals' age (Cammaerts, 2013a). In fact, age-polyethism obviously exists in social insects and has largely been examined (Hölldobler & Wilson, 1990). Personal observations lead to the supposition that individuals' activity also influences their ability in acquiring conditioning: for instance, during nest relocation, conditioning could nearly not be obtained. It has also been shown that diet affects bees' health (Alaux, Ducloz, Crauser, & Le Conte, 2010) and that caffeine boots these insects' memory (Chittka & Peng, 2013).

Ants can easily be maintained in a laboratory throughout the entire year. They are very sensitive and detain plenty of cognitive abilities (Passera & Aron, 2005; Cheng & Wehner, 2002; Dornhaus & Franks, 2008; Chittka & Muller, 2009). They can also easily be conditioned (Dupuy, Sandoz, Giurfa, & Josens, 2006; Cammaerts, 2004 a, 2004b). Such conditioning even helped in studying these insects' sensorial capabilities (Cammaerts, 2007). They are thus excellent and convivial biological models for experimenting on many subjects (chemical communication, navigation systems, social organization), and, among others, on conditioning and its different parameters.

It was decided to use ants as a model, and more specifically the ant *Myrmica ruginodis* Nylander 1846, to examine the impact of three factors (age, activity and diet) on animals' conditioning. In nature, *Myrmica spp* workers are expected to live for three years (Cammaerts, 1977). *Myrmica ruginodis* inhabits borders of forests, nests under branches where the sky is partly visible together with the canopy. It has rather large eyes (169 ommatidia; Rachidi, Cammaerts, & Debeir, 2008) and uses exclusively its vision in navigating (Cammaerts, Rachidi, Beke, & Essaadi, 2012a). It can be visually but not olfactory conditioned under lighting condition. When collectively conditioned, these ants acquired a conditioning score of 80% and kept 75% of it after training ceased (Cammaerts & Nemeghaire, 2012). Our aim was to examine the impact of the three above cited factors (age, activity, diet) on *M. ruginodis* workers' visual conditioning. It was easy to segregate the workers of a colony into groups of differently aged individuals, to increase or decrease the workers' foraging activity and to modify their usual diet. So, we could performed the three planned studies what allowed us deducing information about ants' patterns of life and extending some findings to other animal species. The present paper relates these works.

A

B

C *Left*

C *Middle*

C *Right*

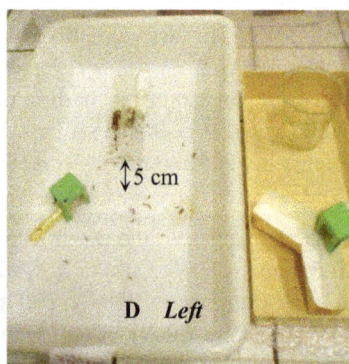
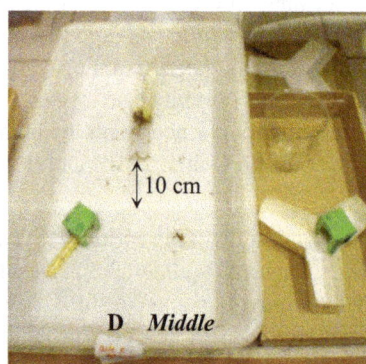
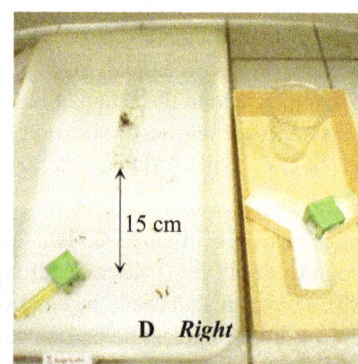

D *Left*

D *Middle*

D *Right*

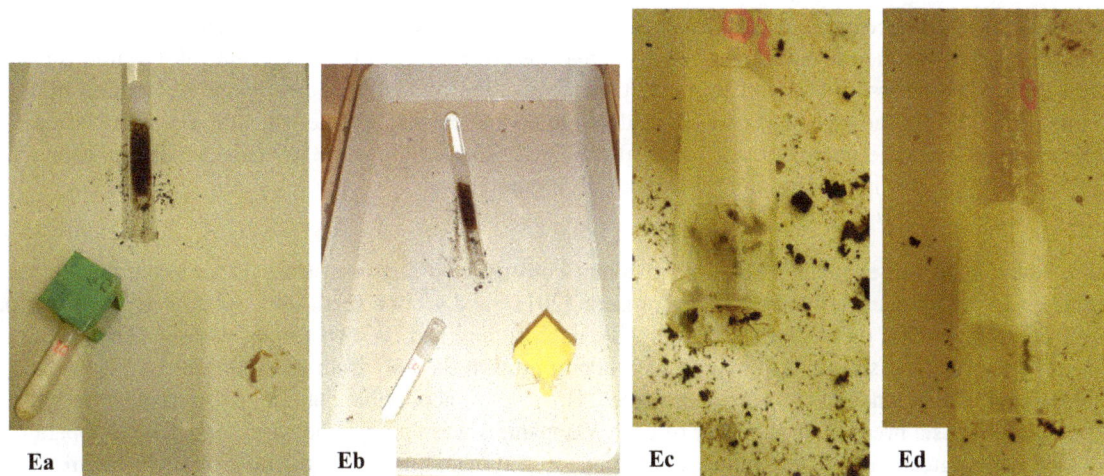

Figure 1. Some views of the experiments. A: ants' training; B: ants' testing; C: very young ants (Left photo), middle-aged ones (Middle photo) and old ants (Right photo) used for examining the impact of age on the conditioning; D: the ants' food was set at 5 cm (Left photo), or at 10 cm (Middle photo), or at 15 cm (Right photo) from the nest, to study the impact of foraging activity on conditioning; E: impact of diet on the conditioning: a: ants are trained to a green cube set on the watered food, b: ants are trained to a yellow cube set on the meat food, c: many ants collected a 20% sugared water, d: few ants collected a 60% sugared water

2. Material and Methods

2.1 Collection and Maintenance of Ants

The present work was conducted on eight colonies of *M. ruginodis*. Two large colonies and a small one were collected, in 2010, in the Aise valley (Ardenne, Belgium); one colony was collected, in 2010, at Petigny (Ardenne, Belgium). A very large colony was collected, in 2013, in the Viroin valley, at Treigne (Ardenne, Belgium) and three colonies could be again collected, in 2013, in the Aise valley (Ardenne, Belgium). In each case, they inhabited the borders of forests, under branches, where the sky was partly visible, the ants often nesting under stones. The large colony collected in the Viroin valley was used to study the impact of the individuals' age on their conditioning; the two large colonies collected in 2010 in the Aise valley and that collected at Petigny were used to study the impact of individuals' activity on the conditioning performance; the colonies collected in 2013 in the Aise valley served to examine the effect of diet on the animals' conditioning. The small colony collected, in 2010, in the Aise valley allowed making control experiments.

Each collected colony contained one to four queens, brood (larvae, nymphs) and about 500 workers. The colonies were maintained in the laboratory in artificial nests made of one to three glass tubes half-filled with water, a cotton-plug separating the ants from the water. During the ants' maintenance, the glass tubes were deposited in trays (34 cm × 23 cm × 3 cm), the sides of which were covered with talc (Figure 1Eb). The trays served as foraging areas, food being delivered in them. The ants were fed with sugar-water provided *ad libitum* in a small glass tube plugged with cotton, as well as with a droplet of tap water set on the foraging area and with cut *Tenebrio molitor* larvae served three times a week on a glass-slide. Some feeding conditions varied in the course of given experiments. Temperature was maintained at 20 ± 2 °C, humidity at about 80%, this remaining constant over the course of the experimentation. The lighting had a constant intensity of 330 lux when caring for the ants (e.g. providing food, renewing nesting tubes) and testing them; during other time periods, the lighting was dimmed to 110 lux. The electromagnetic field had an intensity of 3-5 $\mu W/m^2$.

2.2 Experimental Apparatus

2.2.1 Ants' Conditioning

Ants of each experimental colony were collectively visually trained using an experimental apparatus consisting of a hollow green cube constructed of strong paper (Canson ®). The colors have been analyzed to determine their wavelength reflection (Cammaerts, 2007; M.-C. Cammaerts & D. Cammaerts, 2009). The cubes were constructed according to the instructions given in Cammaerts and Nemeghaire (2012). The ceiling of each cube was filled unlike the four vertical faces, this allowing the ants entering the cubes. The hollow green cube of each experimental colony was set over the opening of the cotton plugged tube filled with sugared water (Figure 1A).

Each cube was relocated once or twice each day, but never according to a regular itinerary (such as every 12 h or 24 h), and the reward was then renewed if necessary. This was done to avoid the establishment of a trail (M.-C. Cammaerts & R. Cammaerts, 1980), and to prevent the acquisition of spatial and/or temporal learning by the ants (Cammaerts, 2004a). The green cube was considered to be the 'correct' choice when the ants were tested as explained below. In one experiment, the ants were also trained using a hollow yellow cube set above the cut *T. molitor* larvae (as explained in the section '3. Results 3.3', Figure 1Eb).

2.2.2 Ants' Testing

Ants were individually tested in a Y-shaped apparatus constructed of strong white paper according to the instructions given in Cammaerts, Rachidi and Cammaerts (2011), and set in a small tray (30 cm × 15 cm × 4 cm), apart from the experimental colony's tray (Figure 1B). Each colony had its own testing design. The apparatus had its own bottom and the sides were covered with talc to prevent the ants from escaping. In the Y-apparatus, the ants deposited no trail since they were not rewarded. However, it is possible that they utilized other chemical secretions as traces. As a precaution, the floor of each Y-apparatus was changed between tests. The Y-apparatus was provided with a green cube - or a yellow one, in one experiment - in one or the other branch (Figure 1B). The branch was randomly chosen, while also ensuring that half of the tests was conducted with the cube in the left branch and the other half with the cube in the right branch of the Y maze. Control experiments were performed using also such Y-apparatus, on never conditioned ants belonging to the small colony recolted, in 2010, in the Aise valley (see 2.1, end of the first alinea).

To conduct a test on an experimental colony, 20 workers of that colony - randomly chosen from the workers of that colony - were transferred one by one to the area at the entrance of the Y-apparatus. Each transferred ant was observed until it turned either to the left or to the right in the Y-maze, and its choice was recorded. Only the first choice of the ant was recorded and this only when the ant was entirely under the cube, i.e. beyond a pencil drawn thin line indicating the entrance of a branch (Figure 1B). Afterwards, the ant was removed and transferred to a polyacetate cup, in which the rim was covered with talc, until 20 ants were so tested, this avoiding testing the same ant twice. All the tested ants were then placed back on their foraging area. To conduct controls, 20 ants of a non-conditioned colony were similarly tested in identicalY-apparatus.

2.3 Assessment of the Ants' Response

The ants' responses were quantified during the control and test experiments. In each experiment, the numbers of ants which turned left or right during the control experiments, or which approached the "correct" green (or the yellow) cube, or went to the "wrong" empty branch of the Y were recorded. The mean number of ants which correctly responded per colony was assessed and the percentage of correct responses for the tested ant population was established (Tables 1, 2 and 3).

For each experiment, the number of correct responses was statistically analyzed using the non-parametric χ^2 test applied to the experimentally obtained values and the theoretical expectation of identical values for the two branches of the Y maze. Since we did not compare the results of each successive assessment to the same control, the Bonferroni correction must not be applied. Responses were considered as being not significant when $P > 0.05$ (Siegel & Castellan, 1989).

2.4 Experimental Protocol

In general, tests were performed in the course of time, at least once per day. The start was given by the colored cube deposit on the ants' tray. As soon as the ants' conditioning score no longer changed (neither increased nor decreased), the training apparatus was removed from the foraging area of the colony and tests were once more started, until the ants' memory (or retention) score showed no further change (neither increase or decrease).

2.5 Experimental Planning

2.5.1 Impact of Individuals' Age

The large colony collected at Treigne was divided in four smaller ones: one contained the older ants foraging on the tray; a second one was made up of ants located in the foraging area as well as near the nest entrances (outside and inside); a third colony consisted of the very young ants, lighter in color than the first and the second groups, located inside the nest; the fourth colony was made up of the queen and the unselected ants e.g. a few old, middle-aged and young ones. Brood elements (eggs, larvae, nymphs) were equally distributed between the first three newly constituted smaller colonies. Each new colony was maintained in a glass tube half filled with water (Figure 1C), in a tray (34 cm × 23 cm × 3 cm) as usual and their meat and sugar food was located on the left and the right respectively of the nest entrance at 7 cm of distance. The fourth colony was not tested. At a precise time, a hollow green cube was set above the opening of the small glass tube containing the sugar water of each of the three experimental colonies, and the training of these colonies began. Tests were then conducted, as explained above, on the course of the ants' conditioning acquisition, and then of their conditioning loss (Table 1).

2.5.2 Impact of Individuals' Foraging Activity

Three colonies of *M. ruginodis* were transferred into somewhat larger trays (42 cm × 27 cm × 7 cm) so that their food could be located as follows. For each colony, the sugar food was located to the right of the nest entrance while the meat food was on the left. For the first colony, the distance between the food sites and the nest entrance was 5 cm; for the second colony, this distance was 10 cm; for the third colony, the distance between food and nest entrance was 15 cm, the food being so nearly as distant as the colony cemeteries which are always established, by the ants, in one, some or all corners of the colony tray (Figure 1D). As explained above, training consisted in setting a hollow green cube above the opening of the small glass tube containing sugared water (sugar food). Ants progressively associated the green cube and the presence of sugar food, thus undergoing operant conditioning. As soon as the training apparatus were set in place, time was continuously checked and ants were tested, as detailed above, in the course of their conditioning acquisition, and then their conditioning loss (or potential loss) (Table 2).

2.5.3 Impact of Individuals' Diet

Three colonies were used to examine the impact of individual diet on conditioning. For each of these colonies, the small tube containing the sugar water was removed and replaced by an identical small tube but filled with a given, experimental diet (Figure 1E).

For one colony, this diet consisted in tap water. The diet of the second colony was diluted sugared water (20% saccharose). For the third colony, the diet was a concentrated aqueous solution of sugar (60% saccharose). The experiments were performed with firstly a hollow green cube above the tube containing the aqueous solution (Figure 1Ea), then, secondly, with a hollow yellow cube above the *T. molitor* larvae (Figure 1Eb). Also, as a precaution, the colony having received tap water during the first experiment received 20% aqueous solution of saccharose in the course of the second experiment; that having received the diluted solution during the first experiment was given 60% solution of saccharose in the course of the second experiment; the colony having first received the more concentrated sugar solution then received only tap water. As soon as the hollow cubes were set in place, time was recorded and the tests performed, as previously until the ants reached their maximum conditioning score (Table 3).

3. Results

3.1 Impact of Individuals' Age on Their Conditioning Performance

Table 1. Number of correct choices *versus* wrong ones - and proportion of correct responses - given by *Myrmica ruginodis* workers of different ages

Experiments	Approximate age of tested ants					
Times	3 - 4 months		1 - 2 years		2 - 3 years	
Control			9 *vs* 10	45%		
Tests 22 hrs	10 *vs* 10	50%	14 *vs* 7	70%	12 *vs* 8	60%
30 hrs	10 *vs* 10	50%	15 *vs* 5	75%	12 *vs* 8	60%
43 hrs	10 *vs* 10	50%	16 *vs* 4	80%	13 *vs* 7	(65%)
58 hrs	→ 10 *vs* 10	50%	17 *vs* 3	85%	12 *vs* 8	60%
70 hrs			18 *vs* 2	(90%)	→ 11 *vs* 9	55%
84 hrs			→ 17 *vs* 3	85%	10 *vs* 10	50%
94 hrs			16 *vs* 4	80%	10 *vs* 10	50%
118 hrs			17 *vs* 3	85%	10 *vs* 10	50%
141 hrs			17 *vs* 3	85%		
170 hrs			16 *vs* 3	80%		
192 hrs			18 *vs* 2	90%		
238 hrs			17 *vs* 3	85%		

The cue at which ants were conditioned was a green hollow cube, set above the ants' sugar water, at 0 hrs for each kind of ants and removed (→) at 72 hrs, 96 hrs, 104 hrs respectively for very young ants, middle aged ants, and rather old ones. A black circle (◯) indicates the acquisition of full conditioning.

During the control experiment, statistically as many ants moved into each branch of the Y apparatus, a hollow green cube being present (9 out of 20 ants) or not (11 out of 20 ants) (Table 1, line 1).

The callow ants never acquired visual conditioning. Even after 58 hrs of training, they continued to hesitate when confronted with a hollow green cube on one hand and an empty branch on the other (Table 1, second column). Middle aged ants very quickly acquired visual conditioning and reached the spectacular score of 90% after 70 hrs. After their training ended, they went on responding to the green cube, presenting a score of 85%, 85%, 80%, 90%, 85%, 85%, 80%, 34 hrs, 57 hrs, 86 hrs, 108 hrs, 154 hrs, 202 hrs, 369 hrs later respectively. Middle aged ants appeared so to have a strong visual memory. Very old ants had difficulty in acquiring visual conditioning. When tested, they often stopped and moved their head upon their arrival at the decision point of the Y-apparatus. After 43 hrs, they acquired a score of only 65% which thereafter decreased to 60% and 55%. They retained nothing of their weak conditioning.

3.2 Impact of Individuals' Foraging Activity on Their Conditioning Performance

A control experiment, conducted using the small colony collected in 2010 in the Aise valley, on non-conditioned ants, showed that such ants chose the empty branch of the Y maze as often as that with a hollow green cube (Table 2, line 1).

Table 2. Number of correct choices *versus* wrong ones - and proportion of correct responses - given by *Myrmica ruginodis* foragers which food (so cue and reward) was set at different distances from the nest

Experiments	Distances between nest and food (so cue and reward)					
Times	5 cm		10 cm		15 cm	
Control			10 *vs* 10	50%		
Tests 24 hrs	15 *vs* 5	75%	12 *vs* 8	60%	10 *vs* 10	50%
32 hrs	16 *vs* 4	80%	14 *vs* 7	70%	11 *vs* 9	55%
48 hrs	17 *vs* 3	(85%)	15 *vs* 5	75%	13 *vs* 7	65%
72 hrs	→ 15 *vs* 5	75%	16 *vs* 4	(80%)	15 *vs* 5	(75%)
96 hrs	13 *vs* 7	65%	→ 15 *vs* 5	75%	→ 14 *vs* 6	70%
119 hrs	14 *vs* 6	70%	13 *vs* 7	65%	11 *vs* 9	55%
144 hrs	15 *vs* 5	75%	13 *vs* 7	65%	12 *vs* 8	60%
168 hrs			13 *vs* 7	65%	12 *vs* 8	60%
192 hrs					12 *vs* 8	60%

The cue at which ants were conditioned was set in place at 0 hrs for each colony and removed (→) at 72 hrs, 96 hrs, 104 hrs respectively for colonies which food was presented at 5cm, 10 cm, 15 cm from the nest, so after full conditioning was achieved (◯).

All the trained ants could acquire operant conditioning but with different performances according to the foraging activity they need for collecting their food (Figure 1D).

When meat and sugar food was located very near the nest entrance (Figure 1D, Left photo), ants presented a very short latency period (about 12 hrs) before acquiring a conditioning score of 85% in 48 hrs. Training then being stopped, the ants slowly lost their conditioning (in about three days) and kept 75% of it (Table 2, second column).

Ants receiving their meat and sugar food at middle distance between their nest entrance and the borders of their foraging area (Figure 1D, middle photo) presented a latency period of about 35 hrs and then acquired an operant conditioning of 80% in about three days ($\approx$ 72 hrs). The training ending at that time, the ants kept their conditioning for a short time (about 10 - 15 hrs), then lost it in about 20 hrs and kept 65% of it (Table 2, third column).

The ants having their meat and sugar food far from their nest entrance (e.g. at nearly the same distance as their cemeteries) presented a relatively long latency period (about 3 days), then acquired conditioning with a score of

75% in about a total of 84 hrs. After the end of the training, these ants soon lost their conditioning (either with no latency period, or with a short barely perceptible one) and kept only 60% of it (Table 2, fourth column).

3.3 Impact of Diet on Individuals' Conditioning Performance

Again, control experiments showed that non-conditioned ants moved at random through a Y maze provided with a hollow green or yellow cube in one of its branches (Table 3, control with each kind of cube).

The results regarding the consumption or the non-consumption of sugar by the ants were unexpected. When the hollow green cubes were set above the aqueous solutions provided, the ants receiving only tap water never acquired conditioning. At the beginning, they often walked quickly all around the provided water, and after one day, they scarcely came to drink the offered water. On the contrary, the ants receiving a 20% aqueous solution of saccharose very quickly acquired visual conditioning, moved quietly on the area and were often numerous in drinking the sugar water offered. The ants having a 60% aqueous solution of saccharose at their disposal very quickly acquired visual conditioning but reached a lower score than the ants consuming a more diluted solution (Table 3, upper part). They were calm but not very numerous in drinking the solution offered. It was also observed that ants receiving only tap water ate an entire *T. molitor* larva in one day, while during that time, those having diluted sugar water ate only half a larva and the ants having a concentrated solution of sugar did not eat the larva which was found intact after one day. The non-acquisition of conditioning by the ants receiving no sugar may be due to the fact that the cube was set above tap water which is not a true reward for the ants. So, the experiment was reproduced using a yellow cube, this cube being set above the meat food. When hollow yellow cubes were set above the meat food, the results were similar to the previous ones: the ants receiving no sugar, though having in this case a true reward under the cues to memorize, never acquired visual conditioning, while the other ants acquired such conditioning perfectly. The latter ants' score was a little lower that the score obtained when rewarding the ants with sugared water, this being usual behavior for ants (M.-C. Cammaert, Rachidi, & D. Cammaerts, 2011). As with the green cube above the sugar food, the ants receiving 20% sugared water reached a better (but not statistically better) score than those receiving 60% sugared water (Table 3, lower part). It was continuously observed that more ants visited the 20% solution of sugar than the 60% one (Figures 1Ec, 1Ed).

Table 3. Number of correct choices *versus* wrong ones - and proportion of correct responses - given by *Myrmica ruginodis* workers which received different diets

Experiments Times	Diet of tested ants and numerical results					
CS on the water	0% of sugar		20% of sugar		60% of sugar	
Control			with a green cube	9 *vs* 11 45%		
Tests 8 hrs	11 *vs* 9	55%	14 *vs* 7	70%	15 *vs* 5	75%
21 hrs	10 *vs* 10	50%	17 *vs* 3	85%	16 *vs* 4	80%
32 hrs	9 *vs* 11	45%	18 *vs* 2	(90%)	16 *vs* 5	(80%)
48 hrs	11 *vs* 9	55%	17 *vs* 3	85%	16 *vs* 4	80%
CS on the meat	20% of sugar		60% of sugar		0% of sugar	
Control			with a yellow cube	11 *vs* 9 55%		
Tests 8 hrs	16 *vs* 4	80%	15 *vs* 5	75%	9 *vs* 11	45%
20 hrs	17 *vs* 3	85%	16 *vs* 4	80%	10 *vs* 10	50%
31 hrs	17 *vs* 3	(85%)	16 *vs* 4	(80%)	10 *vs* 10	50%
47 hrs	17 *vs* 3	85%	16 *vs* 4	80%	10 *vs* 10	50%

E.g. either only tap water, or an aqueous solution of saccharose 20% or such a solution 60%, the element to which the ants should be conditioned (= CS) being set either above the water solution or above the meat continuously provided. A black circle (◯) indicates the acquisition of full conditioning and consequently the end of the training and the experiment.

4. Discussion

The visual conditioning score and memory results obtained here are in agreement with those found during a detailed study of collective operant conditioning in the ant *M. ruginodis* (Cammaerts & Nemeghaire, 2012). The present work shows that conditioning is largely impacted by the age, the activity and the diet of the animals.

4.1 Age

Very young ants could not acquire conditioning. This situation exists while such ants stay inside the nest. However, during this time period, the size of the callow ants' mushroom bodies (a part of the ants' brain) increases (Gronenberg, Heeren, & Hölldobler, 1996). So, during their first weeks of life, young ants may acquire knowledge thanks to mechanisms others than conditioning, such as imprinting and/or habituation. Indeed, it has been shown that they then become imprinted to the odor of their nest entrances (Cammaerts, 2013b) and, a little later, to the specific odor of their foraging area (Cammaerts, submitted). It could be suggested that, for young ants, odors are perhaps more important than visual elements. Even though *M. ruginodis* middle-aged workers exclusively use their vision (Cammaerts et al., 2012a), very young ants are more sensitive to their congeners' odor than to these conspecific ants' visual aspect (Cammaerts & Gosset, submitted). Such an acquisition of knowledge thanks to imprinting and/or habituation, as well as such importance of the odors during the early time of life, may be the case for other animal species, even if later on, learning and visual elements become more important (wasps: Signoretti, Guscelli, Simonelli, d'Ettorre, & Cervo, 2013; human beings: Ward Platt, 2009). Conditioning acquisition ability is maximal when ants are middle-aged. This result is in agreement with the fact that ants of this age then learn their species' trail following behavior (Cammaerts, 2013c) and their specific alarm reaction (Cammaert, in press). Old workers retain nothing of their poor, difficult conditioning. So, an obvious age-polyethism also exists for conditioning abilities as it exists for many social tasks (Hölldobler & Wilson, 1990). In nature, *Myrmica spp* workers live for about three years, the young, middle-aged and old ones being about 1, 2 and 3 years old respectively, and a pronounced age-polyethism becomes automatically established in each colony.

This impact of age on animals' acquisition of long lasting knowledge is understood to occur in many species by ethologists and non-specialists, such as trainers or drillers of dogs or horses (Villers, 2013) and also by psychologists (Prull, Babrieli, & Bunge, 2000).

4.2 Foraging Activity

When the ants had to find their food far from their nest, and were relatively more active in collecting food, they could not acquire efficient conditioning. The ants, which found their food quickly and easily, acquired efficient conditioning. This explains some common observations: for instance, during nest relocation, when ants have to rebuild some part of their nest, when sexuals are erratically moving in the foraging area, ready to fly away, it is practically impossible to obtain workers' conditioning. The ants are preoccupied by transporting larvae, by building or by protecting the alates respectively; they hardly forage and poorly acquire conditioning. This impact of activity, work or state of mind, on an animal's ability to acquire conditioning or knowledge is well known by drillers and trainers of several animal species. These persons select appropriate individuals and/or time periods for training or teaching (Villers, 2013). It is also known that animals must be not perturbed by unusual environment for acquiring efficient conditioning (Pavlov, 1927).

It must be precised that the colored hollow cube was a conditioned stimulus and not a visual cue which allowed ants navigating. To navigate, ants used cues located inside as well as outside of their tray, cues which were rather similar for the three experimented colonies and were constant in the course of the experiment.

From our results regarding the impact of ants' foraging activity on their conditioning (in fact their association between a perceived cue and the presence of food), it can be deduced that the elements located near the nest will be more easily memorized than those located far from the nest. This is logical since ants will encounter more often the cues located near the nest than those farther located. The ants will thus collect firstly any food found very near the nest, and only later that discovered far from the nest. Also, since cues located at greater distances are less memorized than the proximate ones and may change in the course of one trip, ants use chemical trails to help them returning to their nest after having collected food far from the nest or after having gone to cemeteries (Cammaerts, Morel, Martino, & Warzée, 2012b). Such a deduction is valuable in the case of bees. For food sources located very near the nest, a simple eight dance is enough to help bees finding food and nest; when the food source is located far from the hive, a very informative waggle dance is necessary. In the latter case, bees use information about direction , distance and quality of food in order to situate the food and inform congeners about that location (von Frisch, 1946).

4.3 Consumption of Sugar

The ants having received no sugary food could not be conditioned. Moreover, in this situation, four to ten very young ants died in three days. Sugary food appeared so to be essential, vital. This explains the natural diet of ants: they commonly eat aphids' honeydew, a highly sugared food (Leroy, Capella, & Haubruge, 2009); *Myrmica spp* workers were often seen on flowers, near the nectarines (personal observation). We presume that ants eating primarily seeds or meat (Passera & Aron, 2005) also look for some food containing carbohydrates and that leafcutter ants (same reference as above) find sugary elements in their common food.

On the other hand, several studies have shown the importance of glucose for efficient brain function (Lieberman, 2003). Each nerve cell, in fact, uses essentially glucose in order to function. The direct impact of sugar consumption on memory has also been demonstrated (Calvaresi & Bryan, 2001). Our punctual results on ants' conditioning under different sugary food consumption conditions, find in the here above cited works, an obvious explanation.

Acknowledgements

We are very grateful to Mr. T. Sullivan who meticulously corrected the English of our paper and to an anonymous referee whose comments allowed us improving our paper.

References

Alaux, C., Ducloz, F., Crauser, D., & Le Conte, Y. (2010). Diet effects on honeybee immunocompetence. *Biology Letter, 6*, 562-565. http://dx.doi.org/10.1098/rsbl.2009.0986

Calvaresi, E., & Bryan, J. (2001). Vitamins, cognition and aging: a review. *Journal of Gerontology,* 327-339. http://dx.doi.org/10.1093/geronb/56.6.P327

Cammaert, M.-C., & Gosset, G. Ontogenesis of visual and olfactory kin recognition in the ant *Myrmmica sabuleti* (Hymenoptera, Formicidae). *Annales de la Société Entomologique de France,* submitted.

Cammaerts M.-C., & Nemeghaire, S. (2012). Why do workers of *Myrmica ruginodis* (Hymenoptera, Formicidae) navigate by relying mainly on their vision? *Bulletin de la Société Royale Belge d'Entomologie, 148*, 199-208.

Cammaerts, M.-C. (1977). Etude démograpique annuelle des sociétés de *Myrmica rubra* L. des environs de Bruxelles. *Insectes Sociaux, 24*, 147-161. http://dx.doi.org/10.1007/BF02227168

Cammaerts, M.-C. (2004a). Classical conditioning, temporal learning and spatial learning in the ant *Myrmica sabuleti. Biologia, 59*, 243-256.

Cammaerts, M.-C. (2004b). Operant conditioning in the ant *Myrmica sabuleti. Behavioral Processes, 67*, 417-425. http://dx.doi.org/10.1016/j.beproc.2004.07.002

Cammaerts, M.-C. (2007). Colour vision in the ant *Myrmica sabuleti* MEINERT, 1861 (Hymenoptera: Formicidae). *Myrmecological News, 10*, 41-50.

Cammaerts, M.-C. (2013a). Age dependent spatio-temporal learning in the ant *Myrmica sabuleti* (Hymenoptera, Formicidae). *Bulletin de la Société Royale Belge d'Entomologie, 149*, in press.

Cammaerts, M.-C. (2013b). Ants' learning of nest entrance characteristics (Hymenoptera, Formicidae). *Journal of Entomological Research, 6*.

Cammaerts, M.-C. (2013c). Learning of trail following behaviour by young *Myrmica rubra* workers (Hymenoptera, Formicidae). *ISRN Entomology, 6*.

Cammaerts, M.-C. Learning of foraging area specific odor by ants (Hymenoptera, Formicidae). *Research Trends,* submitted.

Cammaerts, M.-C. Learning of the specific alarm pheromone by young workers of the ant *Myrmica sabuleti. Journal of Insect Sciences,* in press.

Cammaerts, M.-C., & Cammaerts, D. (2009). Light thresholds for colour vision in the workers of the ant *Myrmica sabuleti* (Hymenoptera: Formicidae). *Belgian Journal of Zoology, 138*, 40-49.

Cammaerts, M.-C., Morel, F., Martino, F., & Warzée, N. (2012b). An easy and cheap software-based method to assess two-dimensional trajectories parameters. *Belgian Journal of Zoology, 142*, 145-151.

Cammaerts, M.-C., Rachidi, Z., & Cammaerts, D. (2011). Collective operant conditioning and circadian rhythms in the ant *Myrmica sabuleti* (Hymenoptera, Formicidae). *Bulletin de la Société Royale Belge d'Entomologie, 147*, 142-154.

Cammaerts, M.-C., Rachidi, Z., Beke, S., & Essaadi, Y. (2012a). Use of olfactory and visual cues for traveling by the ant *Myrmica ruginodis* (Hymenoptera, Formicidae). *Myrmecological News, 16*, 45-55.

Cammaerts, M.-C., & Cammaerts, R. (1980). Food recruitment strategies of the ants *Myrmica sabuleti* and *Myrmica ruginodis*. *Behavioural Processes, 5*, 251-270. http://dx.doi.org/10.1016/0376-6357(80)90006-6

Cheng, K., & Wehner, R. (2002). Navigating desert ants (*Cataglyphis fortis*) learn to alter their search patterns on their homebound journey. *Physiological Entomology, 27*, 285-290. http://dx.doi.org/10.1046/j.1365-3032.2002.00298.x

Chittka, L., & Muller, H. (2009). Learning, specialization, efficiency and task allocation in social insects. *Communicative & Integrative Biology, 2*, 151-154.

Chittka, L., & Peng, F. (2013). Caffeine boots bees' memories. *Science, 339*, 8. http://dx.doi.org/10.1126/science.1234411

Dornhaus, A., & Franks, N. R. (2008). Individual and collective cognition in ants and other insects (Hymenoptera: Formicidae). *Myrmecological News, 11*, 215-226.

Dupuy, F., Sandoz, J. C., Giurfa, M., & Josens, R. (2006). Individual olfactory learning in *Camponotus* ants. *Animal Behavior, 72*, 1081-1091. http://dx.doi.org/10.1016/j.anbehav.2006.03.011

Giurfa, M. (2003). The amazing mini-brain: lessons from a honey bee. *Bee World, 84*, 5-18.

Gronenberg, W., Heeren, S., & Hölldobler, B. (1996). Age-dependent and task-related morphological changes in the brain and the mushroom bodies of the ant *Camponotus floridanus*. *Journal of Experimental Biology, 199*, 2011-2019.

Guerrieri, F., & d'Ettorre, P. (2010). Associative learning in ants: conditioning of the maxilla-labium extension response in *Camponotus aethiops*. *Journal of Insect Physiology, 56*, 88-92. http://dx.doi.org/10.1016/j.jinsphys.2009.09.007

Hölldobler, B., & Wilson, E. O. (1990). *The ants* (p. 732). Harvard University Press, Springer-Verlag Berlin.

Laloi, D., Sandoz, J. C., Picard-Nizou, A. L., Marchesi, A., Pouvreau, A., Tasei, J. N., … Pham-Delègue, M. H. (1999). Olfactory conditioning of the proboscis extention in bumble bees. *Entomologia Experimentalis et Applicata, 90*, 123-129. http://dx.doi.org/10.1046/j.1570-7458.1999.00430.x

Leroy, P., Capella, Q., & Haubruge, E. (2009). L'impact du miellat de puceron au niveau des relations tritrophiques entre les plantes-hôtes, les insectes ravageurs et leurs ennemis naturels. *Biotechnology, Agronomy, Society and Environment, 13*, 325-334. Retrieved from http://popups.ulg.ac.be/Base/document.php?id=4171.

Lieberman, H. R. (2003). Nutrition, brain function and cognitive performance. *Appetite*, 245-254. http://dx.doi.org/10.1016/S0195-6663(03)00010-2

Matsumoto, Y., Menzel, R., Sandoz, J. C., & Giurfa, M. (2012). Revisiting olfactory classical conditioning of the proboscis extension response in honey bees: a step toward standardized procedures. *Journal of Neuroscience Methods, 211*, 159-167. http://dx.doi.org/10.1016/j.jneumeth.2012.08.018

Passera, L., & Aron, S. (2005). *Les fourmis: comportement, organisation sociale et évolution* (p. 480). Ottawa, Canada: Les Presses Scientifiques du CNRC.

Pavlov, I. P. (1927). *Conditioned reflexes. An investigation of the physiological activity of the cerebral cortex* (p. 142). Translated and edited by G. V. Anrep. London: Oxford University Press.

Pearce, J. M. (1997). *Animal Learning and Cognition: an introduction* (p. 333). Hove: Psychology Press.

Prull, M. W., Babrieli, J. D. E., & Bunge, S. A. (2000). *Ch. 2. Age-related changes in memory: A cognitive neuroscience perspective*. In Craik FIM, Salthouse TA. The handbook of aging and cognition. Erlbaum. ISBN 978-0-8058-2966-2.

Rachidi, Z., Cammaerts, M.-C., & Debeir, O. (2008). Morphometric study of the eye of three species of Myrmica (Formicidae). *Belgian Journal of Entomology, 10*, 81-91.

Schwarz, S., & Cheng, K. (2010). Visual associative learning in two desert ant species. *Behavioral Ecology and*

Sociobiology, 64, 2033-2041. http://dx.doi.org/10.1007/s00265-010-1016-y

Siegel, S., & Castellan, N. J. (1989). *Nonparametric statistics for the behavioural sciences* (p. 396). Singapore: McGraw-Hill Book Company.

Signorotti, L., Guscelli, E, Simonelli, P., d'Ettorre, P., & Cervo, R. (2013). Preimaginal learning and nestmate recognition in the paper wasp *Polistes dominula. Colloque de la section française de l'IUSSI*, Villetaneuse.

Skinner, B. J. (1938). *The behavior of organisms: an experimental analysis* (p. 457). New York: Appleton-Century-Crofts.

Skinner, B. J. (1966). *Operant Behavior*. In W. K. Honig (Ed.) (pp. 1-27). New York: Appleton-Century-Crofts.

Villers, E. (2013). Le chien, le cheval et une seule méthode. *Hebdos du Suroît, première édition. Vaudreuil-Dorion.*

Von Frisch, K. (1946). Die Tänze der Bienen. *Österreichische Zoologische Zeitschrift, 1*, 1-48.

Ward Platt, M. (2009). *Le guide essentiel pour le développement de votre enfant, de 0 à 5 ans.* In B. Wurman & R. Saul (Eds.) (p. 240).

Watson, J. B. (1925). *Behaviorism* (pp. 180-190). New York, NY: W. W. Norton & Company, Inc.

Watson, J. B. (1932). Behaviorism. *American Journal of Psychiatry, 89*, 187-189.

Comparison of the Internal Transcribed Spacer (ITS) Regions Between Hybrid and Their Parents in Scallop

Biao Wu[1], Aiguo Yang[1], Jiakun Yan[1], Wandong Xu[3], Tao Yu[2], Jiteng Tian[1], Zhihong Liu[1], Liqing Zhou[1] & Xiujun Sun[1]

[1] Key Laboratory of Sustainable Development of Marine Fisheries, Ministry of Agriculture, Yellow Sea Fisheries Research Institute, Chinese Academy of Fishery Sciences, Qingdao 266071, PR China

[2] Changdao Enhancement and Experiment Station, Chinese Academy of Fishery Science, Changdao 265800, PR China

[3] Kenli Prefecture Ocean and Fisheries Bureau, Dongying 257500, PR China

Correspondence: Aiguo Yang, Key Laboratory of Sustainable Development of Marine Fisheries, Ministry of Agriculture, Yellow Sea Fisheries Research Institute, Chinese Academy of Fishery Sciences, Qingdao 266071, PR China. E-mail: yangag@ysfri.ac.cn

Abstract

The remarkable heterosis of the heterozygous F_1 derived from Zhikong scallop *Chlamys farreri* (♀) and Japanese scallop *Patinopecten yessoensis* (♂) has been proved during cultured process, and some genetic characteristics analysis have also been conducted using some molecular markers ISSR, SSR, MSAP, and so on. However, no study about the comparison between hybrid offspring and parents based on the nuclear ribosomal DNA internal transcribed spacer (ITS) was explored. In this study, the ITS-1 and ITS-2 of the *C. farreri*, *P. yessoensis* and their hybrid were amplified by PCR using specific primers and characteristics of those ITS regions, genetic regularity were analyzed. The results suggested that ITS-2 (334 bp- 352 bp) were a little longer than ITS-1 (241 bp- 256 bp) in length while the average GC content were lower than AT in most individuals, and they all had high genetic variation. Also, the gene flow between the hybrid and female parents were higher than that of hybrid and male parents. Our data implied a commonality that the ITS regions of offspring shared more similarity with that of female parents. This data in the present investigation will help for further studies in heterosis mechanism and utilization of hybrid scallop in China.

Keywords: *Chlamys farreri*, *Patinopecten yessoensis*, hybrid, internal transcribed spacer (ITS)

1. Introduction

Scallop is enjoyed as a good taste food source by humans in China, and four species are widely cultured. The Zhikong scallop, *Chlamys farreri,* a native species in China, is one of the most economically scallops, whose production had almost reached approximately 80% of the total scallop production before introducing the Bay Scallop (*Argopecten irradians*) and Japanese scallop (*Patinopecten yessoensis*) from overseas. It has a wide distribution along the coasts of North China, Korea, Japan, and Eastern Russia (Zhang et al., 2011). However, in recent years, because of high temperature in summer, deterioration of water quality, and low quality seeds, the mass mortality of *C. farreri* have frequently broken out in summer season in the main cultured areas, and the mortality reached even 80% in some areas (Guo et al., 1999; Zhang & Yang, 1999), which had seriously affected the development of its industry (Xiao et al., 2005). It is imperative for us to actively breed new varieties with high adversity resistance and fast-growing through using traditional and new breeding methods. Hybridization is a very effective means of breeding for improving the germplasm resources. The Japanese scallop *P. yessoensis*, a cold-tolerant species inhabiting coastal waters of the northern islands of Japan, the northern part of the Korean Peninsula, and Russian Primorye, was introduced to China about two decades ago from Japan (Li et al., 2007). The intercross of *C. farreri* (♀) and *P. yessoensis* (♂) has been proved to be a good way to improve the scallop quality, comparing to *C. farreri*, the hybrid F_1 had similar appearance traits, higher survival rate of 95% and growth rate was improved by 23%, while there was large scale death of *C. farreri* in high water temperature season, which revealed that the heterosis of the heterozygous F_1 was very apparent (Yang et al., 2004).

Molecular technique supplies a good tool for effective investigation on genetic analysis. Many researches showed that the molecular basis of heterosis may be attributed to the increased gene expression level in the hybrid or to the altered regulation of gene expression in the hybrid either at the global level or specific classes of genes (Leonardi et al., 1991; Romagnoli et al., 1990; Tsaftaris, 2006). As to the hybridization of C. farreri and P. yessoensis, some analyses were conducted based on some molecular technologies like random amplified polymorphic DNA (RAPD), inter-simple sequence repeat (ISSR), sequence-related amplified polymorphism (SRAP), and methylation-sensitive amplification polymorphism (MSAP) in our previous study and all the results showed that the hybrids have rich genetic diversity level (Cheng et al., 2009; He et al., 2007; Yang et al., 2009; Yu et al., 2010). However, only limited information is available regarding ribosomal DNA in hybridization study.

The internal transcribed spacer (ITS) of nuclear ribosomal DNA (rDNA) is non-coding DNA sequence. It has a high degree of mutation, made it become one of the most extensively sequenced molecular markers (Won & Renner, 2005). This region is part of the rDNA cistron, which consists of 18S, ITS-1, 5.8S, ITS-2 and 28S, and is present in several hundred copies in most eukaryotes (Aktas et al., 2007; Cheng et al., 2006; Fernández-Tajes et al., 2010). To date, the ITS sequences have been successfully employed to distinguish related species and infer phylogenetic relationships from populations to families and even higher taxonomic levels. Genetic approaches based on DNA sequencing are widely used nowadays, but the level of genetic variation between the hybrid scallop offspring and their parents based on ITS sequence rDNA remains few. Thus, it is valuable to study this region for better understanding on the genetic variation of the hybrid offspring and their parents.

The aim of this study was to amplify and sequence the ITS-1 and ITS-2 regions of C. farreri, P. yessoensis and their hybrid to provide basic characteristics data of these sequences, then to assess the genetic variation among the populations. These results are expected to provide molecular biological basis for the protection and utilization of scallop resources in China.

2. Materials and Methods

2.1 Sample and DNA Extraction

Healthy P. yessoensis with the averaging shell length of 100 mm from Changdao Bay (Shandong Province), and C. farreri with 65 mm from Qingdao Bay (Shandong Province), were collected as the parent scallop, respectively. The parent scallop were cultured under the hatchery-reared condition until gonads mature by increasing the water temperature and then hybridization between male P. yessoensis and female C. farreri was implemented randomly by artificial breeding in Changdao Enhancement and Experiment Station, Chinese Academy of Fishery Science. The offspring were cultured to adult in Qingdao Bay. The adductor muscle of parents and hybrid offspring were collected from the fresh bodies and stored at -80 °C for genomic DNA extraction. The DNA extract process was performed as described previously by Aljanabi (Aljanabi & Martinez, 1997) with slight modification. Approximate 100 mg of adductor muscle was digested for 4 h at 56 °C in the lysis buffer (10 mM Tris-Cl, 10 mM EDTA, 0.5 % SDS, 2 % proteinase K). The extracted DNA quantity, purity and integrity were tested by spectrophotometry (A_{260}/A_{280}) and 1.5% agarose gel electrophoresis, and the DNA was stored in the DEPC-treated water at -80 °C for use.

2.2 PCR Amplification and Sequencing

The PCRs were carried out to amplify the complete ITS regions of P. yessoensis, C. farreri and their offspring. Primers, ITS-1-F and ITS-1-R for amplifying ITS-1 region, ITS-2-F and ITS-2-R for ITS-2 region, were synthesized by BGI and used for PCR amplify as described by Hedgecock (Hedgecock et al., 1999) and. PCRs were performed in 30 μl volume reaction mixture composed of 100 ng genomic DNA, 3.0 μl 10 × buffer, 1.6 μl Mg^{2+} (20 mM), 2.0 μl dNTP (10 mM each), 2 μl each primer (10 μM), 1 U DNA polymerase (TianGen). PCR amplifications were carried out in a Master cycler (Eppendorf), using the reaction condition settings as follows: initial denaturation at 94 °C for 5 min, followed by 38 cycles of denaturing for 40 s at 94 °C, annealing for 40 s at 52 °C for ITS-1 and 56 °C for ITS-2, extension for 45 s at 72 °C, and a final extension for 10 min at 72 °C. The target amplified fragments were selected by 1.5 % agarose gel electrophoresis and cloned into pMD18-T vector (TaKaRa) according to the instruction and then transformed into competent cells E. coli. The positive recombinants were detected by PCR and selected for sequencing (Sangon Biotech). The primers used in this study were listed in Table 1.

Table 1. Primers used in this study

Primer name	Sequence (5'-3')
ITS-1-F	GGTTTCTGTAGGTCAACCTGC
ITS-1-R	CTGCGTTCTTCATCGACCC
ITS-2-F	GGGTCGATGAAGAACGCAG
ITS-2-R	GCTCTTCCCGCTTCACTCG

2.3 Data Analysis

Sequences obtained were edited using EditSeq model of DNAStar version 7.1 and revised manually according to the results of Base calling and NCBI blast (http://blast.ncbi.nlm.nih.gov/Blast.cgi). Sequence length and nucleotide composition were computed by MEGA 5.01 (Tamura et al., 2011). Number of Haplotypes, Haplotype (gene) diversity (*HD*), Nucleotide diversity (*PI*), and genetic divergence (F_{st}) were calculated by DNASP V5 (Librado & Rozas, 2009). Gene flow (N_m) was calculated by formula, $N_m = (1- F_{st})/2F_{st}$ (Xu et al., 2011).

Table 2. Summary of length variation (bp) and GC content (%) of the nuclear ribosomal ITS

Parameter	ITS-1						ITS-2					
	T	C	A	G	Length	G+C	T	C	A	G	Length	G+C
C1	20.7	25.9	34.3	19.1	251	45.02	23.1	24.9	30.2	21.9	334	46.7
C2	22.9	29.2	30.0	17.8	253	47.04	24.0	24.9	29.3	21.9	334	46.7
C3	22.1	30.4	28.9	18.6	253	49.01	24.0	24.9	29.3	21.9	334	46.7
C4	24.4	25.2	31.5	18.9	254	44.09	23.4	24.9	30.2	21.6	338	46.4
C5	20.9	25.7	34.0	19.4	253	45.06	21.6	26.6	32.2	19.5	338	46.2
C6	25.8	23.8	31.0	19.4	252	43.25	23.4	24.9	30.2	21.6	338	46.4
C7	20.2	32.5	30.6	16.7	252	49.21	23.4	24.9	30.2	21.6	338	46.4
C8	19.7	33.1	28.0	19.3	254	52.36	24.0	26.3	29.3	20.4	338	46.7
C10	19.4	28.5	34.8	17.4	253	45.85	24.9	25.1	28.7	21.3	334	46.4
C11	22.0	31.5	31.1	15.4	254	46.85	23.4	24.9	30.2	21.6	338	46.4
C12	22.9	28.1	32.0	17.0	253	45.06	24.0	24.9	29.3	21.9	334	46.7
F1	21.0	25.8	33.7	19.4	252	45.24	23.1	25.1	30.2	21.6	338	46.7
F2	21.0	25.8	33.7	19.4	252	45.24	23.4	24.9	30.5	21.3	338	46.2
F3	17.9	31.3	31.3	19.4	252	50.79	23.4	24.9	30.2	21.6	338	46.4
F4	22.7	25.4	34.4	17.6	256	42.97	24.3	25.4	30.2	20.1	334	45.5
F5	20.9	26.0	34.3	18.9	254	44.88	23.1	25.1	30.2	21.6	338	46.7
F6	20.9	27.6	33.1	18.5	254	46.06	23.1	25.1	30.2	21.6	338	46.7
F7	20.9	26.0	33.9	19.3	254	45.28	23.1	25.1	30.2	21.6	338	46.7
F8	22.5	27.3	32.4	17.8	253	45.06	22.8	26.0	31.1	20.1	338	46.2
F10	26.5	24.9	29.6	19.0	253	43.87	23.4	24.9	30.2	21.6	338	46.4
F12	20.9	28.3	31.5	19.3	254	47.64	22.8	26.0	30.5	20.7	338	46.7
P1	23.7	27.8	30.3	18.3	241	46.06	22.6	24.6	30.2	22.6	341	47.2
P2	23.8	25.0	33.2	18.0	244	43.03	22.9	24.9	30.4	21.7	345	46.7
P3	25.3	24.5	32.2	18.0	245	42.45	22.5	25.1	31.1	21.4	351	46.4
P4	24.8	29.8	28.9	16.5	242	46.28	21.4	25.7	33.5	19.4	346	45.1
P5	24.8	29.8	28.9	16.5	242	46.28	22.6	24.6	29.6	23.2	341	47.8
P6	22.7	26.4	33.9	16.9	242	43.39	23.8	26.1	30.5	19.6	341	45.7
P7	23.0	24.7	35.0	17.3	243	41.98	22.9	24.6	30.2	22.3	341	46.9
P8	21.4	28.4	33.3	16.9	243	45.27	23.2	25.8	30.5	20.5	341	46.3
P9	22.2	26.7	32.1	18.9	243	45.68	24.2	26.2	28.9	20.7	343	46.9
P10	28.1	26.4	28.5	16.9	242	43.39	22.7	25.0	31.3	21.0	352	46.0
P11	25.0	24.2	32.4	18.4	244	42.62	23.4	24.3	31.4	20.9	350	45.1
P12	23.0	26.3	33.7	16.9	243	43.20	***	***	***	***	***	***

Note: C, *C. Farreri*; F, hybrids; P, *P. yessoensis*.

3. Results

3.1 Length and GC Content

The ITS regions containing partial sequences of 18S, 5.8S and 28S ribosomal RNA genes were obtained by PCR in at least 10 individuals from each of the three populations. The compete sequence length and GC content of ITS-1 and ITS-2 regions after deleted the 18s, 5.8s and 28s regions were shown in Table 2. It revealed that the length of ITS of the offspring were more similar to their female parents *C. farreri* in both ITS-1 (ranged from 251 bp to 254 bp) and ITS-2 (ranged from 334 bp to 338 bp), while their male parents *P. yessoensis* were shorter in ITS-1 (ranged from 241 bp to 245 bp) and longer in ITS-2 (ranged from 341 bp to 352 bp). Average GC content in ITS-1 of *C. farreri* (43.25 % - 52.36 %) and offspring (42.97 % - 50.79 %) were a little higher than that in *P. yessoensis* (41.98 % - 46.28 %), in comparison, the average GC contents in ITS-2 were nearly equal among the three populations (45.1 % - 47.2 %).

3.2 Gene Polymorphism Analyses

Genetic variation parameters based on the data of ITS-1 and ITS-2 from two parental populations and their offspring were shown as Table 3. As the haplotype number, it displayed that 11/11 ITS-1 from *C. farreri*, 9/10 from hybrids, and 11/12 from *P. yessoensis* were haplotype, compared to that of ITS-2 were 5/11, 6/10, 11/12, which showed that it had higher haplotype's proportion in ITS-1 than that in ITS-2 of all the three populations. The haplotype diversity of ITS-1 from hybrids was lower than that from their parent, however, the nucleotide diversity of ITS-1 from hybrids was the highest among three populations. While in ITS-2, haplotype diversity of hybrids was higher than *C. farreri* but lower than *P. yessoensis*, and nucleotide diversity of hybrids was the lowest.

Table 3. Number of Haplotypes, Haplotype (gene) diversity (*HD*), Nucleotide diversity (*PI*)

Parameter		Sample size	Number of Haplotypes	Haplotype diversity(*HD*)	Nucleotide diversity(*PI*)
C	ITS-1	11	11	1.00	0.39208
	ITS-2	11	5	0.8182	0.41796
F	ITS-1	10	9	0.9449	0.41217
	ITS-2	10	6	0.8444	0.30373
P	ITS-1	12	11	0.9848	0.36904
	ITS-2	11	11	1.00	0.57240

Note: C, *C. Farreri;* F, hybrids; P, *P. yessoensis.*

As shown in Table 4, the gene flow was found to be varied significantly. The gene flow of ITS-2 between *hybrid offspring* and *P. yessoensis* was the minimum (0.2906), and the maximum appeared in the ITS-1 between the hybrid and *C. farreri*. The gene flow between offspring and their female parent *C. farreri*, 7.5645 of ITS-1 and 1.9452 of ITS-2, were higher than other interspecies. Besides, compared to ITS-1, the gene flow of ITS-2 were lower than ITS-1 between offspring and their female and male parents. The genetic divergence of ITS-1 and ITS-2 among different populations had at least significance difference except that of ITS-1 between hybrids and female parents, what's more, two of them reached extreme significant difference ($P < 0.01$). Comparatively speaking, the genetic difference between the hybrid and *C. farreri* were lower than other interspecies in both ITS-1 and ITS-2.

Table 4. Gene flow and genetic divergence

Parameter	C		F		P	
	ITS-1	ITS-2	ITS-1	ITS-2	ITS-1	ITS-2
C	——	——	7.5645	1.9452	0.6543	0.3394
F	0.062	0.044	——	——	0.6430	0.2906
P	0.43314[*]	0.5105[**]	0.43743[*]	0.632[**]	——	——

Note: C, *C. farreri*; F, hybrids; P, *P. yessoensis*. Figures below diagonal represent genetic divergence (F_{st}) corresponding significance test of divergence between populations ($P < 0.05$ means statistical significance; $P < 0.01$ means extreme significance, while figures above diagonal represent gene flow (N_m) between populations.

4. Discussion

In recent years, there has been an increased interest in understanding ITS regions research (Chen et al., 2010; Freire et al., 2011; Shafiei et al., 2013; Sudheer Pamidimarri et al., 2009; Sum et al., 2014). Comparing to the encoding regions, the ITS rDNA sequences belong to the non-coding regions, and they own some special property such as relatively higher variability, more rich divergence information, easily amplification, thus they are commonly applied in many fields and have been proven to be a powerful and useful genetic maker for genetic breeding and infer evolutionary relationships. Many studies were conducted in those fields. For example, the ITS-1 regions of wild giant clam population were amplified, and showed high polymorphism with 29% variation arising from base substitutions (Sudheer Pamidimarri et al., 2009; Yu et al., 2000); the ITS-2 regions of orient clam *Meretrix meretrix* Linnaeus were used for phylogenetic study among clam populations with different stripe color, which ITS were proved to be a good tool for genetic analysis (Li et al., 2006). These reported studies indicated that most previous research focused on variability and phylogenetic and taxonomic relationships, so far, knowledge about the relationships characterization between hybrids and parents based on ITS is not well known.

In the present study, the two internal spacers, ITS-1 and ITS-2 of the scallop hybrid and their parents were obtained successfully. The results showed that the original sequences included some partial sequence of conserved 18S, 5.8S and 28S rRNA genes, indicating the validity of ITS region were obtained. According to the reports about ITS sequences, the length of ITS sequences and GC content were variable in different species. The amplified ITS-2 fragments were about 500 bp in Pearl Oyster, spanning the partial sequences of 5.8S and 28S rRNA genes, whereas rhodnius presented a less variable ITS-1 with around 300 bp, and the species of the Triatoma and Panstrongylus genera presented an amplified ITS-1 fragment between 600 and 1000 bp (He et al., 2005; Tartarotti and Ceron, 2005). Both ITS-1 and ITS-2 of the three populations in our study had normal length of approximately 300 bp, and GC contents were a little lower than AT contents. What's interested was that the ITS length size and GC contents of the hybrid offspring was found to have more similarity with their female parents than male parents. What's more, the same results were also found in the following gene polymorphism analyses in this study. Especially, the gene flow and genetic divergence all obviously showed the higher identical between the hybrid offspring and their female parents in both ITS-1 and ITS-2 regions. In the previous study, we found the hybrid F_1 were similar to the female parents in the genetic analyses via the molecular markers RAPD, SSR, ISSR and SRAP (Cheng et al., 2009; He et al., 2007; Yang et al., 2008; Yang et al., 2009), although the F_1 genome came from male and female parents, which might be the reason for that the hybrid F_1 had similar appearance and higher resistance traits with mothers. The results in this study are considered to be very parallel to those results we previously obtained using other molecular markers. This report about the ITS genetic regularity referring to the *C. farreri*, *P. yessoensis* and their hybrid offspring enhances our knowledge regarding to heterosis in scallop.

Acknowledgements

This work is supported by the grants from Independent Innovation Funds of Shandong Province (2013CXC80202) and Special Scientific Research Funds for Central Non-profit Institutes, Yellow Sea Fisheries Research Institutes (20603022013012).

References

Aktas, M., Bendele, K. G., Altay, K., Dumanli, N., Tsuji, M., & Holman, P. J. (2007). Sequence polymorphism in the ribosomal DNA internal transcribed spacers differs among Theileria species. *Vet Parasitol, 147*, 221-230. http://dx.doi.org/10.1016/j.vetpar.2007.04.007

Aljanabi, S. M., & Martinez, I. (1997). Universal and rapid salt-extraction of high quality genomic DNA for PCR-based techniques. *Nucleic Acids Res, 25*, 4692-4693. http://dx.doi.org/10.1093/nar/25.22.4692

Chen, L. H., Yu, Z., & Jin, H. P. (2010). Comparison of Ribosomal DNA ITS Regions Among Hippophae rhamnoides ssp. sinensis from Different Geographic Areas in China. *Plant Mol Biol Rep, 28*, 635-645. http://dx.doi.org/10.1007/s11105-010-0194-0

Cheng, H. L., Xia, D. Q., Wu, T. T., Meng, X. P., Ji, H. J., & Dong, Z. G. (2006). Study on sequences of ribosomal DNA internal transcribed spacers of clams belonging to the Veneridae family (Mollusca: Bivalvia). *Yi Chuan Xue Bao, 33*, 702-710.

Cheng, N. N., Yang, A. G., Liu, Z. H., Zhou, L. Q., & Wu, B. (2009). Analysis on heterosis of Chlamys farreri × Patinopecten yessoensis by SRAP marker. *Marine Sciences, 33*, 107-111.

Fernández-Tajes, J., Freire, R., & Méndez, J. (2010). A simple one-step PCR method for the identification between European and American razor clams species. *Food Chem, 118*, 995-998. http://dx.doi.org/10.1016/j.foodchem.2008.10.043

Freire, R., Arias, A., Méndez, J., & Insua, A. (2011). Identification of European commercial cockles (Cerastoderma edule and C. glaucum) by species-specific PCR amplification of the ribosomal DNA ITS region. *Eur Food Res Technol, 232*, 83-86. http://dx.doi.org/10.1007/s00217-010-1369-5

Guo, X. M., Ford, S. E., & Zhang, F. S. (1999). Molluscan aquaculture in China. *J Shellfish Res, 18*, 19-31.

He, B., Yang, A. G., Wang, Q. Y., Liu, Z. H., & Zhou, L. Q. (2007). ISSR analysis of the F1 hybrids of scallop Chlamys farreri ♀ × Patinopecten yessoensis ♂. *Journal of Dalian Fisheries University,* 273-277.

He, M. X., Huang, L. M., Shi, J. H., & Jiang, Y. P. (2005). Variability of Ribosomal DNA ITS-2 and Its Utility in Detecting Genetic Relatedness of Pearl Oyster. *Mar Biotechnol, 7*, 40-45. http://dx.doi.org/10.1007/s10126-004-0003-6

Hedgecock, D., Li, G., Banks, M. A., & Kain, Z. (1999). Occurrence of the Kumamoto oyster Crassostrea sikamea in the Ariake Sea, Japan. *Mar Biol, 133*, 65-68. http://dx.doi.org/10.1007/s002270050443

Leonardi, A., Damerval, C., Hebert, Y., Gallais, A., & de Vienne, D. (1991). Association of protein amount polymorphism (PAP) among maize lines with performances of their hybrids. *Theor Appl Genet, 82*, 552-560. http://dx.doi.org/10.1007/BF00226790

Li, Q., Xu, K., & Yu, R. (2007). Genetic variation in Chinese hatchery populations of the Japanese scallop (Patinopecten yessoensis) inferred from microsatellite data. *Aquaculture, 269*, 211-219. http://dx.doi.org/10.1016/j.aquaculture.2007.04.017

Li, T. W., Zhang, A. G., Su, X. R., Li, C. H., Liu, B. Z., Lin, Z. H., & Cai, X. L. (2006). The analysisi of ITS-2 in Meretrix meretrix with different stripes. *Oceanologia Et Limnologia Sinica,* 132-137.

Librado, P., & Rozas, J. (2009). DnaSP v5: a software for comprehensive analysis of DNA polymorphism data. *Bioinformatics, 25*, 1451-1452. http://dx.doi.org/10.1093/bioinformatics/btp187

Romagnoli, S., Maddaloni, M., Livini, C., & Motto, M. (1990). Relationship between gene expression and hybrid vigor in primary root tips of young maize (Zea mays L.) plantlets. *Theor Appl Genet, 80*, 769-775. http://dx.doi.org/10.1007/BF00224190

Shafiei, R., Sarkari, B., & Moshfe, A. (2013). A Consistent PCR-RFLP Assay Based on ITS-2 Ribosomal DNA for Differentiation of Fasciola Species. *Iran J Basic Med Sci, 16*, 1266-1269.

Sudheer Pamidimarri, D. V. N., Chattopadhyay, B., & Reddy, M. P. (2009). Genetic divergence and phylogenetic analysis of genus Jatropha based on nuclear ribosomal DNA ITS sequence. *Mol Biol Rep, 36*, 1929-1935. http://dx.doi.org/10.1007/s11033-008-9401-6

Sum, J. S., Lee, W. C., Amir, A., Braima, K. A., Jeffery, J., Abdul-Aziz, N. M., Fong, M. Y., & Lau, Y. L. (2014). Phylogenetic study of six species of Anopheles mosquitoes in Peninsular Malaysia based on inter-transcribed spacer region 2 (ITS2) of ribosomal DNA. *Parasit Vectors, 7*, 309. http://dx.doi.org/10.1186/1756-3305-7-309

Tamura, K., Peterson, D., Peterson, N., Stecher, G., Nei, M., & Kumar, S. (2011). MEGA5: molecular evolutionary genetics analysis using maximum likelihood, evolutionary distance, and maximum parsimony methods. *Mol Biol Evol, 28*, 2731-2739. http://dx.doi.org/10.1093/molbev/msr121

Tartarotti, E., & Ceron, C. R. (2005). Ribosomal DNA ITS-1 Intergenic Spacer Polymorphism in Triatomines (Triatominae, Heteroptera). *Biochem Genet, 43*, 365-373. http://dx.doi.org/10.1007/s10528-005-6776-0

Tsaftaris, S. (2006). Molecular aspects of heterosis in plants. *Physiol Plantarum, 94*, 362-370. http://dx.doi.org/10.1111/j.1399-3054.1995.tb05324.x

Won, H., & Renner, S. S. (2005). The internal transcribed spacer of nuclear ribosomal DNA in the gymnosperm Gnetum. *Mol Phylogenet Evol, 36*, 581-597. http://dx.doi.org/10.1016/j.ympev.2005.03.011

Xiao, J., Ford, S. E., Yang, H. S., Zhang, G. F., Zhang, F. S., & Guo, X. M. (2005). Studies on mass summer mortality of cultured zhikong scallops (Chlamys farreri Jones et Preston) in China. *Aquaculture, 250*, 602-615. http://dx.doi.org/10.1016/j.aquaculture.2005.05.002

Xu, M. Y., Li, J. J., Guo, B. Y., Lv, Z. M., Zhou, C., & Wu, C. W. (2011). Genetic diversity of seven populations of Octopus variabilis in China's coastal waters based on the 12s rRNA and COIII gene analysis. *Oceanologia Et Limnologia Sinica, 42*, 387-396.

Yang, A. G., Wang, Q. Y., Liu, Z. H., & Zhou, L. Q. (2004). The hybrid between the scallops Chlamys farreri and Patinopecten yessoensis and the inheritance characteristics of its first filial generation. *Marine Fisheries Research, 25*, 1-5.

Yang, P., Yang, A. G., Liu, Z. H., & Zhou, L. Q. (2008). The Selection of Universal Microsatellite Primers between Patinopecten yessoensis and Chlamys farreri and Its Application in Hybrid Identification. *Journal of Anhui Agri. Sci., 36*, 8287-8289.

Yang, P., Yang, A. G., Shan, W. H., Liu, Z. H., & Zhou, L. Q. (2009). Research on genetic variation of hybrid scallop from Chlamys farreri and Patinopecten yessoensis during post-embryo growth. *Progress in Fishery Sciences, 30*, 65-71.

Yu, E. T., Juinio-Menez, M. A., & Monje, V. D. (2000). Sequence variation in the ribosomal DNA internal transcribed spacer of Tridacna crocea. *Mar Biotechnol (NY), 2*, 511-516. http://dx.doi.org/10.1007/s101 260000033

Yu, T., Yang, A. G., Wu, B., & Zhou, L.Q. (2010). Analysis of Chlamys farreri, Patinopecten yessoensis and their offspring using methylation-sensitive amplification polymorphism(MSAP). *Journal of Fisheries of China*, 1335-1342.

Zhang, F. S., & Yang, H. S. (1999). Analysis of the causes of mass mortality of faming Chlamys farreri in summer in costal areas of Shandong, China. *Marine Science*, 44-47.

Zhang, X., Zhao, C., Huang, C., Duan, H., Huan, P., Liu, C., … Xiang, J. (2011). A BAC-based physical map of Zhikong scallop (Chlamys farreri Jones et Preston). *PLoS One, 6*, e27612. http://dx.doi.org/10.1371/journal. pone.0027612

3

Bio-Active Compounds Composition in Edible Stinkbugs Consumed in South-Eastern Districts of Zimbabwe

Robert Musundire[1], Johnson C. Zvidzai[1] & Cathrine Chidewe[2]

[1] Department of Food Science and Postharvest Technology, Chinhoyi University of Technology, Chinhoyi, Zimbabwe

[2] Department of Biochemistry, University of Zimbabwe, Harare, Zimbabwe

Correspondence: Robert Musundire, Department of Food Science and Postharvest Technology, Chinhoyi University of Technology, P. Bag, 7724, Chinhoyi, Zimbabwe. E-mail: rmusundire@cut.ac.zw

Abstract

Encosternum delegorguei Spinola (Hemiptera: Tessaratomidae) are consumed as relish and with traditional claims of having medicinal roles in the South-Eastern districts of Zimbabwe. However, very little has been explored scientifically to validate these claims. The current study was conducted to investigate bio-active compound composition and diversity of stable antibacterial activity from *E. delegorguei* extracts. Methanol, ethanol and aqueous extractions of *E. delegorguei* were performed followed by qualitative, quantitative analyses of phytochemical/bioactive compounds and determination of antibacterial activities using disc diffusion method on ten clinically important microbes. Alkaloids, flavonoids, anthraquinones, tannins, phlobatannins, steroids, triterpenoids and cyanogen glycosides were detected in the insect extracts. Flavonoids were detected in significantly higher concentrations in unprocessed compared to processed insects. Mean DPPH free radical scavenging activities were 78% and 88% for traditionally processed and raw insect extracts respectively. Traditional processing resulted in reduction of bioactive compounds (22.2% total phenolics; 68.4% flavonoids) and free radical scavenging activities by 10%. However, it resulted in an increase of cyanogen glycosides by 65.7%. Methanol extracts produced highest mean inhibition zones of 20 mm while aqueous and ethanol extracts had mean inhibition of 0 to 15 mm as compared to control with 20-40 mm. High flavonoids levels could be beneficial to consumers. However, a potential trade-off from elevated levels of cyanogen glycosides after processing needs further investigation. The free radical scavenging activity displayed by *E. delegorguei* extracts indicate a potential source of natural anti-oxidants that can be formulated into commercial products.

Keywords: antibacterial, *Encosternum delegorguei*, free radical scavenging, phytochemical

1. Introduction

The Edible Stinkbug, *Encosternum delegorguei* Spinola (Hemiptera: Tessaratomidae) is distributed widely in subtropical woodland and bush veldt with occurrences in Zimbabwe and northern provinces of South Africa (Dzerefos, Wtkowski, & Toms, 2009; Picker, Griffiths, & Weaving, 2004). In Zimbabwe, the Edible Stinkbug is restricted to the southern most parts which include the Nerumedzo region in Bikita (approximately 20°1'23.22"S, 31°41'17.65"E) and Zaka (approximately 20°4'28"S, 30°49'58"E) districts in Masvingo province (Chavhunduka, 1975; Mawere, 2012).

Encosternum delegorguei is an important source of income in parts of Zimbabwe and South Africa. It has also immensely contributed towards attainment of food security to rural communities in these two countries during winter season (Mawere, 2012; Kwashirai, 2007; Defoliart, 1995; Chavhunduka, 1975). Teffo, Toms, and Eloff (2007) reported the nutritional importance of the Edible Stinkbug consumed in Limpopo province, South Africa. The insects were reported to have high protein, fat, amino acid, minerals and vitamin contents.

Although much of the chemical analytical research work involving *E. delegorguei* has been focussed on its nutritional composition (Teffo et al., 2007), an emerging dimension in the research on *E. delegorguei* is on medicinal value of compounds derived from this insect species. A number of insect species are known to sequester compounds from host plants and store them as defence mechanisms. This ability of insects to sequester and produce allelochemicals and phytochemicals has raised a lot of interest in the research for new drugs and search for alternatives to synthetic pesticides (Elemo, 2011; Silberbush, Markman, Lewinsohn, Bar, & Cohen, 2010; Teffo, Aderogba, & Eloff, 2010; Zaku, Abdulrahaman, Onyeyili, Aguzue, & Thomas, 2009; Moraes et al., 2008).

In the Zimbabwean communities that consume *E. delegorguei*, traditional claims asserts medicinal roles of these insects that include cure for asthmatic and heart diseases, aiding digestive systems, acting as appetizers and enhancing sexual desires. Traditional preparation methods of consumption for these insects are therefore aimed to achieve a quality that will deliver medicinal properties. However, very little has been explored scientifically to validate these practices.

In addition, the Edible Stinkbug is well known for releasing a very offensive smell which is associated with unknown volatile defensive compounds. However, some of these compounds are removed from the insects using a traditional aqueous processing method. Depending on the degree to which the traditional processing steps are followed, insects can either be well-prepared, where the requisite taste and flavour is acquired or can be spoiled. In the latter case, volatile compounds accumulate on the thoracic segments and impart a very bitter taste to the insects. Traditional beliefs also assert stomach currant properties from these spoiled and bitter insects.

This study was born out of the need to investigate the bioactive compound composition and diversity of stable antibacterial activities from *E. delegorguei* extracts as a first step in the validation of traditional claims on medicinal properties associated with this insect species. In addition, the study was also aimed at determining the effect of traditional processing on the quality and quantity of bio-active compounds in order to predict potential implications on health benefits and risks to consumers.

2. Materials & Methods

2.1 Origin and Collection of E. delegorguei

Samples of *E. delegorguei* were collected from Bikita (approximately 20°1'23.22"S, 31°41'17.65"E), during winter season (month of June). Live insects were collected using the jarring and knock down approach from tree branches and transported for laboratory analyses in perforated polypropylene sacks kept in cool boxes.

2.2 Preparation of Insects for Extraction of Phytochemicals

A portion of 500 g of live insects were gradually killed with 5 litres of lukewarm water (~37 °C) and stirred for 5 minutes until the insects were dead. Water was drained from the dead insects giving rise to an aqueous liquid (supernatant) which was kept as the traditional extract while the insects were dried by heating and stirring in traditionally prepared clay pots for 3 minutes. Meanwhile during the heating and drying process, a flame was used to burn off volatile compounds from the dead insects. Heating was stopped when insects changed colour from green to golden brown. The insects killed using this method were kept in perforated plastic bags that were kept at ambient temperatures until use for experimental purposes.

Raw and traditionally prepared insect samples were oven-dried at 60 °C overnight followed by grinding separately using the pestle and motor.

2.3 Extraction of Phytochemicals From Raw E. delegorguei

An amount of 10 g (28 insects), was weighed and homogenised using a pestle and motor in 60 ml of methanol, and this extraction product was centrifuged using bench centrifuge at 3 000 rpm for 10 min (Harbone, 1973). A second extraction was performed using ethanol as the solvent. A third and final extraction was performed on whole organisms with water as a solvent. The extracts were all kept under refrigeration at 4 °C.

2.4 Microbial Strains Tested

Test microorganisms used in this study were *Bacillus subtilis*, *Escherichia coli*, *Faecal streptococcus*, *Lactobacillus* species, *Proteus vulgaris*, *Pseudomonas aeruginosa*, *Salmonella enteritidis*, *Salmonella typhi*, *Shigella* species and *Staphylococcus aureus* all obtained from Medical Microbiology Department, University of Zimbabwe. The bacterial isolates were incubated at 37 °C for 18 h during testing.

2.5 Determination of Phytochemical Composition From the Raw and Traditionally Processed E. delegorguei

2.5.1 Assaying for Oxalates

Oxalates were extracted and determined by titration (Amoo & Agunbiade, 2010). Raw and traditionally processed samples (2 g) were digested with 50 ml of 0.75 M H_2SO_4 for 2 hours, stirred and filtered using Whatman No. 1 filter paper. An aliquot of 125 ml of the filtrate was heated until it was close to boiling point (80-90 °C) and titrated against standardised 0.5 M $KMnO_4$ solution to a faint pink colour.

2.5.2 Assaying of Phytates

Phytic acid content was determined by a colorimetric method (Vaintraub & Laptewa, 1988). Raw and traditionally processed samples (1 g) were extracted with 10 ml of 0.5 M HCl for 1 hour at room temperature, diluted, centrifuged and analysed for phytic acid by addition of 1 ml of Wade reagent (0.3 g of ferric chloride ($FeCl_3.6H_2O$) and 3 g of sulphosalycylic acid dissolved in 1L of distilled water) to 3 ml of sample extract. Absorbance was read at 500 nm on a Jenway 6405 UV/VIS spectrophotometer (Jenway Ltd., Essex, UK). The concentration of phytic acid was calculated by comparison with standard solutions of calcium phytate (Sigma –Aldrich Chemie, Steinheim, Germany, www.sigmaaldrich.com).

2.5.3 Assaying for Cyanogen Glycosides

Cyanogen glycosides were quantified as total cyanide (Makkah, 2003). Raw and traditionally processed samples (4 g) were added to 125 ml water followed by 2.5 ml chloroform in a Kjeldahl flask and then distilled. HCN released was absorbed in 2% (w/v) potassium hydroxide (total volume after extraction was 20 ml). An aliquot (5 ml) of the solution was mixed with 5 ml of alkaline picrate and heated in a boiling water bath for 5 minutes. After cooling the absorbance was read at 520 nm. Potassium cyanide (240 mg KCN/ L, Sigma-Aldrich Chemie, Steinheim, Germany, www.sigmaaldrich.com) was used as a standard.

2.5.4 Assaying for Alkaloids

Alkaloids were determined by a gravimetric method (Harbone, 1973). Raw and traditionally processed samples (5 g) were weighed and dispersed into 50 ml of 10% (v/v) acetic acid solution in ethanol. The mixture was vortexed and allowed to stand for 4 hours before it was filtered. The filtrate was evaporated to one quarter the original volume on a hot plate. Concentrated ammonium hydroxide was added drop wise in order to precipitate the alkaloids. A pre-weighed filter paper was used to filter off the precipitate and was then washed with 1% (v/v) ammonium hydroxide solution. The filter paper containing the precipitate was dried in an oven at 60 °C for 30 minutes, followed by transferring to a dessicator to cool and re-weighing until a constant weight. The weight of alkaloid was determined by weight difference of filter paper and expressed as g /100 g fresh weight of sample.

2.5.5 Assaying for Total Phenolics and Flavonoids

For the determination of total phenolics and flavonoids, raw and traditionally processed samples (1 g) were dissolved in 20 ml 50% methanol. The mixture was vortexed for 1 minute and then sonicated for 20 minutes. The mixture was centrifuged at 3000 g for 10 minutes and the supernatant was used for the analysis of total phenolics and flavonoids.

Total phenolics were determined by the Folin-Ciocalteu method using gallic acid (0.5 mg/ml, Sigma-Aldrich Chemie, Steinheim, Germany, www.sigmaaldrich.com) as the standard (Penarrieta, Alvaradoa, Bergenstahlc, & Akkesonb, 2007). The Folin-Ciocalteu reagent, diluted 10 times (2.5 ml), 2 ml of saturated sodium carbonate (75 g/L) and 50 μl of sample (diluted ten times) were mixed and homogenized for 10 seconds and heated for 30 minutes at 45 °C. The absorbance at 765 nm was read after cooling to room temperature.

Total flavonoids were determined according to Jimo, Adedapo, Aliero, Koduru, and Afolayani (2010). To 0.5 ml sample, 0.5 ml of 2% $AlCl_3$ in ethanol solution was added. After 1 hour at room temperature, the absorbance was measured at 420 nm. Total flavonoids content was evaluated as catechin equivalence CE g/100g dry weight of extract.

2.5.6 Assaying for Tannins

Tannins were extracted using methanol and determined by spectrophotometric method (Price, Scoyoc, & Butler, 1978). Raw and traditionally processed samples (1 g) were defatted using diethyl ether and transferred to a 100 ml glass beaker. To the defatted material, 10 ml methanol was added and the beaker placed in an ultrasonic ice-water bath for 30 minutes at room temperature. The contents of the beaker were transferred to centrifuge tubes and centrifuged for 10 minutes at 3000 g. Tannins were determined by the modified Vanillin-HCl method using catechin (5 mg/ml, Sigma-Aldrich Chemie, Steinheim, Germany, www.sigmaaldrich.com) as the standard stock solution (Price et al., 1978).

2.6 Bio-Assaying for DPPH Radical Scavenging Activity

The free radical scavenging activities of methanolic extracts were measured by decrease in the absorbance of methanol solution of DPPH (P. V. Sharma, Paliwal, & S. Sharma, 2011). A stock solution of DPPH (33 mg in 1 L, Sigma-Aldrich Chemie, Steinheim, Germany, www.sigmaaldrich.com) was prepared in methanol, which gave initial absorbance of 0.430, and 5 ml of this stock solution was added to 1 ml of *E. delegorguei* extract solution at different concentrations (100 – 1000 μg/ml). After 30 minutes, absorbance was measured at 517 nm and compared with standards (100 – 1000 µg/ml). The radical scavenging activity (%) was obtained by expressing the difference between the absorbance of control mixture and that of the test compounds over the absorbance of the control mixture (Anwar, Qayyum, Hussain, & Iqbal, 2010) from the equation:

$$\% \text{ activity} = [(OD_{control} - OD_{sample}) \times 100] \,/\, OD_{control} \qquad (1)$$

2.7 Disc Diffusion Assaying for Antibacterial Activity

Each organism was maintained on nutrient agar plates and recovered for testing by growth in nutrient broth for 18 hours (Taylor, Manandhar, Hudson, & Towers, 1995). Before use each bacterial culture was diluted 1:100 with fresh sterile nutrient broth and incubated for 3-5 hours to standardise the culture using 0.5 McFarland standard solutions. Test organisms were streaked in a radial pattern on sterile nutrient agar plates. Sterile 6 mm diameter filter paper discs were impregnated with the sterile test materials and placed onto nutrient agar. Negative controls were prepared using the same solvents used to dissolve the plant extracts. Gentamycin (10 µg/ml, Sigma-Aldrich Chemie, Steinheim, Germany, www.sigmaaldrich.com) was used as positive reference standards. The innoculated plates were incubated at 37 °C for 24 hours. The antibacterial activity was measured as the diameter in (mm) of clear zone of growth inhibition. Each test was run in triplicates.

2.8 Statistical Analysis

Statistical analyses of the data were done using GraphPad Prism 5.03 software. The biochemical assays were done in triplicate using three samples. Data obtained were expressed as mean $\pm$ standard deviation, and were subjected to one way analysis of variance (ANOVA) and means were separated by Turkey's pair wise and multiple comparison tests at $P < 0.05$ where appropriate.

3. Results

3.1 Qualitative Analysis of Phytochemicals/Bio-Active Compounds From Raw E. delegorguei Extracts

The initial qualitative analysis of *E. delegorguei* extracts are shown in Table 1. Prior to quantification, alkaloids, flavonoids, cardiac glycosides, steroids, triterpenoids and free reducing sugars were found in varying concentrations as observed from the colour intensities produced. Compounds that were detected in the screening stage were confirmed by the quantitative tests shown in Table 2. Low levels of anthraquinones and phlobatannins were observed in this study. Amino acids were not detected in aqueous and organic solvent extracts.

Table 1. Qualitative amounts of phytochemical compounds from methanol and aqueous extracts of *E. delegorguei*

Phytochemical	Test Done	Methanol extract from raw insects	Aqueous extracts (supernatant after traditional processing)
Alkaloids	Dragendorffs test	+++++	++
	Mayers test	+++	-ve
	Tannic acid test	++++	+++
Flavonoids	NH₄OH test	+	-ve
	NaOH test	+++++	+++
Reducing sugars		+++++	+++++
Tannins	Ferric chloride test	-ve	-ve
	Tannin alkaline reagent test	+	-ve
	Vanillin-Hydrochloride test	+	-ve
Phlobatannins		+	-ve
Steroids		++++	++++
Steroids & Triterpenoids		+++++	+++++
Amino acid using ninhydrin		-ve	-ve
Cardiac glycosides	Keddes test	++++	++++
	Keller- Killan test	+++++	+++++
Anthraquinones		+	+

* + trace amount, ++/+++ moderate amount, ++++/ +++++ appreciable amount, - completely absent.

3.2 Composition of Phytochemical/Bioactive Compounds From E. delegorguei

There were higher quantities of phytochemicals/bio-active compounds in raw *E. delegorguei* for all tests done compared to the traditionally processed samples except for cyanogens which got elevated by approximately three times in processed insects (Table 2). Traditional processing of *E. delegorguei* was noted to reduce significantly the levels of total phenolics, tannins and oxalates through the aid of aqueous system. The extents of reductions in the respective compounds were 22.2% for total phenolics, 29.7% for alkaloids, 30.2% oxalates and 67.7% for tannins (Table 2). However, there was a significant increase (65.7%) in the quantities of cyanogen glycosides after raw insects were processed according to traditional practices.

Phytates were not recorded from both raw and traditionally processed insect extracts while flavonoids occurred in relatively high levels. The traditional processing procedure also significantly reduced the quantity of flavonoids in the processed insects by 68.4% (Table 2).

Table 2. Phytochemical composition of raw and traditionally processed *E. delegorguei*

Phytochemical	Raw insects	Traditionally processed	Percent (%) increase (+)/decrease (-) between raw & traditionally processed insects
Total Phenolics[d] (g GAE/ 100 g)	3.6 ± 0.4^a	2.8 ± 0.5^a	(-) 22.2
Tannins[e] (g CE/ 100 g)	0.31 ± 0.01^a	0.10 ± 0.04^b	(-) 67.7
Flavonoids[e] (g CE/ 100 g)	15.20 ± 1.00^a	4.80 ± 0.40^b	(-) 68.4
Alkaloids (g/ 100g)	7.4 ± 0.6^a	5.2 ± 0.2^b	(-) 29.7
Oxalates (g/ 100g)	1.26 ± 0.07^a	0.88 ± 0.15^b	(-) 30.2
Cyanogen glycosides[g] (μg/ 100g)	23 ± 3.1^a	67.0 ± 3.4^b	(+) 65.7
Phytates (g/ 100g)	Not detected	Not detected	No change

[*1] Values are means ± standard deviation. In the same row means with different superscripts are significantly different (P < 0.05).

[*2] CE- Catechin equivalents, GAE – Gallic acid equivalents.

3.3 DPPH Free Radical Scavenging Assaying From E. delegorguei Extracts

The values of DPPH free radical scavenging activity for *E. delegorguei* extracts were not significantly different for raw and traditionally processed *E. delegorguei* extracts. The free radical scavenging activities were comparable to values of known compounds used as controls in this study (ascorbic acid, catechin and butylated hydroxyanisole) whose values ranged between 87 to 98% (Table 3).

Table 3. DPPH radical scavenging activity (%) produced from methanolic extracts of raw and traditionally processed *E. delegorguei* in comparison to activities of known compounds

Sample (1 mg/ml of each test compound was used)	% DPPH radical scavenging activity
Raw *E. delegorguei* methanolic extract	88 ± 3^a
Traditionally processed *E. delegorguei* extract	78 ± 7^{ab}
Ascorbic acid	92 ± 1^c
BHA	97 ± 1^d
Catechin	92 ± 3^a

[* 1] Values are means ± standard deviation.

[* 2] Means with different superscripts are significantly different (P < 0.05).

3.4 Antibacterial Activity Determination of Ethanol, Methanol and Water Extracts

Antibacterial bioactivity of extracts from *E. delegorguei* using disc diffusion assay realised either bactericidal or bacteriostatic action to the following microbes that are Gram positive: *B. subtilis*, *Lactobacillus* sp.,

Staphylococcus aureus, *Salmonella enteritidis*, *Salmonella typhi*, and Gram negative species which included *E. coli* and *P. aeruginosa*, as shown in Table 4.

Table 4. Antimicrobial activity (inhibition diameter (mm)) of *E. delegorguei* extracts using disc diffusion assay on several microbial isolates

Microorganism	Type of extract				Gentamycin
	Methanol extract	Traditional extract	Ethanol extract	Aqueous extract from thoracic glands	
Escherichia coli	20.0 ± 0	9.0 ± 0	10.3 ± 0.6	9.0 ± 0	30.0 ± 0
Salmonella typhi	17.3 ± 0.6	NI	6.3 ± 0.6	NI	30.0 ± 0
Salmonella enteritidis	20.3 ± 0.6	NI	NI	10.0 ± 0	20.0 ± 0
Proteus vulgaris	20.0 ± 0	15.3 ± 0.6	10.0 ± 0	10.0 ± 0	35.0 ± 0
Shigella sp.	20.3 ± 0.6	10.3 ± 0.6	6.0 ± 0	15.3 ± 0.6	35.3 ± 0.6
Pseudomonas aeruginosa	14.0 ± 0	12.3 ± 0.6	10.3 ± 0.6	14.0 ± 0	35.0 ± 0
Bacillus subtilis	10.7 ± 0.6	NI	NI	NI	35.3 ± 0.6
Staphylococcus aureus	12.0 ± 0	10.3 ± 0.6	15.0 ± 0	6.0 ± 0	40.0 ± 0
Faecal streptococcus	25.0 ± 0	$12.3 \pm 0.$	NI	10.0 ± 0	35.0 ± 0
Lactobacillus sp.	20.3 ± 0.6	10.0 ± 0	7.0 ± 0	9.0 ± 0	25.0 ± 0

*[1] NI = no inhibition.

*[2] Gentamycin antibacterial activity was referenced as 100% inhibition and its inhibition zone was highest at 40 mm diameter.

The methanol extract produced a relatively high comparable zone of inhibition of about 20 mm on *E. coli*, *S. enteritidis*, *P. vulgaris*, *Lactobacillus* sp., *Shigella* sp., and *F. streptococcus* strains to that of the positive control. In general, the aqueous and ethanol extracts had subdued inhibition effect on all the microbial strains and in some instances not having an effect at all. On average, the overall inhibition from the insect extracts was about 10 mm, which was 72% less antibacterial compared to the positive control of 10 µg/ml gentamycin (Table 4).

4. Discussion

Methanol and aqueous extracts of *E. delegorguei* were noted to comprise of alkaloids, flavonoids, cyanogen glycosides, steroids, triterpenoids and free reducing sugars. The defence mechanism employed *E. delegorguei* could possibly make use of some of these chemicals that they can synthesise from plants that they feed on in a manner as suggested by Teffo et al. (2007). In addition, compounds acquired from plants during foraging can function as precursor molecules for the biosynthesis of other compounds that are found in the insects.

Quite a considerable amount of attention has been given on medically applicable phytochemicals from plants (De Britto, Gracelin, & Sebadastian, 2011; Olusesan, Ebele, Onwuegbuchulan, & Olorunmola, 2010; Yusha'u, Hamza, & Abdullahi, 2010; O. O. Igbinosa, E. O Igbinosa, & Aiyegoro, 2009; Teffo et al., 2007; Prabuseenivasan, Jayakumar, & Ignacimuthu, 2006; Akinyemi, Oladapo, Okwara, Ibe, & Fasure, 2005; Zaidan, et al., 2005). However, few studies have focussed on phytochemicals from insects. Much of the studies done so far were centred on the nutritional compositions of edible insects that included the African palm weevil, *Rhychophorus phoenicis* (Elemo, 2011), Edible Stinkbug, *E. delegorguei* (Teffo et al., 2007) and a wide range of other insects (Defoliart, 1995; Banjo, Lawal, & Songonuga, 2006). Our study makes an effort to profile phytochemicals/bio-active compounds from this edible insect species and this could be the foundation of more refined experimental protocols for profiling these compounds in all edible insect species.

The traditional processing procedure which involved the use of warm water and subsequent drying by heating could realise a reduction in the quantities of phytochemicals/bio-active compounds due to chemical degradation. Similar findings on reduction of phytochemicals after aqueous extraction have been noted with plant extracts (Soetan & Oyewole, 2009).

Boiling and sun drying of insects have been shown to eliminate potentially harmful compounds such as neurotoxins and in some cases improve nutritional quality through partial inactivation of protease inhibitors (Akinnawo, Abatan, & Ketiku, 2002; Marickar & Paltabiraman, 1988). Our study showed similar results for most phytochemicals except for cyanogen gylcosides which have been shown to cause cyanogenesis (liberation of hydrogen cyanide leading to poisoning) if consumed by humans (Lechtenberg, 2011). The accentuation in quantities of cyanogen glycosides due to the traditional processing procedure could indicate a complex degradation and unbinding process of the major chemical constituents in *E. delegorguei* during heating. This observation from our study is quite interesting and contrary to traditional beliefs by consumers that assume that traditional processing assists in removing all harmful compounds from the insects.

Despite the potential threat from increased levels of cyanogen glycosides due to traditional processing, based on analyses of bio-active compound composition, *E. delegorguei* could still be a vital source of flavonoids. These compounds have been shown to have anti-allergic, anti-diarrheal, antiulcer, and anti-inflammatory agents and are considered essential nutrients (Bravo, 1998; Middleton, 2000). From a nutritional perspective, dietary intake of flavonoids has been estimated to range from 23 mg/day in the Netherlands, 28 mg/day in Denmark and 170 mg/day in the US (Cook & Samman, 1996; Leth & Justesen, 1998). Our study shows that the quantities detected even after traditional processing could be enough to meet the minimum acceptable daily intake by consumers. However, bio-availabilities of these flavonoids from *E. delegorguei* need to be determined.

A number of studies have revealed the physiological effects associated with alkaloids, flavonoids, cardiac glycosides, steroids, and triterpenoids. Saponins, oxalates and tannins for example at high levels are known to interfere with digestive processes and thus their presence in food is undesirable (Ijeh, Ejike, Nkwonta, & Njoku, 2010). Reduction in potentially harmful compounds of these groups due to the traditional processing procedure could therefore be beneficial to *E. delegorguei* consumers.

DPPH free radical scavenging activity displayed by *E. delegorguei* methanolic extracts is high, that is, within a range of 13% of activities of currently known scavenging bioorganic molecules that include ascorbic acid (vitamin C), butylated hydroxyanisole (BHA) and catechin (Sharma et al., 2011; Umaru, Adamu, Dahiru, & Nadro, 2007). However, it was noted from our study that subjecting the insects to heat processing reduces the DPPH free radical scavenging activity by 10%. It was also observed that the raw insect extracts expressed the same level of antioxidant activity compared to the known biological antioxidants (that include ascorbic acid) as well as the *Dodonae viscosa* var. *angustifolia* leaf extracts which is one of the insect host plant in South Africa as was recorded by Teffo et al. (2010). Naturally antioxidants have a fundamental physiological role in the human body by reducing tissue damaging free radicals (Tapiero, Tew, Nguyen, & Mathe, 2002). The provision of these possibly free radical quenching agents from the diet can be envisaged to greatly contribute towards the antioxidant preventive measures. In addition, they can also improve the digestion system; can function on reduction of coronary heart diseases, and some types of cancer and inflammation (Jayasri, Mathew, & Radha, 2009; Mattson & Cheng, 2006; Uddin, Akond, Mubassara, & Yesmin, 2008).

Aqueous extracts with antibacterial activity comprising of reducing sugars have been recorded in *A. digitata* (Yusha'u et al., 2010). Ethanol extracts had the least antibacterial inhibition effect followed by aqueous extracts. The poor antibacterial activities expression from ethanol solvents have been observed in several studies (Zaidan et al., 2005; Bhakuni, Dhar, Dhawan, & Mehrotra, 1969; Ahmad, Mehmood, & Mohammad, 1998). Ethanol can therefore be considered to be inappropriate chemical as a solvent for the efficient extraction of a number of phytochemicals from either plants or insects.

Methanol based extracts have been shown to have a broad spectrum of antibacterial activities although these were of plant origin: *Dodonae viscosa* (Teffo et al., 2010). Teffo and co-workers (2010) studied the antimicrobial activities of *E. delegorguei* and its plant food source *D. viscosa* var. *angustifolia*. Based on the bioautography technique, their results indicated antimicrobial activities of methanol extracts on *Staphylococcus aureus* and *Escherichia coli*. Our results using the disc diffusion method show that methanol extracts have some negative effects on some bacteria of clinical interest. These findings lay a foundation for subsequent work towards elucidating the individual components from *E. delegorguei* insects. Further studies are underway to determine minimum inhibitory concentrations and bactericidal concentrations using methanol extracts of whole insects.

5. Conclusions

Encosternum delegorguei has beneficial and harmful bio-active components which are rapidly degraded or extracted (except for cyanogen glycosides) from the insects through a traditional warm water and heating procedure in preparation for consumption. High levels of flavonoids could be beneficial to consumers. However, a potential trade-off resulting from elevated levels of cyanogen glycosides due to processing needs further investigation. The DPPH free radical scavenging activity displayed by *E. delegorguei* methanol extracts is very comparable to the currently known scavenging bioorganic molecules and could indicate a potential source of natural anti-oxidants that can be formulated into commercial products. However, it was noted that subjecting the insects to heat processing reduces the DPPH free radical scavenging activity by 10%. This could be less beneficial to consumers of this edible insect species.

Currently, further work is in progress to perform quantification and elucidation of chemical structures of phytochemicals from *E. delegorguei*. Research work orientated towards such depth will enable us to determine the discrete compounds that are in *E. delegorguei* using methanol as solvents and come up with a standardized chemical process to harness and concentrate the relevant phytochemicals for specific nutritional and medicinal applications.

References

Ahmad, I., Mehmood, Z., & Mohammad, F. (1998). Screening of some Indian medicinal plants for their antimicrobial properties. *Journal of Ethnopharmacology, 62*, 183-193. http://dx.doi.org/10.1016/S0378-8741(98)00055-5

Akinnawo, O. O., Abatan, M. O., & Ketiku, A. O. (2002). Toxicological study on the edible larva of *Cirina forda* (Westwood). *African Journal of Biomedical Research, 5*, 43-46.

Akinyemi, K. O., Oladapo, O., Okwara, C. E., Ibe, C. C., & Fasure, K. A. (2005). Screening of crude extracts of six medicinal plants used in South-West Nigerian unorthodox medicine for antimethicillin resistant *Staphylococcus aureus* activity. *BMC Complementary and Alternative Medicine, 5*, 6-15. http://dx.doi.org/10.1186/1472-6882-5-6

Amoo, A., & Agunbiade, F. O. (2010). Some nutrients and antinutrient components of Ptergota macrocarpa seed flour. *Journal of Environmental, Agricultural and Food Chemistry, 9*, 293-300.

Anwar, F., Qayyum, H. M. A., Hussain, A. I., & Iqbal, S. (2010). Antioxidant activity of 100% and 80% methanol extracts from barley seeds (*Hordeum vulgare* L.) stabilization of sunflower oil. *Grasas y Aceites, 61*(3), 237-243. http://dx.doi.org/10.3989/gya.087409

Banjo, A. D, Lawal, O. A., & Songonuga, E. A. (2006). The nutritional value of fourteen species of edible insects in southwestern Nigeria. *African Journal of Biotechnology, 5*, 298-301.

Bárt, I., Ala, A. T., Socha, R., Šimek, P., & Kodrík, D. (2010). Analysis of the lipids by adipokinetic hormones in the firebug *Pyrrhocoris apterus* (Hepteroptera: Pyrrhocoridae*). European Journal of Entomology, 107*, 509-520. http://dx.doi.org/10.14411/eje.2010.058

Bhakuni, D. S., Dhar, M. L., Dhawan, B. N., & Mehrotra, B. N. (1969). Screening of Indian plants for biological activity, part II. *Indian Journal of Experimental Biology, 7*, 250-262.

Bravo, L. B. (1998). Polyphenols: chemistry, dietary sources, metabolism, and nutritional significance. *Nutrition Reviews, 56*, 317-33. http://dx.doi.org/10.1111/j.1753-4887.1998.tb01670.x

Chavhunduka, D. M. (1975). Insects as a source of food to the African. *Rhodesian Science News, 9*, 217-220.

Cook, N. C., & Samman, S. (1996). Flavonoids-Chemistry, metabolism, cardioprotective effects, and dietary sources. *Nutritional Biochemistry, 7*, 66-76. http://dx.doi.org/10.1016/0955-2863(95)00168-9

De Britto, A. J, Gracelin, D. H. S, & Sebadastian, S. R. (2011). Antibacterial activity of a few medicinal plamts against *Xenthomonas campestris* and *Aeromonas hydrophilia*. *Journal of Biopestcides, 4*, 57-60.

Defoliart, G. R. (1995). Edible insects as mini-livestock. *Biodiversity and Conservation, 4*, 306-321. http://dx.doi.org/10.1007/BF00055976

Dzerefos, C. M., Witkowski, E. T. F., & Toms, R. (2009). Life-history traits of the edible stinkbug, *Encosternum delegorguei* (Hem., Tessaratomidae), a traditional food in southern Africa. *Journal of Applied Entomology, 133*, 749-759. http://dx.doi.org/10.1111/j.1439-0418.2009.01425.x

Ekhaise, F. O., Soroh, A. E., & Falodun, A. (2010). Antibacterial properties and preliminary phytochemmical analysis of methanolic extrcat of *Ocimum gratissium* (scent leaves). *Bayero Journal Pure Applied Sciences, 3*, 81-83.

Elemo, B. O., Elemo, G. N., Makinde, M., & Erukainure, O. L. (2011). Chemical Evaluation of African Palm Weevil, *Rhychophorus phoenicis*, Larvae as a food source. *Journal of Insect Science, 11*, 146-154. http://dx.doi.org/10.1673/031.011.14601

Gade, G., & Simek, P. (2010). A novel member of the adipokinetic peptide family in a "living fossil", the ice crawler *Galloisiana yuasai*, is the first identified nueropetide from the order Grylloblactodea. *Peptides, 31*, 372-376. http://dx.doi.org/10.1016/j.peptides.2009.10.016

Harbone, J. B. (1973). *Phytochemical methods: A guide to modern technique of plant analysis* (2nd ed.). London: Chapman A. & Hall.

Igbinosa, O. O., Igbinosa, E. O., & Aiyegoro, O. A. (2009). Antimicrobial activity and phytochemical screening of stem bark extracts from *Jatropha curcas* (Linn). *African Journal of Pharmacy and Pharmacology, 3*, 058-062.

Ijeh, I. I., Ejike, C. E., Nkwonta, O. M., & Njoku, K. B. (2010). Effect of traditional processing techniques on the nutritional and phytochemical Composition of African Bread-Fruit (*Treculia africana*) Seeds. *Journal of Applied Science and Environmental Management, 14*, 169-173.

Jayasri, M. A., Mathew, L., & Radha, A. (2009). A report on the anti-oxidant activities of leaves and rhizomes of *Costus pictus* D. Don. *International Journal of Integrative Biology, 5*, 20-26.

Jimo, F. O., Adedapo, A. A., Aliero, A. A., Koduru, S., & Afolayani, A. J. (2010). Evaluation of the polyphenolic, nutritive, biological activities of the acetone, methanol and water extracts of *Amaranthus asper*. *The Open Complementary Medicine Journal, 2*, 7-14. http://dx.doi.org/10.2174/1876391X01002010007

Kwashirai, C. V. (2007). Indigenous management of teak woodland in Zimbabwe, 1850-1900. *Journal of Historical Geography, 3*, 816-832. http://dx.doi.org/10.1016/j.jhg.2006.10.023

Lechtenberg, M. (2011). *Cyanogenesis in higher plants and animals*. Chichester: eLS. John Wiley & Sons Ltd.. http://dx.doi.org/10.1002/9780470015902.a0001921.pub2

Leth, T., & Justesen, U. (1998). Analysis of flavonoids in fruits, vegetables and beverages by HPCL-UV and LC-MS and estimation of the total daily flavonoid intake in Denmark. In R. Amado, H. Andersson, S. Bardocz, & F. Serra. (Eds.), *Polyphenols in Food Luxembourg* (pp. 39-40). Official Publications European Communication.

Mahesh, B., & Satish, S. (2008). Antimicrobial activity of some important medicinal plant against plant and human pathogens. *World Journal of Agricultural Sciences, 4*, 839-843.

Makkah, H. P. S. (2003). *Chemical and biological assays for quantification of major plant secondary metabolites*. Vienna, Australia: International Atomic Energy Agency.

Marco, H. G., Simek, P., Clark, D. K., & Gade, G. (2013). Novel adipokinetic hormones in the kissing bugs *Rhodnius prolixus, Triatoma infestans, Dipetalogaster maxima* and *Panastrongylus megistus. Peptides, 41*, 21-30. http://dx.doi.org/10.1016/j.peptides.2012.09.032

Marickar, Y., & Paltabiraman, T. N. (1988). Changes in protease inhibitory activity in plant seeds on heating. *Journal of Food Science and Technology, 25*, 59-62.

Mattson, M. P., & Cheng, A. (2006). Neurohormetic phytochemicals: low-dose toxins that induce adaptive neuronal stress responses. *Trends in Neurosciences, 29*, 632-639. http://dx.doi.org/10.1016/j.tins.2006.09.001

Mawere, M. (2012). Buried and forgotten but not dead': Reflections on '*Ubuntu*' in Environmental Conseravtion in Souther eastern Zimbabwe. *Global Journal of Human Social Science, 12*, 1-9.

Moraes, T. M., Rodrigues, C. M., Kushima, H., Bauab, T. M., Villegas, W., Pellizzon, C. H., ... Hiruma-Lima, C. A. (2008). *Hancornia speciosa*: indications of gastroprotective, healing and anti-Helicobacter pylori actions. *Journal of Ethnopharmacology, 120*, 161-168. http://dx.doi.org/10.1016/j.jep.2008.08.001

Olusesan, A. G., Ebele, O. C., Onwuegbuchulan, O. N., & Olorunmola, E. J. (2010). Prelimnary in vitro antibacterial activities of ethanolic extracts of *Fiscus sycomorus* and *Ficus platyphylla* Del. (Moraceae). *African Journal of Microbiology Research, 4*, 598-601.

Penarrieta, J. M., Alvaradoa, A., Bergenstahlc, B., & Akkesonb, B. (2007). Spectrophotometric methods for the measurement of total phenolic compounds and total flavonoids in foods. *Revista Boliviana de Quimica, 24*, 1-8.

Picker, M., Griffiths, C., & Weaving, A. (2004). *Field guide to insects of South Africa*. Struik Publishers, South Africa.

Prabuseenivasan, S., Jayakumar, M., & Ignacimuthu, S. (2006). *In vitro* antibacterial activity of some plant essential oils. *BMC Complementary and Alternative Medicine, 6,* 39-45. http://dx.doi.org/10.1186/1472-6882-6-39

Price, M. L., Scoyoc, S. V., & Butler, L. G. (1978). A critical evaluation of the vanillin reaction as an assay for tannin in sorghum grain. *Journal of Agriculture and Food Chemistry, 26,* 1214-1218. http://dx.doi.org/10.1021/jf60219a031

Sharma, P. V., Paliwal, R., & Sharma, S. (2011). Preliminary phytochemical screening and in vitro antioxidant potential of hydro-ethanolic extract of *Euphorbia neriifolia* Linn. *International Journal of Pharmtech Research, 56,* 200-207.

Silberbush, A., Markman, S., Lewinsohn, E., Bar, E., & Cohen, J. (2010). Predator-released hydrocarbons repel oviposition by a mosquito. *Ecology Letters, 13,* 1129-1138. http://dx.doi.org/10.1111/j.1461-0248.2010.01501.x

Soetan, K., & Oyewole, O. E. (2009). The need for adequate processing to reduce antinutritional factors in plants used as human foods and animal feeds: A Review. *African Journal of Food Science, 3,* 223-232.

Tapiero, H., Tew, K. D., Nguyen, B. G., & Mathe, G. (2002). Polyphenols: do they play a role in the prevention of human pathologies? *Biomedical and Pharmacology, 56,* 200-207. http://dx.doi.org/10.1016/S0753-3322(02)00178-6

Taylor, R. S. L., Manandhar, N. P, Hudson, J. B., & Towers, G. H. N. (1995). Screening of selected medicinal plants of Nepal for antimicrobial activities. *Journal of Ethnopharmacology, 546,* 153-159. http://dx.doi.org/10.1016/0378-8741(95)01242-6

Teffo, L. S., Aderogba, M. A., & Eloff, J. N. (2010). Antibacterial and antioxidant activities of four kaempferol methyl ethers isolated from *Dodonae viscosa* var. *angustifolia* leaf extracts. *South African Journal of Botany, 76,* 25-29. http://dx.doi.org/10.1016/j.sajb.2009.06.010

Teffo, L. S., Toms, R. B., & Eloff, J. N. (2007). Preliminary data on the nutritional composition of the edible stink-bug, *Encosternum delegorguei* Spinola, consumed in Limpopo province, South Africa. *South African Journal of Science, 103,* 434-436.

Uddin, S. N., Akond, M. A., Mubassara, S., & Yesmin, M. N. (2008). Antioxidant and antibacterial activities of *Trema cannabina. Middle-East Journal of Scientific Research, 3,* 105-108.

Umaru, H. A, Adamu, R., Dahiru, D., & Nadro, M. S. (2007). Levels of antinutritional factors in some wild edible fruits of Northern Nigeria. *African Journal of Biotechnology, 6,* 1935-1938.

Vaintraub, I. A., & Laptewa, N. A. (1988). Colorimetric determination of phytate in unpurified extracts of seeds and the products of their processing. *Analytical Biochemistry, 175,* 227-230. http://dx.doi.org/10.1016/0003-2697(88)90382-X

Yusha'u, M., Hamza, M. M., & Abdullahi, N. (2010). Antibacterial activity of *Adansonia digitata* stem bark extracts on some clinical bacterial isolates. *International Journal of Biomedical and Health Sciences, 6,* 129-135.

Zaidan, M. R. S., Noor-Rain, A., Badrul, A. R., Adlin, A., Norazah, A., & Zakaiah, I. (2005). *In vitro* screening of five local medicinal plants for antibacterial activity using disc diffusion method. *Tropical Biomedicine, 22,* 165-170.

Zaku, S. G., Abdulrahaman, F. A., Onyeyili, P. A., Aguzue, O. C., & Thomas, S. A. (2009). Phytochemical constituents and effects of aqueous root-bark extracts of *Ficus sycomorus* L. (Moracaea) on muscular relaxation, anaesthetic and sleeping time on laboratory animals. *African Journal of Biotechnology, 8,* 6004-6006.

4

Major Metabolic Diseases Affecting Cows in Transition Period

Leilson R. Bezerra[1], Cezario B. de Oliveira Neto[1], Marcos J. de Araújo[1], Ricardo L. Edvan[1],
Wagner D. C. de Oliveira[1] & Fabrício B. Pereira[1]

[1] School of Zootecnia, Federal University of Piauí, Bom Jesus, Piauí, Brazil

Correspondence: Leilson Rocha Bezerra, Campus Professora Cinobelina Elvas, University Federal of Piauí, BR 135, km03, Bairro Planalto Horizonte, Bom Jesus, Piauí State, Brazi.Email: leilson@ufpi.edu.br

Abstract

The aim of this study was to perform a literary review on the main metabolic diseases affecting cows during the transition period. The heat increment promoted by higher energy demand that occurs during the transition period between the end of lactation and early management, combined with low dry matter intake due to fetal growth major and consequent reduced ability of the rumen, make the dairy cow highly susceptible to the metabolic diseases ketosis, milk fever and hepatic lipidosis. The increase in blood concentrations of non-esterifies fatty acids during this transition period appear to be linked to the onset of these disorders and this can be explained by the high energy mobilization because of the negative energy balance. Diets with high energy density during this period are extremely necessary to minimize the effects of negative energy balance. In addition it is recommended to feed the animals with foods smaller particles because the capacity of the rumen is reduced. It also recommends the provision of lipid sources such as protected fat and vegetable oils that do not degrade the ruminal microorganisms and even precursors of glucose, such as propylene glycol or starchy concentrates, and not by lipid.

Keywords: diagnosis, fatty liver syndrome, ketosis, milk fever, postpartum, ruminant

1. Introduction

The success of the production cycle of a cow is determined by its production level, the postpartum recovery reproductive function and absence of pathology. Undoubtedly, the achievement of these objectives depends largely on the state of animal in its early days postpartum. So much so, that the level of production, the level of intake and blood parameters (NEFA, ketone bodies) in the first week postpartum are good indicators of the quality of initiating lactation (Grummer, Mashek, & Hayirli, 2004).

The energy balance is the result of the difference between the needs of the animal and food contributions. During last 2-4 weeks of gestation there is an increase substantial energy requirements due to fetal development and the needs of colostrum synthesis. This is accompanied by a decrease in the ingestion of materials dry. These two circumstances are often responsible for the development of a negative energy balance that initiates A few weeks before delivery. Cattle have the ability to compensate for deficits food energy through the mobilization of body fat. However, an excess mobilization of fat leads to disease and reproductive problems

The metabolic diseases or disorders of production are caused by an imbalance of nutrients that enter the animal organism (glicídeos, proteins, minerals, and water), your metabolism and graduates through feces, urine, milk and fetus. Nutritional imbalances affecting livestock are produced because the supply or use of foods is not meet nutritional requirements for maintenance, growth, production, reproduction (Martinez et al., 2014). When these imbalances are of short duration and are not too severe, the metabolism of the animal can compensate by using their body reserves. However, if the imbalance is severe or moderate but persistent animal body depletes its reserves and disease occurs (Wittwer, 2000). Unfortunately, most of these diseases has an effect difficult to perceive and act by limiting the production of the species of a persistently causing a decrease in the profitability of livestock enterprise.

The transition period consists of two phases, the first being formed by last three weeks before calving and the second by the first three weeks postpartum. This period is marked by changes, some of these are related to

Alterations Increases in energy requirements driven by both fetal needs and lactogenesis, endocrine and metabolic preparing cows for childbirth and lactation (Morgante et al., 2012; Piccione et al., 2012).

According to most recent National Animal Health Monitoring System for dairy cattle (National Animal Health Monitoring System, 2008), leading causes of morbidity in dairy cattle are clinical mastitis, lameness, infertility, retained placenta, milk fever, reproductive problems, and displaced abomasum. Of cows removed from herds, about 53% leave for one or more of the above reasons. Additionally, the rate of mortality of cows in U.S. dairy herds is nearly 6%, with 43% of these related to periparturient health issues, and likely a large portion of those classified as "unknown" (25%) occurring as a result of complications from the above. Overall, 16.2% of the cows that are permanently removed from a dairy herd are removed before 50 days in milk. These cows represent losses before the most profitable period of lactation. The relationship of the above disorders to excess prepartum body condition score (BCS) has been documented by numerous researchers and extensively reviewed (Bewley & Schultz, 2008). Briefly, cows with excessive body condition at calving, or excessive weight loss after calving, demonstrate overall decreased reproductive performance and increased likelihood of dystocia, retained placenta, metritis, milk fever, cystic ovaries, lameness, and mastitis as well as metabolic disorders, fatty liver, and ketosis.

2. Development

2.1 Ketosis and Fatty Liver Syndrome

Ketosis and Fatty Liver Syndrom is a metabolic disorder which often goes undiagnosed and leads to constricted performance and an impairment of general condition. It´s primarily occurs 2-7 weeks after calving (Gillund, Reksen, Grohn, & Karlberg, 2001) and occurs in consequence of negative energy balance, which will promote fat ana mobilizaçlão form of triacylglycerides (TGL). Before reaching the liver, TGL lose ester molecule in order to facilitate the conduction blood. Upon arriving at the liver TGL will be transformed into ketone bodies ketone bodies, acetoacetate (AcAc) and acetone (Ac) can be reduced to β-hydroxybutyrate (BHB) in an enzymatic reaction or decarboxylated to generate energy (Nielsen & Ingvartsen, 2004).

In early lactation, dairy cows high-yielding production suffer a variety of metabolic changes derived from energy deficiency, due to poor dry matter ingestion and high milk production, which are factors that predispose to ketosis (Chapinal et al., 2011) . A study to describe the prevalence of primary subclinical ketosis in New Zealand demonstrated age and calving interval are predisposing factors to the onset of ketosis (Compton, McDougall, Young & Bryan, 2014). Already Fatty Liver represents the major factor predisposing body condition score because fat and very fat cows have to have the disorder in the postpartum period. Furthermore, the Fatty liver may be a secondary complication to the cow to have a negative energy balance. Once developed fatty liver, and due to the low rate of export triglyceride in the form of lipoproteins, fatty liver persist for a long period. Reduction of fatty deposits in the liver is usually initiated when the cow enters positive energy balance, and end exhausted in the course of several weeks (Chapinal et al., 2012).

In the period between the end of late gestation and early lactation, there is a marked change from the dry matter intake by females. In cows reduction in consumption in this period is even more pronounced than for the heifers as shown in Figure 1. Over the first 30 days postpartum, when there is the peak of lactation, also observed consumption capacity suboptimal mobilization of body reserves for milk production, and often weight loss. The observed negative energy balance is due to insufficient nutrients necessary for lactation by dietary intake, because at this stage there is a priority of nutrients by the mammary gland (Leroy, Vanholde, Van Knegsel, Garcia-Ispierto & Bols, 2008) mediated by limiting the consumption capacity, causing a serious change in blood metabolites and hematological profile of animals (Piccione et al., 2012; Bezerra et al., 2013). With the advance of lactation (60 days postpartum), known as an intermediary period, milk production begins to decrease in the order of 2.5 % per week, while the voluntary intake increases gradually. Finally, at 90 days post partum, post- peak DM intake milk production continues to decrease gradually (NRC, 2008).

These responses, exaggerated by moderate under nutrition status in pregnant animals, are mediated by reduced tissue sensitivity and responsiveness to insulin, associated with decreased tissue expression of the insulin-responsive facilitative glucose transporter (GLUT4) (Bell & Bauman 1997). Peripheal tissue responses to insulin remain severely attenuated during early lactation but recover as the animal progresses through mid-lactation (Guesnet, Massoud & Demarne 1991, Bell & Bauman 1997, Sasaki 2002). Thus, during the period of NEB, key hormone expression and tissue responsiveness alter to increase lipolysis and decrease lipogenesis, causing high levels of non-esterified fatty acids (NEFA) and β-hydroxybutyrate (BHB) concentration which are indicative of lipid mobilization and fatty acid oxidation (Sakha, Ameri & Rohbakhsh, 2006, Wathes et al., 2009). Excessive fat mobilization can induce an imbalance in he patic carbohydrate and fat metabolism, which may result in ketosis (Goff & Horst, 1997).

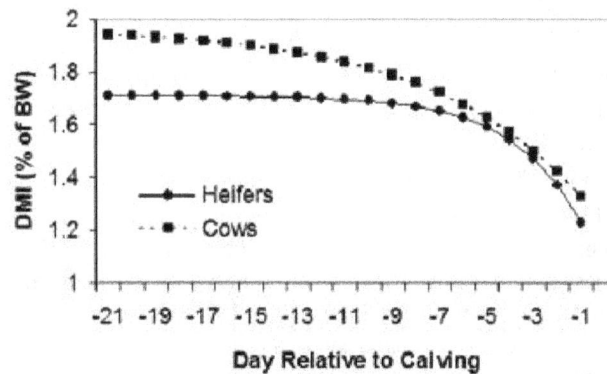

Figure 1. Dry matter intake comparative between cows and heifers during the transition period (Grummer, Mashek & Hayirli, 2004)

Fatty liver occurs when blood levels are elevated NEFA. This is dramatically raising birth. The liver's ability to grasp the NEFA is proportional to their concentration in the blood. The NEFA taken up by the liver are esterified to triglycerides or oxidized in mitochondria or peroxisomes (microsomes). TGL can be stored or exported as part of a low density lipoprotein. Compared with other species, the export capacity of liver triglycerides in ruminants is low, not knowing the cause. Under circumstances where an increase is produced in the liver uptake of NEFA (for example, when there are low levels of glucose and insulin in the blood), the esterification of fatty acids and accumulation of triglycerides in the liver occurs (Andresen, 2001). Complete oxidation of NEFA forms CO_2. Incomplete oxidation produces ketone bodies (especially acetoacetate and beta- hydroxy - butyrate. Formation of ketone bodies is also favored when blood levels of glucose and insulin are low, partly due to increased mobilization of fatty acids from tissue adipose. Low levels of insulin may increase the oxidation of fatty acids by reduction in malonyl -CoA concentration in hepatocytes and reduced sensitivity of the -1 carnitine palmitoiltrasferasa action of malonyl -CoA. Carnitine palmitoyltransferase - 1 is responsible for the translocation of fatty acids from the cytosol to the mitochondria for oxidation; action is inhibited by malonyl - CoA. Propionic acid is anti - ketogenic, probably due to its indirect effects in promoting insulin secretion and its direct effects on hepatic metabolism.

In the case of Fatty Liver Syndrome the deposition of TAG in liver is the consequence of mobilization of NEFA from adipose tissue exceeding capabilities of liver for oxidation and secretion of lipids (Gross, Schwarz, Eder, Van Dorland & Bruckmaier, 2013). Cows with fatty liver have greater adipose stores and mobilize more TAG, which leads to greater plasma NEFA concentrations, because adipose tissue from cows with fatty liver is less responsive to lipogenic substances and more responsive to lipolytic substances. Furthermore, cows with fatty liver have decreased fatty acid oxidation, hepatic apolipoprotein synthesis and lipid secretion, as indicated by decreased plasma apolipoprotein and lipid concentrations and decreased serum lecithin: cholesterol acyltransferase (LCAT) activity (Bobe, Young & Beitz, 2004). Besides disturbances in lipid metabolism, cows with fatty liver also have disturbances in glucose metabolism: Cows with fatty liver are either hyperinsulinemic-hyperglycemic or hypoinsulinemic-hypoglycemic (Holtenius, 1991), because either peripheral glucose uptake is decreased, indicating insulin resistance, or insulin and glucagon secretion and, therefore, hepatic gluconeogenesis are decreased. Furthermore, plasma amino acids are decreased. In summary, the availability of glucose, amino acids, and lipids for peripheral tissues is decreased in cows with fatty liver.

The diagnosis is made by clinical history, clinical signs and laboratory tests. The confirmation must be made by measurement of serum blood glucose (<50 mg/100 ml) b-hydroxybutyrate (> 14.4 mg / dl). The presence of ketones in the urine ("Ketostick") also assists in confirmation. The subclinical ketosis is characterized by a β-hydroxybutyrate (BHB) concentration in blood serum and milk (Figure 2; Herdt, Dart & Neuder, 2001). To generate this metabolic situation, an animal model was created. The model, based on group-specific interaction of dietary energy supply and body condition, is appropriate for testing the medical effectiveness of treating this kind of ketosis and its concomitants (Denis-Robichaud, Dubuc, Lefebvre & DesCôteaux, 2014; Campos, González, Coldebella & Lacerda, 2005). Research has been tested other forms of diagnosis as a magnetic resonance-based metabolomics. Sun et al. (2014) observed that plasma [1]H-nuclear magnetic resonance-based metabolomics, coupled with pattern recognition analytical methods, not only has the sensitivity and specificity to distinguish cows with clinical and subclinical ketosis from healthy controls, but also has the potential to be developed into a clinically useful diagnostic tool that could contribute to a further understanding of the disease mechanisms.

According Mahrt, Burfeind & Heuwieser (2014) to measure BHBA, blood samples of continuously fed dairy cows can be drawn at any time of the day. A single measurement provides very good test characteristics for on-farm conditions. Blood samples for BHBA measurement should be drawn from the jugular vein or tail vessels; the mammary vein should not be used for this purpose.

In search for accuracy of milk ketone bodies from flow-injection analysis for the diagnosis of hyperketonemia in dairy cows. Denis-Robichaud, Dubuc, Lefebvre, and DesCôteaux (2014) observed that accounted for milk BHBA and milk acetone values simultaneously had the highest accuracy of all tested models for predicting hyperketonemia. These results support that milk BHBA and milk acetone values from flow-injection analysis are accurate diagnostic tools for hyperketonemia in dairy cows and could potentially be used for herd-level hyperketonemia surveillance programs.

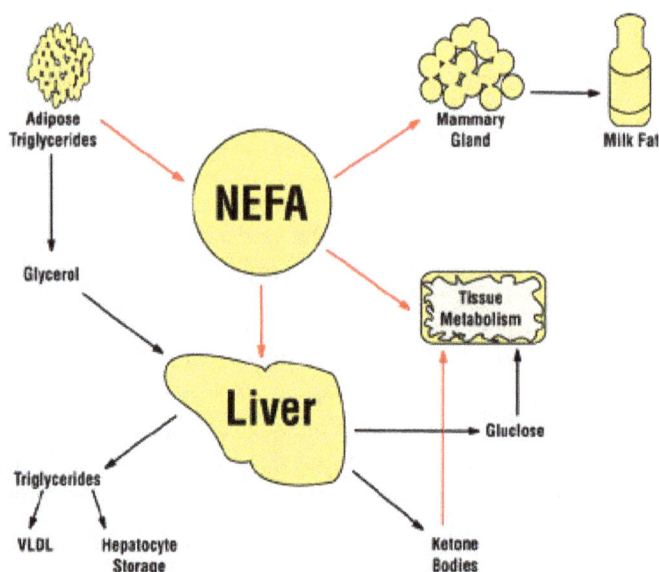

Figure 2: Non-esterfied fatty acids can be used as a sensitive indicator of energy balance in blood and milk. (Herdt, Dart & Neuder, 2001)

The evaluation of NEFA and BHB represents a strategy for the monitoring of subclinical ketosis and prepartum negative energy balance in dairy cows (Contreras, O'Boyle, Herdt, & Sordillo, 2010). The intravenous Glucose tolerance test (GTT) was useful o studying the physiological adaptation of animals to transition period since it produced a specific insulin response path (Morgante et al., 2012)

Treatment should be carried out with the 40% glucose, glucocorticoids, propylene glycol, sodium propionate and B12 (greater propionate production). Should avoid excessive weight loss at calving (management body condition score). The transition of diets should be performed with caution, always carefully to ensure gradual changes in the types of crops, gradual changes in the amounts of concentrates to supply adequate energy levels at different stages of production. Supplementation with niacin (6-12g / d) works best when forages and grains are supplied separately (largest fluctuation of glucose, insulin, NEFA and ketones in the blood) (Wittwer, 2000).

Recent research has demonstrated a reduction in lipoprotein assembly and secretion of TGL can promote deficits choline in non-ruminants. The addition of other methyl donors such as methionine serves to prevent the accumulation of lipids in the liver in mice, perhaps as substrates for the synthesis of choline. Currently there is considerable interest in the use of choline and related compounds to reduce fatty liver associated with the onset of labor. Rumen on the hill is a promise to modulate the metabolism of cows in transition period and reduce the incidence and severity of fatty liver at calving (Figure 3). The frequency response of a significant positive milk production of rumen-protected choline is observed in 50% of studies. Metabolic responses to rumen-protected choline have been mistaken. The predictable response to feeding rumen protected choline may depend on the basal diet, the supply of other B vitamins and related factors, and other management factors, including body condition score of cows entering the transition period (Donkin, 2011). Moreover, the formation of lipoproteins in the liver is not only in dependence of nutrients such as choline but a number of related factors to fat utilization by ruminants that are sensitive to these nutrients because of the deleterious effect that it promotes rumen. Thus, despite advances, little can be said about reducing the incidence of fatty liver by the use of this substance.

The prevention is determined that intervention with glucagon as a treatment/prevention of fatty liver is most effective within 14 days after parturition. The results demonstrated that subcutaneous injections of glucagon of 7.5 and 15 mg/d starting at 2 d postpartum are sufficient for fatty liver prevention; however, some cows developed fatty liver already at d 2 postpartum. Previous results confirm (Osman et al., 2008) showing that prenatally and subcutaneously injected glucagon will decrease markedly the accumulation of lipid in the liver of the post parturient dairy cow. Daily administration of the same amount (15 mg/day) of glucagon for several days prenatally in a limited number of cows was effective in preventing fatty liver during the early post parturient period.

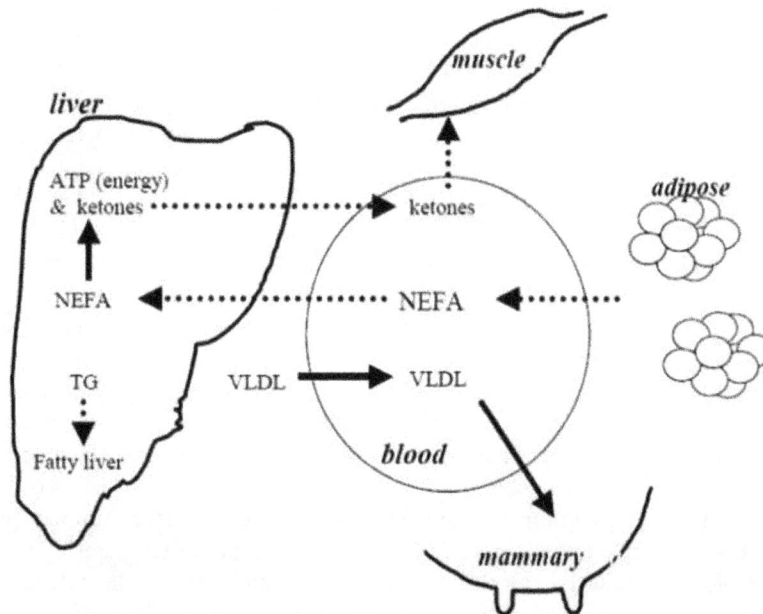

Figure 3. Fat mobilization occurred in cases of fatty liver syndrome or hepatic lipidosis and formation of fat globules in the mammary gland

2.2 Hypocalcemia Dairy Cow or Milk Fever

Hypocalcemia is particularly amenable to strategies tailored to individual cows or targeted groups of cows. First, a substantial proportion of cows are affected by hypocalcemia. The birth and early lactation are periods of too much stress for dairy cows due to the large metabolic challenges that occur in this period. During the last 2 weeks pre-calving, dairy cows are usually in negative energy balance and calcium and, in the last days before calving, the balance of other nutrients such as protein, vitamins and minerals may also be compromised. This reduction in serum calcium concentrations usually occurs about 12 to 24 hours after calving (Figure 4, Kimura, Reinhardt & Goff, 2006; Goff, 2008), where cases of hypocalcemia are more frequent and the most predisposing cows heifers.

Also known as milk fever or puerperal paresis, hypocalcemia is a metabolic-nutritional disease caused by the organism insufficiencies in maintaining serum calcium this period constant transformation in cows. It is a metabolic disease that affects cattle, especially animals with high milk production, which is the lowering of serum calcium and a subsequent number of other problems, such as decubitus and paresis, among others. The etiology is varied but the disease is associated with the delivery and early lactation, causing an exponential increase in needs for calcium (Ca). Calcium is a macromineral that has important functions in the body, among them are the bone matrix, the process of muscle contractor and transmission of nerve impulses. The level of calcium in plasma is well regulated, and when the level decreases, the parathyroid gland will excrete parathyroid hormone (PTH) (Oetzel & Miller, 2012). This increases the mobilization of calcium from the skeleton and also raises the renal threshold for calcium in the kidneys (Goff, Littledike & Horst, 1986). During the dry period, the supply of calcium through the diet is usually more than adequate to maintain homoeostasis without activating the calcium mobilization system (Ramberg, Mayer, Kronfeld, Phang & Berman, 1970), which is thus usually not activated until parturition. It is the phase most important in the development of milk fever (Kronqvist, Ferneborg, Emanuelson & Holtenius, 2014; DeGaris & Lean, 2008).

One of the major complications in the occurrence of hypocalcemia is that as the Ca is responsible for impulses transmission of nerve and muscle contractions, since the problem set and depending on its severity ie the severity of paresis may happen several other secondary disorders of order productive or reproductive due to this muscle inactivity, can cite among others retained placenta, metritis, ruminal acidosis, ketosis, due also to the reduction of dry matter intake and negative energy balance (Goff, 2008; Oetzel, 2011).

The cause is the deficiency of Ca in the early period of lactation. The concentration of plasma calcium is coordinated by action of calciotropic parathyroid hormone (PTH) and 1.25-dihydroxyvitamin D3 [1.25 (OH) 2D3] which are produced in response to hypocalcemia, acting to increase the intake of calcium in the pool plasma. Any decrease in serum calcium stimulates parathyroid gland to secrete PTH which, within minutes, increase renal calcium reabsorption from the glomerular filtrate. If the decreased plasma calcium is small, calcium returns to normal levels and PTH returned to their baseline levels. However, if the calcium drained plasma pool is in large quantities, the continued secretion of PTH stimulates resorption of calcium from bone For a long time it was considered that the tail of the disorder was given due to a failure to respond to release PTH. However, further studies demonstrated that these glands were able to respond to increased demand for calcium, even in cows with puerperal hypocalcemia. Currently it is known that the pathogenesis of the disease is much more associated with the action of PTH on cells responsible for demineralization (osteoclasts), cells of the intestine responsible for absorption and kidney cells responsible for the reabsorption of calcium in the tubules. Factors such as the world Production of milk, age and race are predisposing cows to have the disturbance, since cows for producing more secreted more calcium and should have efficient metabolism to meet increased demand. The great demand of calcium in early lactation to produce 10 liters of colostrum (colostrum formation begins 30 days before calving) cow loses 23 g of calcium in a single milking (2.3 g/L) which is about nine times more present in the plasma compartment. The calcium lost from the plasma compartment should be replaced by intestinal calcium absorbed and bone reabsorption. During the dry season these mechanisms are inactive, and all cows undergo hypocalcemia in the first days after birth until the intestines and bones are adapted. The adaptation starts with increased PTH and 1,25 - (OH) 2D at the beginning of hypocalcemia. About 24 hours of stimulation of 1,25 - (OH) 2D is required for intestinal calcium transport increase significantly. Bone reabsorption (osteoclast recruitment and activation) is not increased until 48 hours after the stimulation of PTH. In cows with parturient paresis this adaptation to these procedures may be extended. In these animals the decline of plasma and extracellular calcium determine the death of the animal before intestinal and bone adaptation (Radostits, Gay, Blood & Hinchcliff, 2002).

Hypocalcemia may or may not show clinical signs. Studies show that many cases of milk fever, animals do not externalize the clinical signs. The most severe hypocalcemia, said clinic (Ca <5 mg / dl), has very serious economic point of view because if not rapidly controlled may lead to loss of the affected animal; On the other hand, subclinical hypocalcemia assumes a more insidious role leading to loss of production and fertility. Subclinical hypocalcemia in which concentrations of calcium in the blood does not decline as severely affects about 50% of lactating dairy cows. If the animals are supplemented with minerals to reduce the risk of milk fever the hypocalcemic percentage of cows is reduced to about 15 to 25% (Oetzel, 2011).

Regarding the clinical signs of the disease in the stage I, the cow is not yet paresis. At this stage of puerperal hypocalcemia, hypersensitivity conductors nerves and muscles occurs, may cause excitement, muscle tremors, anorexia, ataxia, and general debility. The animal does not want to move, do not feed, but often body temperature is normal and may remain in this stage for hours. The stage II is the prodromica phase of the disease and is characterized by prostration and external decumbency. The tetania observed in the first phase is replaced by prolonged external decumbency with the animal unable to stand, displaying paresis, dry muzzle, cold extremities and temperature above normal (36.5 to 38 °C) . The artery pulse is weak, barely audible heart sounds and moderate heart frequency (to 80/min). It is observed absence of ruminal stasis and movements which can lead to a secondary bloat. The stage III is the most advanced, the animal enters and completes lateral decubitus sagging. Cardiac depression is severe and irregular and almost imperceptible pulse, breathing is shallow and diminished. Untreated animals die peacefully with shock in a state of complete collapse (Goff, 1999; Oetzel, 2011).

The diagnosis of milk fever is given based on the history of the animal at birth, age of dam and in the concentrations of calcium in the blood. The decrease in serum levels of magnesium and phosphorus may also be associated. Blood cell count may be some changes as eosinopenia, neutrophilia and lymphopenia suggestive of adrenocortical hyperactivity, but are nonspecific changes. Because of the symptoms is necessary to perform diagnosis differs in relation to hepatic steatosis, septic endometritis, mastitis and acute rumen acidosis. Calcium levels may appear below 5 mg/dl, but with less than 7 mg/dl levels, the animals have demonstrated clinical signs (Figure 4).

The treatment should be carried out as quickly as possible. Animals should be treated as soon as possible. The calcium treated by oral route is the best approach to hypocalcemia cows that are still standing, and the absorption into your bloodstream in about 30 minutes supplementation (Goff & Horst, 1993). The intravenous administration (IV) calcium is not recommended for the treatment of cows that are still standing (Oetzel, 2011), since this application if not done correctly can result in dead animal by cardiac complication. Cows treated with calcium IV often suffer a relapse hypocalcemic 12 to 18 hours later (Curtis, Cote, McLennan, Smart & Rowe, 1978; Thilsing – Hansen, Jørgensen, & Østergaard, 2002). For cows in stage II and III of milk fever should be treated immediately with a slow IV administration of 500 ml of a solution of calcium gluconate 23 %. This gives 10.8 g of elemental calcium, which is more than sufficient to correct the deficit whole cow's calcium (about 4 to 6 grams).

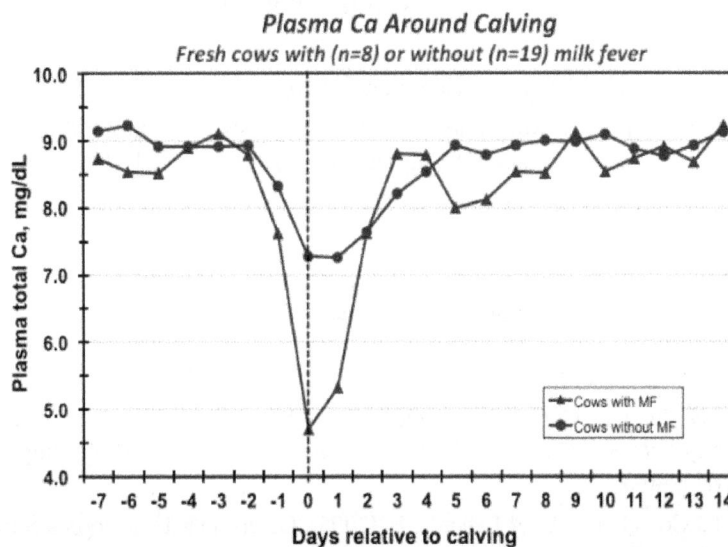

Plasma Ca Around Calving
Fresh cows with (n=8) or without (n=19) milk fever

Figure 4. Period of greatest clinical occurrence of milk fever in cows post calving (Adapted Kimura et al. 2006)

Prevention should be performed with the mineral at ease and even forced supplementation during the transition period. In addition to the findings of a larger number of micro minerals essential to functioning of the animal organism , studies have also addressed the relationship between cations and anions present in a certain diet aiding in the metabolic processes of the animal (e.g., acid-base balance) in a particular production phase, thus the anionic and cationic diets have been widely researched and used in animal production, mainly in feeding cows pre - calving, animal class often neglected because the producers are not producing milk and consequently does not contributed directly in net income from property (Oetzel, 2004). One of the main aims of using anionic diets in cows during transition period is control subclinical hypocalcemia, milk fever or puerperal paresis. Hypocalcemia is characterized by rapid depletion of blood calcium levels due to the large demand for calcium to the mammary gland in early lactation. The hormones responsible the absorption of calcium in the intestine so as bones, is low in activity due to small calcium requirement during the dry period. From the moment in which animal has one hypocalcemia, increase the incidence of other metabolic disorders such as mastitis, metritis, uterine prolapsed, retained placenta and ketosis, since calcium is a major responsible for muscle contraction and hence the mineral uterine atony and disposal the placenta.

Sakha, Mahmoudi & Nadalian (2014) in study to determine the effects of varying dietary cation-anion differences (DCAD) in prepartum period on milk fever, subclinical hypocalcemia and negative energy balance in dairy cows showed that use of anionic diets during three weeks before calving can protect dairy cows from clinical and subclinical hypocalcemia by increasing the calcium level in serum. To reduce the postpartum negative energy balance, replacement of anionic diet by cationic ions soon after calving is suggested (Sakha, Mahmoudi, & Nadalian, 2014).

3. Conclusions

Metabolic diseases are of great economic impact; it usually affects the animals about to reach their maximum potential production. Food consumption cannot be harmed in the coming days to calving and early lactation, since this is a critical period in the nutrition of females. Is any factor that restricts food intake at this stage (such as milk fever or ketosis) increases the metabolism of body fat, when the animal has in order to obtain energy, with consequent accumulation of fat in the liver immobilized, directly affecting the deficit power of females.

It is accepted that reproduction is important for the profitability of dairy farms, and nutritional status and metabolic health are both associated with successful reproduction. Cows that experience periparturient problems have delayed return to ovulation, lower pregnancy per insemination, and increased pregnancy loss. Therefore, implementing nutritional and health programs that reduce the risk of metabolic disturbances are expected to not only improve cow health, but also enhance fertility. Low nutrient intake coupled with high energy demand during the transition period will increase the risk of occurrence of metabolic disorders. Strategies to manipulate peripartum metabolic health involve dietary formulation to minimize the degree and extent of negative nutrient balance, improve Ca homeostasis, and minimize the severity of negative energy balance.

References

Andresen, S. H. (2001). Vacas secas y en Transición. *Rev. investig. vet. Perú, 12*(2), 38-46.

Bell, A. W., & Bauman, D. E. (1997). Adaptations of Glucose Metabolism During Pregnancy and Lactation. *J Mammary Gland Biol Neoplasia, 2*, 265-278. http://dx.doi.org/10.1023/A:1026336505343

Bewley, J. M., & Schutz, M. M. (2008). Review: An interdisciplinary review of body condition scoring for dairy cattle. *Prof. Anim. Sci, 24*, 507-529.

Bezerra, L. R., Torreão, J. N. C., Marques, C. A. T., Machado, L. P., Araújo, M. J., & Veiga A. M. S. (2013). Influence of concentrate supplementation and the animal category in the hemogram of Morada Nova sheep. *Arquivo Brasileiro de Medicina Veterinária Zootecnia, 65*(6), 1738-1744. http://dx.doi.org/10.1590/S0102-09352013000600022

Bobe, G., Young, J. W., & Beitz, D. C. (2004). Invited review: Pathology, etiology, prevention, and treatment of fatty liver in dairy cows. *Journal of Dairy Science, 87*, 3105-3124. http://dx.doi.org/10.3168/jds.S0022-0302(04)73446-3

Campos, R., González, F., Coldebella, A., & Lacerda, L. (2005). Determinação de corpos cetônicos na urina como ferramenta para o diagnóstico rápido de cetose subclínica bovina e relação com a composição do leite. *Archives of Veterinary Science, 10*(2), 49-54.

Chapinal, N., Carson, M. E., LeBlanc, S. J., Leslie, K. E., Godden, S., Capel, M., ... Duffield, T. F. (2012). The association of serum metabolites in the transition period with milk production and early lactation reproductive performance. *Journal of Dairy Science, 95*, 1301-1309. http://dx.doi.org/10.3168/jds.2011-4724

Chapinal, N., Carson, M., Duffield, T. F., Capel, M., Godden, S., Overton, M., ... LeBlanc, S. J. (2011). The association of serum metabolites with clinical disease during the transition period. *Journal of Dairy Science, 94*, 4897-4903. http://dx.doi.org/10.3168/jds.2010-4075

Compton, C. W. R., McDougall, S., Young, L., & Bryan, M. A. (2014). Prevalence of subclinical ketosis in mainly pasture-grazed dairy cows in New Zealand in early lactation. *New Zealand Veterinary Journal, 62*(1), http://dx.doi.org/10.1080/00480169.2013.823829

Contreras, G. A., O'Boyle, N. J., Herdt, T. H., & Sordillo, L. M. (2010). Lipomobilization in periparturient dairy cows influences the composition of plasma nonesterified fatty acids and leukocyte phospholipid fatty acids. *Journal of Dairy Science, 93*, 2508-2516. http://dx.doi.org/10.3168/jds.2009-2876

Curtis, R. A., Cote, J. F., McLennan, M. C., Smart, J. F., & Rowe, R. C. (1978). Relationship of methods of treatment to relapse rate and serum levels of calcium and phosphorous in parturient hypocalcaemia. *Canadian Veterinary Journal, 19*, 155-158.

DeGaris, P. J., & Lean, I. J. (2008). Milk fever in dairy cows: A review of pathophysiology and control principles. *Veterinary Journal, 176*, 58-69. http://dx.doi.org/10.1016/j.tvjl.2007.12.029

Denis-Robichaud, J., Dubuc, J., Lefebvre, D., & DesCôteaux, L. (2014) Accuracy of milk ketone bodies from flow-injection analysis for the diagnosis of hyperketonemia in dairy cows. *Journal of Dairy Science.* http://dx.doi.org/10.3168/jds.2013-6744

Donkin, S. S. (2011). Extension Foundation. America's Research-based Learning Network. *Rumen-Protected Cholin.* May 2011. Retrieved 14 april, 2014, from http://www.extension.org:80/pages/26158/rumen-protected-choline

Gillund, P., Reksen O., Grohn Y. T., & Karlberg, K. (2001). Body condition related to ketosis and reproductive performance in Norwegian dairy cows. *Journal of Dairy Science, 84,* 1390-1396. http://dx.doi.org/10.3168/jds.S0022-0302(01)70170-1

Goff, J. P. (1999). Treatment of calcium, phosphorus, and magnesium balance disorders. *Veterinary. Clin. North Am. Food Animal. Pract., 15,* 619-639.

Goff, J. P. (2008). The monitoring, prevention, and treatment of milk fever and subclinical hypocalcemia in dairy cows. *Veterinary Journal, 176,* 50-57. http://dx.doi.org/10.1016/j.tvjl.2007.12.020

Goff, J. P., & Horst, R. L. (1993). Oral administration of calcium salts for treatment of hypocalcemia in cattle. *Journal of Dairy Science, 76,* 101-108. http://dx.doi.org/10.3168/jds.S0022-0302(93)77328-2

Goff, J. P., & Horst, R. L. (1997). Physiological changes at parturition and their relationship to metabolic disorders. *Journal of Dairy Science, 80,* 1260-1268. http://dx.doi.org/10.3168/jds.S0022-0302(97)76055-7

Goff, J. P., Littledike, E. T., & Horst, R. L. (1986). Effect of synthetic bovine parathyroid hormone in dairy cows: prevention of hypocalcemic parturient paresis. *Journal of Dairy Science, 69,* 2278-2289. http://dx.doi.org/10.3168/jds.S0022-0302(86)80666-X

Gross, J. J., Schwarz, F. J., Eder, K., van Dorland, H. A., & Bruckmaier, R. M. (2013). Liver fat content and lipid metabolism in dairy cows during early lactation and during a mid-lactation feed restriction. *Journal of Dairy Science, 96,* 5008-5017. http://dx.doi.org/10.3168/jds.2012-6245

Grummer, R. R., Mashek, D. G., & Hayirli, A. (2004). Dry matter intake and energy balance in the transition period. *Vet Clin North Am Food, Anim Pract, 20,* 447-470.

Guesnet, P. M., Massoud, M. J., & Demarne, Y. (1991). Regulation of adipose tissue metabolism during pregnancy and lactation in the ewe: the role of insulin. *Journal Animal Science, 69,* 2057-2065.

Herdt, T. H., Dart, B., & Neuder, L. (2001). Will large dairy herds lead to the revival of metabolic profile testing? *Proc Am Assoc Bov Pract, 34,* 27-34.

Holtenius, P. (1991). Disturbances in the regulation of energy metabolism around parturition in cows. *Mh. Vet.-Med, 46,* 795-797.

Kimura, K., Reinhardt, T. A., & Goff, J. P. (2006). Parturition and hypocalcemia blunts calcium signals in immune cells of dairy cattle. *Journal of Dairy Science, 89,* 2588-2595. http://dx.doi.org/10.3168/jds.S0022-0302(06)72335-9

Kronqvist, C., Ferneborg, S., Emanuelson, U., & Holtenius, K. (2014). Effects of pre-partum milking of dairy cows on calcium metabolism at start of milking and at calving. *Journal of Animal Physiology and Animal Nutrition, 98*(1), 191-196. http://dx.doi.org/10.1111/jpn.12038

Leroy, J. L. M. R., Vanholder, T., Van Knegsel, A.T. M., Garcia-Ispierto, I., & Bols, P. E. J. (2008). Nutrient Prioritization in Dairy Cows Early Postpartum: Mismatch Between Metabolism and Fertility? *Reprod Domest Anim, 43*(Suppl.), 96-103. http://dx.doi.org/10.1111/j.1439-0531.2008.01148.x

Mahrt, A., Burfeind, O., & Heuwieser, W. (2014). Effects of time and sampling location on concentrations of β-hydroxybutyric acid in dairy cows. *Journal of Dairy Science, 97*(1), 291-298. http://dx.doi.org/10.3168/jds.2013-7099

Martinez, N., Sinedino, L. D. P., Bisinotto, R. S., Ribeiro, E. S., Gomes, G. C., Lima, F. S., ... Santos J. E. P. (2014). Effect of induced subclinical hypocalcemia on physiological responses and neutrophil function in dairy cows. *Journal of Dairy Science, 97*(2), 874–887. http://dx.doi.org/10.3168/jds.2013-7408

Morgante, M., Gianesella, M., Casella, S., Stelletta, C., Cannizzo, C., Giudice, E., & Piccione, G. (2012). Response to glucose infusion in pregnant and nonpregnant ewes: changes in plasma glucose and insulin concentrations. *Comp Clin Pathol, 21,* 961-965. http://dx.doi.org/10.1007/s00580-011-1208-5

National Animal Health Monitoring System. (2008). *Dairy 2007 Part I: Reference of Dairy Cattle Health and Management Practices in the United States.* 2007. USDA

Nielsen, N. I., & Ingvartsen, K. L. (2004). Propylene glycol for dairy cows: A review of the metabolism of propylene glycol and its effects on physiological parameters, feed intake, milk production and risk of ketosis. *Anim Feed Sci Technol, 115*, 191-213. http://dx.doi.org/10.1016/j.anifeedsci.2004.03.008

Oetzel, G. R. (2004). Monitoring and testing dairy herds for metabolic disease. *Veterinary Clin. North Am. Food Animal. Pract, 20*, 651-674.

Oetzel, G. R. (2011). Non-infectious diseases: Milk fever. In J. W. Fuquay & P. L. H. McSweeney (Eds.), *Encyclopedia of Dairy Sciences* (Vol. 2, pp. 239-245). San Diego: Academic Press.

Oetzel, G. R., & Miller, B. E. (2012). Effect of oral calcium bolus supplementation on early lactation health and milk yield in commercial dairy herds. *Journal of Dairy Science, 95*, 7051-7065. http://dx.doi.org/10.3168/jds.2012-5510

Osman, M. A., Allen, P. S., Mehyar, N. A., Bobe, G., Coetzee, J. F., Koehler, K. J., & Beitz, D. C. (2008). Acute metabolic responses of postpartal dairy cows to subcutaneous glucagon injections, oral glucagon, or both. *Journal of Dairy Science, 91*, 3311-3322. http://dx.doi.org/10.3168/jds.2008-0997

Piccione, G., Messina, V., Marafioti, S., Casella, S., Giannetto, C., & Fazio, F. (2012). Changes of some haematochemical parameters in dairy cows during late gestation, post partum, lactation and dry periods. *Vet Med Zoot, 58*, 59-64.

Radostits, O. M., Gay, C. C., Blood, D. C., & Hinchcliff, K. W. (2002). *Clínica veterinária. Um tratado de doenças dos bovinos, ovinos, suínos, caprinos e eqüinos* (Vol. 9, pp. 1737). Rio de Janeiro: Guanabara Koogan.

Ramberg, C. F., Mayer, G. P., Kronfeld, D. S., Phang, J. M., & Berman, M. (1970) Calcium kinetics in cows during late pregnancy, parturition, and early lactation. *The American Journal of Physiology, 219*, 1166-1177.

Sakha, M., Ameri, M., & Rohbakhsh, A. (2006). Changes in blood β-hydroxybutyrate and glucose concentrations during dry and lactation periods in Iranian Holstein cows. *Comp Clin Pathol, 15*, 221-226. http://dx.doi.org/10.1007/s00580-006-0650-2

Sakha, M., Mahmoudi, M., & Nadalian, M. G. (2014). Effects of dietary cation-anion difference on milk fever, subclinical hypocalcemia and negative energy balance in transition dairy cows. *Journal Research Opinions in Animal and Veterinary Sciences, 4*(2), 69-73.

Sasaki, S. (2002). Mechanism of insulin action on glucose metabolism in ruminants. *Anim Sci J, 73*, 423-433. http://dx.doi.org/10.1046/j.1344-3941.2002.00059.x

Sun, L. W., Zhang, H. Y., Wu, L., Shu, S., Xia, C., Xu, C., & Zheng, J.S. (2014). H-Nuclear magnetic resonance-based plasma metabolic profiling of dairy cows with clinical and subclinical ketosis. *Journal of Dairy Science, 97*(3), 1552-1562. http://dx.doi.org/10.3168/jds.2013-6757

Thilsing-Hansen, T., Jørgensen R. J., & Østergaard, S. (2002). Milk fever control principles: areview. *Acta. Vet. Scand, 43*, 1-19.

Wathes, D. C., Cheng, Z., Chowdhury, W., Fenwick, M. A., Fitzpatrick, R., Morris, D. G., ... Murphy, J. J. (2009). Negative energy balance alters global gene expression and immune responses in the uterus of postpartum dairy cows. *Physiol Genomics, 39*, 1-13. http://dx.doi.org/10.1152/physiolgenomics.00064.2009

Wittwer, F. (2000). Diagnóstico dos desequilíbrios metabólicos de energia em rebanhos bovinos. In F. H. D. González, J. O. Barcellos, H. Ospina, & L. A. O. Ribeiro (Eds.), *Perfil metabólico em ruminantes: seu uso em nutrição e doenças nuricionais*. Porto Alegre, Brasil, Gráfica da Universidade Federal do Rio Grande do Sul.

5

Effects of Delayed First Feeding on Larval Growth and Survival of Yesso Scallop (*Patinopecten yessoensis*)

Zhongqiang Cai[1], Xiujun Sun[2] & Aiguo Yang[2]

[1] Changdao Enhancement and Experiment Station, Chinese Academy of Fishery Sciences, Changdao, China

[2] Key Laboratory of Sustainable Development of Marine Fisheries, Ministry of Agriculture, Yellow Sea Fisheries Research Institute, Chinese Academy of Fishery Sciences, Qingdao 266071, China

Correspondence: Xiujun Sun, Key Laboratory of Sustainable Development of Marine Fisheries, Ministry of Agriculture, Yellow Sea Fisheries Research Institute, Chinese Academy of Fishery Sciences, Qingdao, China. E-mail: xjsun@ysfri.ac.cn

Abstract

The time of first feeding is an important factor for establishing successful initial feeding in molluscan hatcheries. The effects of delayed first feeding on larval growth and survival in the larvae of Yesso Scallop *Patinopecten yessoensis* were investigated in this study. Groups of larvae were fed at early stage of D-shaped larvae or delayed for 24, 48, 72, 96, 120 and 144 h. When first feeding was delayed for less than 72 h, the *P. yessoensis* larvae grew as rapidly as those non-starved larvae. When delayed first feeding for more than 72 h, the growth rates of larvae were significantly reduced (less than 2 μm day^{-1}), half or less than those of the non-starved group. The survival rates decreased sharply with the prolonged starvation period and the extremely low survival rates were observed in the groups that were delayed first feeding for more than 72 h. The results indicate that the prolonged starvation had a deleterious impact on the larval growth and survival when delayed first feeding for 96 h or more. To avoid mass mortality and obtain adequate growth and survival, food availability within 72 h after early D-shaped larvae is critical important for the successful initial feeding in commercial culture of this species.

Keywords: *Patinopecten yessoensis*, delayed first feeding, growth, survival rate

1. Introduction

For marine invertebrates, factors influencing larval survival are predation, starvation, and oceanographic conditions that may transport larvae into unfavorable environments by advection (Sewell et al., 2004). During planktonic period, larvae acquire their food from seawater to get energy for growth and development, and starvation has long been considered to be an important source of mortality of marine invertebrate larvae (Thorson, 1950; Manahan, 1990; Wehrtmann, 1991). Newly hatched planktotrophic larvae have very little nutrient reserves, so the availability of food in the first few days after hatching is critical for their survival (Pawlik, 1992). The amount of phytoplankton in the ocean is rarely present at concentrations necessary to sustain maximum larval growth, which indicates planktonic larvae have the potential to be food-limited in nature and experience high mortality when encountering adverse feeding conditions (Olson & Olson, 1989; Fenaux et al., 1994). In order to better understand the effects of starvation or food limitation on larval growth, survival and development in marine invertebrates, laboratory experiments were widely performed in a variety of species (His & Seaman, 1992; Pedrotti & Fenaux, 1993; Fenaux et al., 1994; McEdward & Qian, 2001; Moran & Manahan, 2004; Yang et al., 2008).

The Yesso scallop *Patinopecten yessoensis* is a cold water bivalve and naturally distributes along the coastline of the northern islands of Japan, northern Korean Peninsula, and Russian Primorye, Sakhalin and Kurile islands (Ito, 1991). Because *P. yessoensis* is larger in size and has higher market values than the native Zhikong scallop *Chlamys farreri*, it has become one of the most important maricultural shellfish in northern China since it was introduced in 1982. Artificial seed production is critically important for the aquaculture industry for the reason that *P. yessoensis* seeds are produced exclusively in hatcheries in China. In commercial shellfish hatcheries, cultured live microalgae are mainly used as the diets for bivalve larvae. However, microalgal production is time consuming and limited by culture conditions and apparatus, and the culture of microalgae can be subject to various types of contamination, resulting in sudden mortality or unsuitability as a food source (Robert & Trintignac, 1997). For that reason, the supply of microalgae in hatcheries is often insufficient to meet the requirements of a large-scale, commercial shellfish breeding. Larvae are potentially at risk of starvation especially when several batches of larvae are reared in hatcheries. Therefore, a greater understanding of the effects of starvation on larvae will assist in the development of adequate feeding management for batches of larvae in a food-limited situation.

Previous studies on *P. yessoensis* have investigated in the aspects of artificial seed production and culture (Zhang et al., 2000), polyploidy induction (Yang et al., 2001), interspecific hybridization (Yang et al., 2004), and population genetics (Li et al., 2007). However, little information is available at present to understand the effects of starvation on the scallop larvae. We therefore investigated the effects of delayed first feeding on growth and survival of *P. yessoensis* larvae, in order to better understand starvation tolerance of this species, and provide information for the successful establishment of exogenous feeding schedules for hatcheries.

2. Materials and Methods

2.1 Larval Culture

Adults of Yesso Scallop *P. yessoensis* were collected from the coast of Zhangzidao Island, Liaoning Province, China. Animals were maintained at 6-8 °C in 10m³ concrete tanks by changing seawater daily and fed *Nitzschia closterium* f. minutissima and *Spirulina Platensis* for about 30 days to speed up their gonad development. Before spawning, the males were removed from the culture tank and transferred to 100-L containers. Broodstock scallops were dried for 1-2 h, and then moved to higher temperature sand-filtered seawater (10 °C) to induce spawning. After the release of eggs, sperm were added to the spawning tank to assure that the sperm/egg ratio was about 5:1. During the incubation, continuous aeration was accomplished by air flow through a series of air stones distributed evenly in the bottom of the tank. When the zygotes developed to early D-shaped larvae (about 72 h after fertilization) at which time they were ready to feed on algae, the larvae were siphoned and isolated on 60-μm sieve for counting and used in starvation experiments.

2.2 Starvation Experiments

Eight different levels of starvation were tested (larvae were starved for 0, 24, 48, 72, 96, 120, 144 h, and unfed group). Three replicates were used for each treatment. In 0 h group, larvae were fed immediately after the occurrence of early D-shaped larvae. Early D-shaped larvae were stocked at a uniform density of 6 nos. ml⁻¹ in plastic circular aquarium barrels containing 3-L sand-filtered seawater. The experiment was conducted at a temperature of 10 ± 2 °C under an ambient photoperiod regime of 14 h of light and 10 h of dark. Culture water was exchanged for 1/2 twice a day. The algal diet of *Isochrysis galbana* was provided at an initial density of 5 000 cells ml⁻¹ day⁻¹, progressively increased to 20 000 cells ml⁻¹ day⁻¹, and *Platymonas subcordiformis* at a density of 2 000 cells ml⁻¹ day⁻¹ were used for the late stage of larval development.

2.3 Larval Growth of Shell Length and Shell Width, Larval Survival

The shell length and shell width of larvae in different treatments were measured using a microscope with a pre-calibrated micrometer at 100× magnification. Larval measurements were made on days 5, 8, 12, 16, 23 after early D-shaped larvae. The growth rates (n = 24 larvae/treatment) were calculated by regressing direct measurements of larval length and width over the culture days. To determine the survival rate of larvae, three 20-ml or 40-ml subsamples were randomly taken from each replicate, and the numbers of larvae were counted using a microscope on days 6, 9, 13, 17, and 24.

2.4 Statistical Analysis

One-way ANOVA was performed to test the effects of starvation on shell length and width of larvae in different treatments. Post-hoc tests (Tukey HSD) were performed in order to test the differences among groups. The significant levels for the growth and survival rates among groups were also determined by One-way ANOVA followed by Tukey HSD test. All statistical analyzes were conducted with the SPSS 17.0 program. Statements of significant differences were based on accepting $P < 0.05$.

3. Results

3.1 Effects of Delayed First Feeding on the Growth of Shell Length

The growth of larval shell length in different treatments during culture days are shown in Figure 2. For all treatment groups, the obvious increases in shell length were observed from day 0 to day 5. On day 5, the larvae in the different treatment groups had the same value of shell length (120.98 ± 1.62 μm; $F = 1.218$, df = 7, 12, $P > 0.05$). In contrast, the significant differences in shell length were observed among groups on day 8 ($F = 14.46$, df = 7, 11, $P < 0.01$). The relatively higher values in shell length were observed in 0 and 24 h groups, with the maximum value of 127.60 ± 0.63 μm in 24 h group, which are significantly higher than those of 48, 72, 96, 120 h and unfed groups (Tukey's HSD test, $P < 0.05$). The significant differences in shell length among groups were also observed on day 12 ($F = 6.40$, df = 7, 10, $P < 0.01$), while no significant difference were detected on day 16 ($F = 2.51$, df = 7, 11, $P > 0.05$). On day 12, the maximum value of shell length was also observed in 24 h group (145.98 ± 0.66 μm), which was significantly higher than that in 96 h, 120 h, and unfed groups. From day 16 to day 23, the remarkable growth in shell length were observed in 0, 24, 48, 72 h groups, but no survived larvae were detected in the groups that were starved for 96 h or more on day 23.

After feeding, larvae that had been starved for 24 h had mean growth rates of 3.71 ± 0.06 μm day^{-1} by day 12, growing as rapidly as larvae in the 0, 48 and 72 h group, with the growth rates of 2.56 ± 0.18 μm day^{-1}, 2.69 ± 0.18 μm day^{-1}, and 2.39 ± 0.01 μm day^{-1}, respectively (Table 1). The growth rates for 96, 120 h and unfed groups were less than 2 μm day^{-1}, which were significantly affected by delayed first feeding. From day 12 to day 23, the growth rates of larvae in the groups that were delayed first feeding for less than 72 h were significantly higher than those of groups that were fed after periods of starvation for 96 h and more (Tukey's HSD test, $P < 0.05$).

Table 1. Mean growth rates of shell length and shell width in different treatments (0, 24, 48, 72, 96, 120, 144 h and unfed) during culture days. Values with different letters are significantly different from each other ($P < 0.05$)

Growth rate	Shell length (μm day^{-1})		Shell width (μm day^{-1})	
	day0-12	day12-23	day0-12	day12-23
0 h	2.56 ± 0.18^{ab}	5.04 ± 0.84^{a}	2.10 ± 0.07^{ab}	4.74 ± 0.85^{a}
24 h	3.71 ± 0.06^{a}	2.55 ± 0.20^{ab}	2.85 ± 0.23^{a}	2.07 ± 0.93^{ab}
48 h	2.69 ± 0.18^{ab}	3.50 ± 0.06^{a}	2.52 ± 0.15^{a}	2.95 ± 0.84^{ab}
72 h	2.39 ± 0.01^{ab}	4.25 ± 0.03^{a}	1.84 ± 0.25^{bc}	4.08 ± 0.13^{ab}
96 h	1.66 ± 0.20^{b}	0.30 ± 0.00^{b}	1.13 ± 0.01^{c}	1.80 ± 0.03^{ab}
120 h	1.91 ± 0.09^{b}	0.52 ± 0.20^{b}	1.58 ± 0.00^{bc}	0.54 ± 0.00^{b}
144 h	2.38 ± 0.88^{ab}	0.39 ± 0.53^{b}	1.57 ± 0.35^{bc}	2.63 ± 1.02^{ab}
unfed	1.23 ± 0.33^{b}	1.02 ± 0.03^{b}	1.84 ± 0.18^{bc}	-

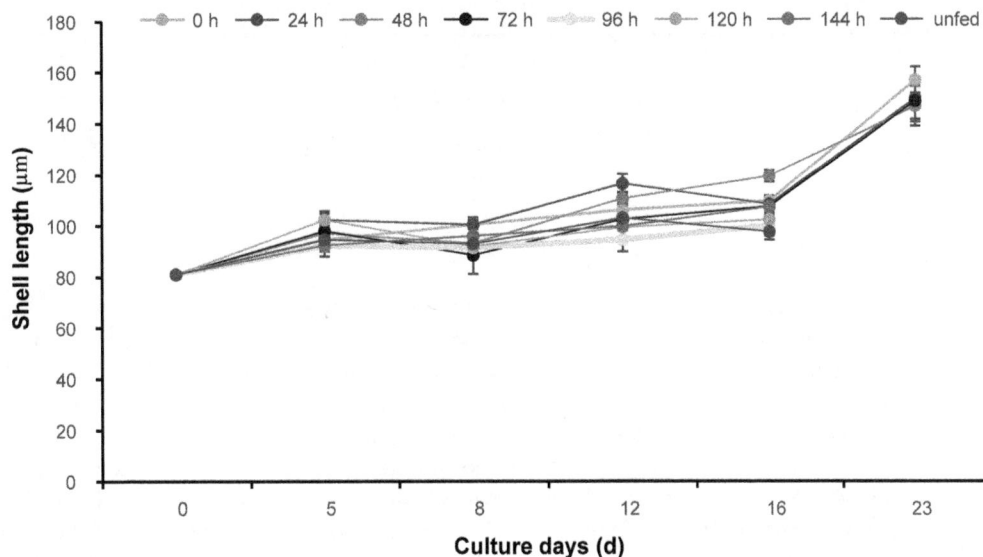

Figure 1. Mean shell length of *Patinopecten yessoensis* in different groups (delayed first feeding for 0, 24, 48, 72, 96, 120, and 144 h, and unfed group) during culture days. All error bars represent standard error of the mean

3.2 Effects of Delayed First Feeding on the Growth of Shell Width

The growth of larval shell width in different treatments during culture days are shown in Figure 2. On day 5 and day 8, there was no significant difference in shell width among groups ($F = 0.63$, df = 7, 12, $P > 0.05$; $F = 1.63$, df = 7, 11, $P > 0.05$), while the significant differences were detected on day 12 ($F = 8.04$, df = 7, 10, $P < 0.01$). On day 12, the maximum shell length was observed in 24 h group, 115.19 ± 2.77 µm, which was not significant but higher than that of 0 h group. The significantly lower values in shell width were observed in groups that were starved for 96 h and more, having 94.58 ± 0.17 µm in 96 h group, 100.04 ± 0.03 µm in 120 h group and 99.86 ± 4.23 µm in 144 h group, respectively (Tukey's HSD test, $P < 0.05$). On day 16, there were significant differences observed in shell width among groups ($F = 4.58$, df = 7, 11, $P < 0.05$). The maximum shell width was detected in 48 h group, with the value of 119.51 ± 2.33 µm, which was significantly higher than that of 96, 120 h and unfed groups. The remarkable growth in shell width were observed in 0, 24, 48, 72 h groups, however, no data was recorded in 96, 120, 144 and unfed groups because no survived larvae were found in those groups on day 23.

For the growth rates of shell width by day 12, the growth advantage was observed in the groups that were starved for 48 h and less (Table 1). In contrast, the growth rates were significantly negatively affected by delayed first feeding for 72 h or more, with less than 2 µm day^{-1} (Tukey's HSD test, $P < 0.05$). Although the highest growth rate of 4.74 ± 0.85 µm day^{-1} occurred in 0 h group, the slight differences had been observed among other groups except for 120 h group (Tukey's HSD test, $P > 0.05$).

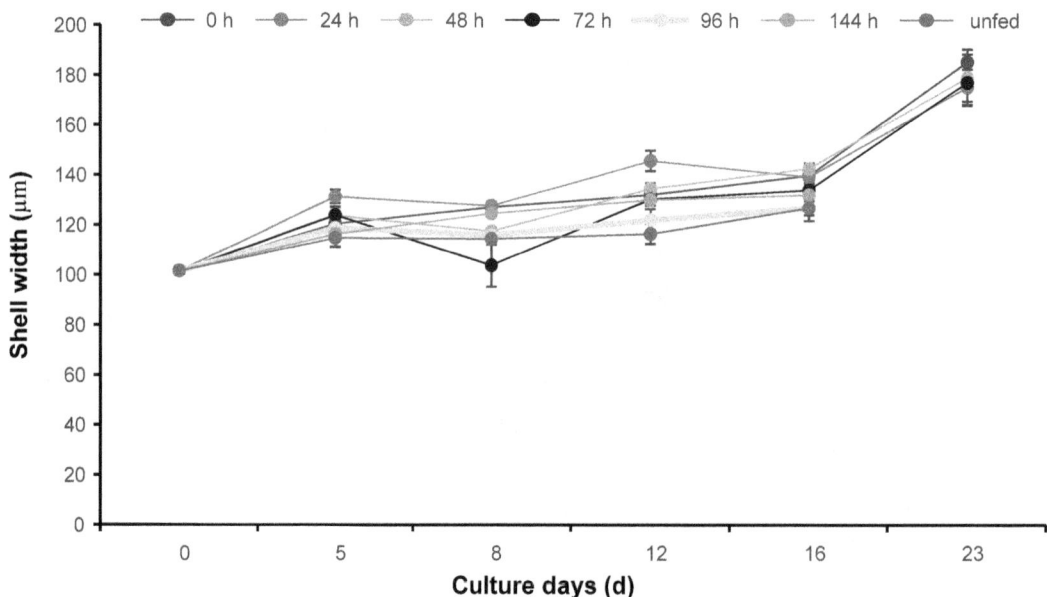

Figure 2. Mean shell width of *P. yessoensis* in different groups (delayed first feeding for 0, 24, 48, 72, 96, 120 and 144 h, and unfed group) during culture days. All error bars represent standard error of the mean

3.3 Effects of Delayed First Feeding on Larval Survival

The survival rates of larvae in different groups are shown in Figure 3. The significantly lower survival rates were observed on all counting days except for day 17, showing steeping declines over the time period (Tukey's HSD test, $P < 0.05$). By day 6, the survival rates decreased sharply with the prolonged starvation period especially in the groups that were starved for 96 h and more, lower than 20.0%, compared to the maximum survival rate of 58.3% in 0 h group. This suggests that delayed first feeding in *P. yessoensis* larvae had a deleterious impact on the larvae when delayed first feeding for 96 h and more. In contrast, no significant difference was detected among 0, 24, 48, and 72 h groups (Tukey's HSD test, $P > 0.05$). On day 9, the maximum survival rate was observed in 24 h group, which was significantly higher than those of the groups that were starved for 72 h and more (Tukey's HSD test, $P < 0.05$). On day 13, no significant difference in survival rate was detected among 0, 24, 48, and 72 h groups, significantly higher than those of 96, 120, 144 h and unfed groups. On day 24, there was no significant difference in survival rates among 24, 48, and 72 h groups, and the highest survival rate (4.17%) was observed in 0 h group. Moreover, most larvae starved for 72 h and more were delayed in development, subsequently becoming deformed before they reached umbo larvae stages.

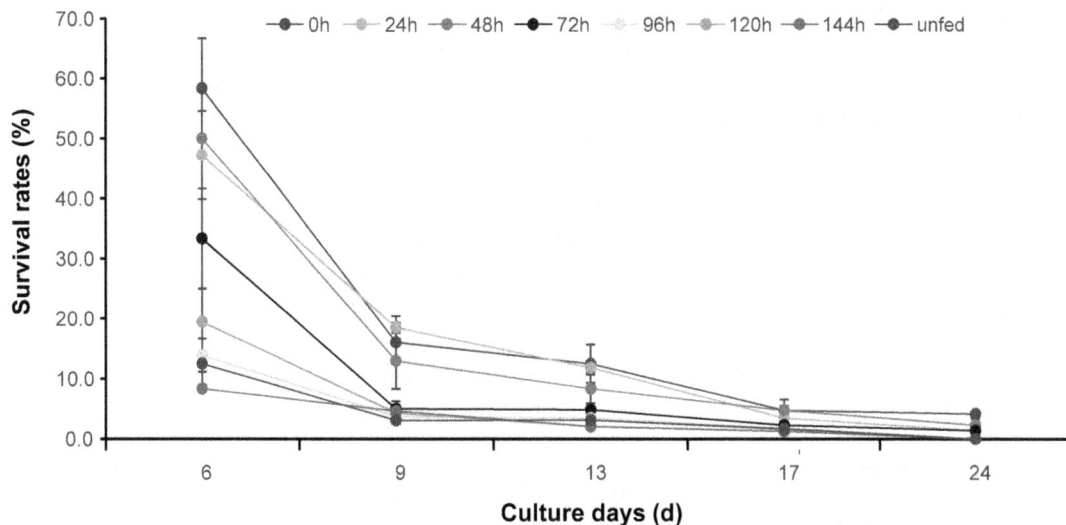

Figure 3. Mean survival rates of *P. yessoensis* larvae in different groups (delayed first feeding for 0, 24, 48, 72, 96, 120, and 144 h, and unfed group) during culture days

4. Discussion

The different responses of larval growth and survival when larvae were exposed to different durations of starvation prior to first feeding indicate that the first feeding time had significant effects on *P. yessoensis* larvae. For marine invertebrates, larvae depend largely on their endogenous yolk reserves to support the energy for developing the feeding ability and physiological mechanisms, which will help to start the normal first feeding, growth and survival (Olson & Olson, 1989). Previous studies on delayed first feeding or starvation in early larval development may have profound effects on subsequent larval growth, development, and survival (Zheng et al., 2005; Tang et al., 2006; Yan et al., 2009). In the present study, when delayed first feeding for 24, 48 and 72 h, the growth rates of *P. yessoensis* larvae following short periods of starvation recovered to the level of the non-starved larvae. However, when larvae were fed after being starved for more than 72 h, growth rates were significantly smaller, half or less than the growth of the control group. Similarly, growth recovery of larvae for short-duration starvation was observed in Ivory shell *Babylonia formosae habei* (Zheng et al., 2005), Manila clam *Ruditapes philippinarum* (Yan et al., 2009), and hard clam *Meretrix meretrix* (Tang et al., 2006). The possible explanations for the ability of larvae to achieve normal growth after short periods of starvation are mainly attributed into three parts. First, larvae might respond to starvation by feeding more vigorously at higher rates or more continuously. Second, larvae might respond to food concentration by altering the allocation of resources between growth and development (Bertram & Strathmann, 1998). As reported, starved larvae of sea cucumber increase the allocation of growth to the ciliated band when feeding is resumed, and therefore they are capable of acquiring more food (Sun & Li, 2014). In this study, the greater sizes in shell length of larvae when delayed first feeding for 24 h suggest that the starved larvae might increase the ability of larval cilia to collect and ingest food particles. Third, starvation early in life might cause significant mortality and select for very hardy larvae that have inherently greater capacity for feeding and growth (McEdward & Qian, 2001).

However, for the larvae starved for more than 72 h, growth became significantly slower and never recovered to the level of non-starved larvae. It is suggested that starvation reduced larval growth rates largely by increasing the duration of the larval period relative to the opportunities for acquiring food (McEdward & Qian, 2001). The prolonged starvation may cause permanent detrimental damage on larval feeding ability, such as the ability of larval cilia to collect and ingest food particles, and further inhibit energy generation for continuing development (Zheng et al., 2005). The irreversible damage on the starved larvae in this study was likely to be responsible for the low survival rates in those groups that were delayed first feeding for 96 h or more. For the crab *Hyas araneus*, when an early larva is starved beyond the critical point, its feeding ability will diminish and eventually experience irreversible damage to the mitochondria and hepatopancreas (Storch & Anger, 1983). It is therefore assumed that extended starvation has a significant impact on larval feeding and digestion, resulting in a rapid degeneration progress of tissues, slow growth and mass mortality (Sun & Li, 2014). Thus, the optimum starvation period for *P. yessoensis* larvae was 72 h or less after early D-shaped larvae, in order to avoid permanent detrimental effects on larvae due to starvation.

For larvae of marine invertebrates, endogenous reserves are always insufficient to account for metabolic demand, and energy for larval development and metamorphosis are mainly from exogenous food (Bertram & Strathmann, 1998). Therefore, the ability to withstand starvation in early larval development varies from species to species mostly depending on their endogenous reserves. For instance, sea star larvae could survive more than 70 days of starvation in early development (Allison, 1994). In contrast, the survival rate of sea cucumber larvae decreased markedly with prolonged starvation, and most larvae were dead after 13 days of starvation (Sun & Li, 2014). Moderate effects on the survival rate of clam larvae (70.83%) were observed during starvation for up to 9 days (Tang et al., 2006). Planktotrophic larvae of sea urchin could survive 14 days when a source of exogenous energy was not present (Meyer et al., 2007). In this study, the survival rates decreased sharply with the prolonged starvation period and eventually more than 98% of starved larvae were dead on day 19 after hatching. Those differences in starvation tolerance among species are probably attributed to nutrition content inside their egg capsules, such as nutritive fluids and yolk material (Pechenik, 1986; Chaparro et al., 1999). Besides the initial endogenous energy reserves, the variability of starvation tolerance among species may also be correlated with alternative sources of exogenous resources, such as dissolved organic material (DOM) or detritus in seawater. It is suggested that those materials are potentially important sources of energy for marine invertebrate larvae during starvation (Gomme, 2001; Moran & Manahan, 2004; Sun & Li, 2014). In this study, because the larvae were reared in sand-filtered seawater, DOM, probably not detritus, was more likely to provide a supplementary source of energy to maintain basic metabolism for the starved larvae, extending the larval period to 19 days after hatching. However, little is known about whether DOM is used differentially by marine benthic invertebrates (McEdward 1997; Moran & Manahan, 2004).

In conclusion, the growth and survival rates of *P. yessoensis* larvae could recover to the level of the non-starved larvae when delayed first feeding for 72 h or less, while it had a deleterious impact on the larval growth and survival when delayed first feeding for 96 h or more. Food availability within 72 h after early D-shaped larvae plays a key role in the successful initial feeding for this species and establishment of exogenous feeding schedules for hatcheries.

Acknowledgements

This work is supported by The Key Laboratory of Mariculture (KLM), Ministry of Education, OUC, and the grants from Independent Innovation Funds of Shandong Province (2013CXC80202), and Emerging Strategic Industries Project of Qingdao City (13-4-1-60hy), and Special Scientific Research Funds for Central Non-profit Institutes, Yellow Sea Fisheries Research Institutes (20603022013012).

References

Allison, G. W. (1994). Effects of temporary starvation on larvae of the sea star *Asterina miniata*. *Marine Biology, 118*, 256-261. http://dx.doi.org/10.1007/BF00349792

Bertram, D. F., & Strathmann, R. R. (1998). Effects of maternal and larval nutrition on growth and form of planktotrophic larvae. *Ecology, 79*, 315-327. http://dx.doi.org/10.1890/0012-9658(1998)079%5B0315: EOMALN%5D2.0.CO;2

Chaparro, O. R., Oyarzun, R. F., Vergara, A. M., & Thompson, R. J. (1999). Energy investment in nurse eggs and egg capsules in *Crepidula dilatata* Lamarck (Gastropoda, Calyptraeidae) and its influence on the hatching size of juvenile. *Journal of Experimental Marine Biology and Ecology, 232*, 261-277. http://dx.doi.org/10.1016/S0022-0981(98)00115-4

Fenaux, L., Strathmann, M. F., & Strathmann, R. R. (1994). Five tests of food-limited growth of larvae in coastal waters by comparisons of rates of development and form of echinoplutei. *Limnology and Oceanography, 39*, 84-89. http://dx.doi.org/10.4319/lo.1994.39.1.0084

Gomme, J. (2001). Transport of exogenous organic substances by invertebrate in teguments: the field revisited. *Journal of Experimental Zoology, 289*, 254-265. http://dx.doi.org/10.1002/1097-010X(20010401/30)289:4% 3C254::AID-JEZ6%3E3.0.CO;2-F

His, E., & Seaman, M. N. L. (1992). Effects of temporary starvation on the survival, and on subsequent feeding and growth, of oyster (*Crassostrea gigas*) larvae. *Marine Biology, 114*, 277-279. http://dx.doi.org/10.1007/B F00349530

Ito, H. (1991). *Patinopecten* (Mizuhopecten) *yessoensis*. In S. E. Shumway (Ed.), *Scallops: Biology, Ecology and Aquaculture. Elsevier, Amsterdam* (pp. 1024-1055).

Li, Q., Xu, K. F., & Yu, R. H. (2007). Genetic variation in Chinese hatchery populations of the Japanese scallop (*Patinopecten yessoensis*) inferred from microsatellite data. *Aquaculture, 269*, 211-219. http://dx.doi.org/10.1016/j.aquaculture.2007.04.017

Manahan, D. T. (1990). Adaptations by invertebrate larvae for nutrient acquisition from seawater. *American Zoologist, 30*, 147-160.

McEdward, L. R., & Qian, P. Y. (2001). Effects of the duration and timing of starvation during life on the metamorphosis and initial juvenile size of the polychaete *Hydroides elegans* (Haswell). *Journal of Experimental Marine Biology and Ecology, 261*, 185-197.

McEdward, L. R. (1997). Reproductive strategies of marine benthic invertebrates revisited: facultative feeding by planktotrophic larvae. *American Zoologist, 150*, 48-72.

Meyer, E., Green, A. J., Moore, M., & Manahan, D. T. (2007). Food availability and physiological state of sea urchin larvae (*Strongylocentrotus purpuratus*). *Marine Biology, 152*, 179-191. http://dx.doi.org/10.1007/s00227-007-0672-6

Moran, A. L., & Manahan, D. T. (2004). Physiological recovery from prolonged 'starvation' in larvae of the Pacific oyster *Crassostrea gigas*. *Journal of Experimental Marine Biology and Ecology, 306*, 17-36. http://dx.doi.org/10.1016/j.jembe.2003.12.021

Olson, R. R., & Olson, M. H. (1989). Food limitation of planktrophic marine invertebrate larvae: does it control recruitment success? *Annual Review of Ecology and Systematics, 20*, 225-247.

Pawlik, J. R. (1992). Chemical ecology of the settlement of benthic marine invertebrate. *Oceanography and Marine Biology: an annual review, 30*, 273-335.

Pechenik, J. A. (1986). The encapsulation of eggs and embryos by molluscs: an overview. *American Malacological Bulletin, 4*, 165-172.

Pedrotti, M. L., & Fenaux, L. (1993). Effects of food diet on the survival, development and growth rates of two cultured echinoplutei (*Paracentrotus lividus* and *Arbacia lixula*). *Invertebrate Reproduction & Development, 24*, 59-70. http://dx.doi.org/10.1080/07924259.1993.9672332

Robert, R., & Trintignac, P. (1997). Substitutes for live microalgae in mariculture. *Aquatic Living Resources, 10*, 315-327. http://dx.doi.org/10.1051/alr:1997035

Sewell, M. A., Cameron, M. J., & McArdle, B. H. (2004). Developmental plasticity in larval development in the echinometrid sea urchin *Evechinus chloroticus* with varying food ration. *Journal of Experimental Marine Biology and Ecology, 309*, 219-237. http://dx.doi.org/10.1016/j.jembe.2004.03.016

Storch, V., & Anger, K. K. (1983). Influence of starvation and feeding on the hepatopancreas of larval *Hyas araneus* (Decapoda, Majidae). *Helgoländer Meeresuntersuchungen, 36*, 67-75. http://dx.doi.org/10.1007/BF01995796

Sun, X. J., & Li Q. (2014). Effects of delayed first feeding on larval growth, survival and development of the sea cucumber *Apostichopus japonicus* (Holothuroidea). *Aquaculture Research, 45*, 278-288. http://dx.doi.org/10.1111/j.1365-2109.2012.03224.x

Tang, B. J., Liu, B. Z., Wang, G. D., Zhang, T., & Xiang, J. H. (2006). Effects of various algal diets and starvation on larval growth and survival of *Meretrix meretrix*. *Aquaculture, 254*, 526-533. http://dx.doi.org/10.1016/j.aquaculture.2005.11.012

Thorson, G. (1950). Reproductive and larval ecology of marine bottom invertebrates. *Biological Reviews, 25*, 1-45. http://dx.doi.org/10.1111/j.1469-185X.1950.tb00585.x

Wehrtmann, I. S. (1991). How important are starvation periods in early larval development for survival of *Crangon septemspinosa* larvae? *Marine Ecology Progress Series, 73*, 183-190. http://dx.doi.org/10.3354/meps073183

Yan, X. W., Zhang, Y. H., Huo, Z. M., Yang, F., & Zhang, G. F. (2009). Effects of starvation on larval growth, survival, and metamorphosis of Manila clam *Ruditapes philippinarum*. *Acta Ecologica Sinica, 29*, 327-334. http://dx.doi.org/10.1016/j.chnaes.2009.09.012

Yang, A. G., Wang, Q. Y., Liu, Z. H., & Zhou, L. Q. (2004). The hybrid between the scallops *Chlamys farreri* and *Patinopecten yessoensis* and the inheritance characteristics of its first filial generation. *Marine Fisheries Research, 25*, 1-5.

Yang, F., Zhang, Y. H., Yan, X. W., & Zhang, G. F. (2008). Effects of starvation and refeeding on larval growth, survival, and metamorphosis of clam *Cyclina sinensis*. *Acta Ecologica Sinica, 28*, 2052-2059.

Yang, H., Li, L., & Guo, X. M. (2001). Preliminary study on inducing polyploidy in Japanese scallop (*Patinopecten yessoensis*) by cytochalasin B. *Acta Zoologica Sinica, 47*, 459-464.

Zhang, Q., Zhang, Q., Li, W., & Wang, X. (2000). Study on high yield technology of raft culture of *Patinopecten yessoensis*. *Marine Science, 24*, 14-16.

Zheng, H. P., Zheng, C. H., Zhou, S. L., & Li, F. X. (2005). Effects of starvation on larval growth, survival and metamorphosis of Ivory shell *Babylonia formosae habei* Altena et al., 1981 (Neogastropoda: Buccinidae). *Aquaculture, 243*, 357-366. http://dx.doi.org/10.1016/j.aquaculture.2004.10.010

6

Efficacy of Two Commercial Systems for Identification of Clinical and Environmental *Escherichia coli*

Hussein H. Abulreesh[1]

[1] Department of Biology, Faculty of Applied Science, Umm Al-Qura University, Saudi Arabia

Correspondence: Department of Biology, Faculty of Applied Science, Umm Al-Qura University, P.O. Box 7388, Makkah 21955, Saudi Arabia. E-mail: hhabulreesh@uqu.edu.sa

Abstract

The aim of this study was to test the efficacy of API 20E and Fluorocult® LMX broth in identifying a collection of 200 *E. coli* isolates. A total of 100 isolates originated from clinical samples (UTI) and 100 isolates from environmental water receiving faecal contamination. Randomly selected isolates that were identified by API 20E and Fluorocult® LMX broth were further identified by PCR targeting a fragment of the *E. coli* 16S rRNA gene. The results showed that overall 95% and 100% of the clinical and environmental isolates respectively were identified with various degrees of accuracy as *E. coli* by API 20E. However, only 86% of the clinical isolates and 32% of environmental isolates were identified with high level of discrimination (90% and above). Identification by Fluorocult® LMX broth successfully identified 90% and 96% of the clinical and environmental isolates respectively as *E. coli*. Further identification by PCR showed that 70% (n = 20) and 55% (n = 20) of the isolates that were previously identified by the two commercial systems were successfully identified by PCR. Identification of *E. coli* isolates of clinical and environmental origins by rapid commercial systems should be interpreted with care, PCR might be used to further confirm the result of rapid identification systems.

Keywords: API 20E, *Escherichia coli*, Fluorocult® LMX broth, PCR

1. Introduction

Escherichia coli is a common inhabitant of the intestinal tract of humans and warm blooded animals (Ashbolt et al., 2004). Generally, *E. coli* is regarded as a commensal in the intestinal tract, yet there are a number of strains that cause diarrheal diseases through contaminated drinking water and food (Chen & Frankel, 2005; Bettelheim, 2007), commensal or non-pathogenic strains of *E. coli* can cause a variety of infections in humans such as urinary tract infection (UTI), bacteremia and meningitis (Janda & Abbott, 2006). Given the exclusive faecal origin of *E. coli*, their presence in the environment, particularly in water is a strong indication of faecal contamination and represents a risk of disease (Leclerc et al., 2001). Thus *E. coli* was suggested to be the best biological faecal indicator and has been recommended by World Health Organization for the microbiological assessment of safe drinking water supplies (Edberg et al., 2000; Tallon et al., 2005).

Escherichia coli is exogenous to aquatic environments, and it is expected that the bacterium would not be able to grow in water due to unfavourable environmental conditions. However, a number of studies have shown that *E. coli* can persists for prolonged periods in various types of aquatic environments, surviving the hostile environmental abiotic and biotic factors (Davies & Evison, 1991; Alkan et al., 1995; Janakiraman & Leff, 1999; Whitman et al., 2004; Wcisto & Chróst, 2000; Wanjugi & Harwood, 2013), being able to survive in water, *E. coli* may undergo physiological and morphological changes probably as a survival strategy. These changes may, in part, enable the bacteria to enter the debatable viable-but-nonculturable (VBNC) stage, which make these bacterial cells undetectable by conventional culture media, yet still viable and may retain their virulence factors and pathogenicity (Xu et al., 1982; Barer & Harwood, 1999; Pommepuy et al., 1996; Oliver, 2000; Pinto et al., 2011).

Different commercial miniaturized systems were developed to identify presumptive *E. coli* isolates of clinical and environmental origins. These identification systems include the widely used API 20E (bioMerieux, France) and the Fluorocult® LMX Broth (Merck, Germany). The API 20E is one of the original miniaturized systems, still in wide spread use for the identification of members of the *Enterobacteriaceae* since the 1970s. This system is based on 20 different biochemical tests such as the production of indole; citrate utilization; voges-proskauer reaction; carbohydrate fermentation and other tests (Smith et al., 1972). Fluorocult® LMX broth is a modified lauryl sulfate tryptic broth for simultaneous detection of coliforms and *E. coli*. This selective enrichment broth contains the chromogenic substrate 5-bromo-4-chloro-3 indolyl-β-D-galactopyranoside, cleaved by coliforms' β-D-galactosidase, and the fluorogenic substrate, 4-methylumbelliferyl-β-D-glucuronide (MUG), which is highly specific for *E. coli* β-D-glucuronidase (Ossmer, 1993).

The current study aims to evaluate the efficacy of the API 20E and Fluorocult® LMX broth in identifying a collection of *E. coli* isolates that were recovered from environmental water and clinical specimens. Further confirmation was performed by PCR amplifying a target fragment of the *E. coli* 16S rRNA gene.

2. Materials and Methods

2.1 Bacterial Isolates

A collection of 200 *E. coli* isolates were used to evaluate the efficacy of two widely used commercial identification systems; the API 20E and the Fluorocult® LMX broth. A total of 100 *E. coli* isolates were obtained from clinical laboratories mainly derived from urine samples of patients suffering a urinary tract infection (UTI), these isolates will be referred to as clinical *E. coli* (CEC). Another 100 *E. coli* isolates were recovered from fresh water receiving faecal contamination and rain-related run-off, these isolates will be referred to as environmental *E. coli* (EEC). Detection of environmental *E. coli* isolates was carried out as described by Abulreesh et al. (2004). Briefly, water samples were collected into sterile polypropylene bottles. A volume of 100 ml subsample was assayed by colony counts on 0.45 μm membrane filters. Membrane filters were incubated on eosin methylene blue agar (EMB) (Oxoid, Basingstoke, UK), plates were initially incubated at 30 °C for 4 hours, followed by 14 hours at 44 °C. Randomly six typical *E. coli* colonies (colonies with 2-3 diameter, exhibiting a greenish metallic sheen by reflected light and dark purple centers by transmitted light), were picked up from each plate (total 100 colonies) and used for further tests. The *E. coli* K-12 strain was used as a positive control for all identification methods including PCR.

2.2 Identification of E. coli by API 20E and Fluorocult® LMX Broth

Before the start of identification tests, all 200 isolates were purified on EMB agar plates to ensure that there is no mixed culture that could influence the reactions of the identification systems. A bacterial suspension approximating a 0.5 McFarland standard was prepared from each of the 200 purified isolates and was used for inoculation of the API 20E strips (bioMerieux, France), incubation was at 37 °C for 24 hours, after incubation the addition of reagents and interpretation of reactions were done according to manufacturers' directions. The 20 biochemical test reactions were converted into an octal profile number. Each profile number was then decoded by using the Analytical Profile Index for the completion of identification.

All 200 previously purified isolates were identified by culture in Fluorocult® LMX broth (Merck, Germany), aliquots of 1.0 ml of the broth were aseptically transferred into a U-bottom sterile polystyrene well cluster (Coaster, Coring Inc., NY, USA). Each well was inoculated with a single purified *E. coli* colony using an inoculation loop. Incubation of the wells was at 37 °C for 24 hours. The formation of blue-green colour (β-D-galactosidase) together with the observation of fluorescence (β-D-glucuronidase) under UV transillumination (366 nm) confirmed that an isolate was *E. coli* (Ossmer 1993).

2.3 Identification by PCR Detecting Fragment of the E. coli 16S rRNA Gene

Identification by PCR was carried out on a randomly selected 40 isolates (20 clinical and 20 environmental), some of these isolates were identified as *E. coli* by API 20E and Fluorocult® LMX broth, and other isolates were not identified by these two systems. The primers used for PCR assay targeted fragment of *E. coli* 16S rRNA gene, the primers RW01 ('5-AACTGGAGGAAGGTGGGGAT-3') and DG74 ('5-AGGAGGTGATCCAAGCA-3') (Invitrogen, Paisley, UK), is expected to give a PCR product of 371 bp as described by (Tsai & Olson, 1992). Bacterial rRNA was extracted by suspending a loopful of *E. coli* colonies in 100 μl of sterile, pure water and boiling for 5 minutes. The suspension was then centrifuged for 5 minutes at 1260 g and 10 μl of the supernatent were used as target DNA. The PCR reaction mixture (50 μl total volume) consisted of the following: 25 μl of 2 x PCR master mix (ABgene, Surry, UK) contains the following: 75 mmol l^{-1} Tris-HCL; 20 mmol l^{-1} (NH$_4$)$_2$SO$_4$; 20 mmol l^{-1} MgCl$_2$; 0.01% (v/v) Tween® 20; 0.2 mmol l^{-1} each of dATP, aGTP, dCTP and dTTP; 1.25 units of Theroprime Plus DNA Polymerase; 0.5 μl of each primer (0.25 μmol l^{-1}); 10 μl bacterial rRNA extract and 14 μl sterile pure water. The PCR program consisted of initial denaturation at 95 °C for 2 minutes, followed by 40 PCR cycles (each cycle is a 1.5 minutes at 95 °C for denaturation and 1 minute at 62 °C for annealing and extension) and a 7 minutes final extension at 62 °C (Tsai & Olson, 1992). The PCR products were analyzed by electrophoresis on a 2.0% agarose gel (BioLine, London, UK), and made visible by ethidium bromide (1.0 μg ml^{-1}) staining and UV transillumination.

2.4 Statistical Analysis

Chi-square test (χ^2) was used to determine the accuracy of API 20E and Fluorocult LMX in identifying all *E. coli* isolates from clinical and environmental sources. The null hypothesis that was tested by Chi-squared statistics reads: Both API 20E and Flurocult LMX broth can accurately identify all *E. coli* isolates regardless of their origin.

3. Results

This study examined the efficacy of the API 20E and the Fluorocult® LMX broth for the identification of a collection of 200 *E. coli* isolates from clinical and environmental origins. Overall 95% of clinical *E. coli* were successfully identified by API 20E with various levels of discrimination (Table 1). All 100 environmental isolates were successfully identified as *E. coli* by API 20E (100%) with various degrees of discrimination (Table 2).

Table 1. Identification of 100 clinical isolates of *E. coli* by API 20E

Level of discrimination	Number of isolates identified	Isolate reference number
Not identified as *E. coli*	5	CEC 1, CEC 12, CEC 22, CEC 27, CEC 35
Low discrimination	8	CEC 3, CEC 5, CEC 26, CEC 40, ECE 77-79, CEC 99
Acceptable discrimination	1	CEC 31
Good discrimination	68	CEC 2, CEC 7-10, CEC 16, CEC 18, CEC 20, CEC 21, CEC 23, CEC 25,CEC 28, CEC 29, CEC 32-34, CEC 37, CEC 38, CEC 41-48, CEC 50, CEC 51, CEC 53, CEC 54, CEC 56-59, CEC 61-76, CEC 81-97, CEC 100
Very good discrimination	12	ECE 6, ECE 13, ECE 14, ECE 17, ECE 24, ECE 30, ECE 36. ECE 39, ECE 49, ECE 52, ECE 55, ECE60
Excellent discrimination	6	CEC 4, CEC 11, CEC 15, CEC 19, CEC 80, CEC 98
Total number of isolates	100	

Table 2. Identification of 100 environmental *E. coli* isolates by API 20E

Level of discrimination	Number of isolates identified	Isolate reference number
Not identified as *E. coli*	0	
Low discrimination	66	EEC 2, EEC 3, EEC 5, EEC 6, EEC 9, EEC 11, EEC 12, EEC 15-17, EEC 19-25, EEC 28, EEC 30-42, EEC 44, EEC 45, EEC 44, EEC 47, EEC 48, EEC 52-55, EEC 58, EEC 59, EEC 61-69, EEC 72-76, EEC 81, EEC 83, EEC 84, EEC 87, EEC 89, EEC 91, EEC 96-100
Acceptable discrimination	2	EEC 43, EEC 70,
Good discrimination	18	EEC 1, EEC 4, EEC 7, EEC 8, EEC 26, EEC 29, EEC 46, EEC 49, EEC 56, EEC 57, EEC 77-80, EEC 90, EEC 93-95
Very good discrimination	10	EEC 10, EEC 13, EEC 14, EEC 18, EEC 50, EEC 51, EEC 60, EEC 82, EEC 85, EEC 86
Excellent discrimination	4	EEC 27, EEC 71, EEC 88, EEC 92
Total number of isolates	100	

Given the importance of accurate identification of these isolates, it can be seen from Table 3 that 86 isolates of the clinical *E. coli* were identified by API 20E with high levels of discrimination (90% and above), in contrast, 68 isolates of the environmental *E. coli* isolates were identified with low levels of discrimination (75% and below). Statistical significant difference ($P < 0.01$, χ^2) showed that API 20E accurately identified more clinical *E. coli* isolates that those of environmental origin (Table 3).

Table 3. Levels of identification for 200 clinical and environmental *E. coli* isolates by API 20E

Level of identification	Clinical *E. coli* (n = 100)	Environmental *E. coli* (n = 100)	P
High level of discrimination (90% and above)	86	32	< 0.01
Low level of discrimination (72% and below)	9	68	
Not identified as *E. coli*	5	0	

P = the probability that API 20E should accurately identify all *E. coli* isolates regardless of their origin (χ^2).

The result of identification of the 200 *E. coli* isolates by Fluorocult® LMX broth is summarized in Table 4. Of the 100 clinical isolates examined, only 90 isolates were identified as *E. coli*, while 96 out of 100 environmental isolates were identified as *E. coli*. No statistical significant difference ($P > 0.5$) was found in the accuracy of identification of *E. coli* isolates from clinical and environmental sources as tested by Chi-squared statistics (Table 4).

Table 4. Identification of 200 *E. coli* isolates by Fluorocult® LMX Broth

Isolates origin	N/P	Isolates that were not identified
Clinical *E. coli*	100/90	CEC 12, CEC 18, CEC 22, CEC 27, CEC 35, CEC 58, CEC 66, CEC 72, CEC 90, CEC 92
Environmental *E. coli*	100/96	EEC 38, EEC 40, EEC 70, EEC 71
P	NS	

N = Total number of isolates tested;

P = Total number of isolates identified as *E. coli*;

P = the probability that Fluorocult LMX broth should accurately identify all *E. coli* isolates regardless of their origin (χ^2);

NS = not significant > 0.5.

It can be noted that of the 10 clinical isolates that were not identified by Fluorocult® LMX broth, four isolates were also not identified as *E. coli* by API 20E (CEC 12, 22, 27, 35), while six of these 10 isolates were identified as *E. coli* by API 20E with good level of discrimination (CEC 18, 58, 66, 72, 90, 92) (Table 5). As far as the environmental isolates are concerns, all of the four isolates that were not identified by Fluorocult® LMX broth, they were identified by API 20E with various degrees of discrimination as follows: one isolate was identified with excellent level of discrimination (EEC 71), one isolate was identified with acceptable level of discrimination (EEC 70) and two isolates with low level of discrimination (EEC 38 and 40) (Table 5).

Table 5. API 20E identification of clinical and environmental *E. coli* isolates that gave negative reaction with Fluorocult® LMX Broth

Isolate reference number	API 20E Identification	Level of discrimination
CEC 12	*Enterobacter agglomerans*	74.75
	Kliebsiella oxytoca	23.2% low discrimination
CEC 18	*E. coli*	96.5% good
CEC22	*Klebsiella oxytoca*	97.4% doubtful
CEC 27	*Citrobacter freundii*	99.9% excellent
CEC35	*Citrobacter freundii*	99.9% excellent
CEC58	*E. coli*	96.5% good
CEC66	*E. coli*	96.5% good
CEC 72	*E. coli*	96.5% good
CEC 90	*E. coli*	96.5% good
CEC 92	*E. coli*	96.5% good
EEC 28	*E. coli*	72% low
	Yersinia aldovae	25%
EEC 40	*E. coli*	72% low
	Yersinia aldovae	25%
EEC 70	*E. coli*	77% acceptable
EEC 71	*E. coli*	99.9% excellent

A total of 40 (20 clinical and 20 environmental) isolate were randomly selected and further identified by PCR targeting a fragment of the *E. coli* 16S rRNA gene. On the whole, PCR successfully identified 14 (70%) out of the 20 clinical isolates that were examined, of which, 13 isolates identified as *E. coli* by API 20E, Fluorocult® LMX broth and PCR, and three isolates (CEC 12, 41 and 35) were not identified as *E. coli* by all three methods (Table 6). A total of four isolates (CEC 2, 9, 20 and 41) were successfully identified by API 20E and Fluorocult® LMX broth, however they gave no PCR products with primers RW01 and DG74 (Table 6).

Table 6. Identification of randomly selected clinical *E. coli* isolates by PCR

Isolate reference number	Identification by API 20E	Identification by Fluorocult® LMX broth	Identification by PCR
CEC 2	good discrimination	+	-
CEC 9	good discrimination	+	-
CEC 11	excellent discrimination	+	+
CEC 12	not identified as *E. coli*	-	-
CEC 15	Excellent discrimination	+	+
CEC 18	good discrimination	-	+
CEC 20	good discrimination	+	-
CEC 28	good discrimination	+	+
CEC 34	good discrimination	+	+
CEC 35	not identified as *E. coli*	-	-
CEC 39	Very good discrimination	+	+
CEC 41	Good discrimination	+	-
CEC 44	Good discrimination	+	+
CEC 49	Very good discrimination	+	+
CEC 55	Very good discrimination	+	+
CEC 58	Good discrimination	+	+
CEC 62	Good discrimination	+	+
CEC 69	Good discrimination	+	+
CEC 75	Good discrimination	+	+
CEC 88	Good discrimination	+	+

Identification of environmental isolates by PCR showed that only 11 isolates out of 20 (55%) produced the expected PCR products. These isolates were also successfully identified by API 20E and Fluorocult® LMX broth (Table 7). Three environmental isolates (EEC 38, 40 and 70) were not identified by PCR and by Fluorocult® LMX broth as well. Five isolates (EEC 5, 12, 41, 42, 43) were not identified by PCR although they were successfully identified by API 20E and Fluorocult® LMX broth (Table 7).

Table 7. Identification of randomly selected environmental *E. coli* isolates by PCR

Isolate reference number	Identification by API 20E	Identification by Fluorocult® LMX broth	Identification by PCR
EEC 3	Low discrimination	+	+
EEC 5	Low discrimination	+	-
EEC 12	Low discrimination	+	-
EEC 18	Very good discrimination	+	+
EEC 25	Low discrimination	+	+
EEC 38	Low discrimination	-	-
EEC 40	Low discrimination	-	-
EEC 41	Low discrimination	+	-
EEC 42	Low discrimination	+	-
EEC 43	Acceptable discrimination	+	-
EEC 44	Low discrimination	+	+
EEC 45	Low discrimination	+	+
EEC 46	Good discrimination	+	+
EEC 47	Low discrimination	+	+
EEC 50	Very good discrimination	+	+
EEC 56	Good discrimination	+	+
EEC 66	Low discrimination	+	+
EEC 70	Acceptable discrimination	-	-
EEC 88	Excellent discrimination	+	+
EEC 92	Excellent discrimination	+	+

4. Discussion

The rapid commercial identification systems were developed to provide fast and accurate identification of bacteria in clinical settings and when investigating health hazards associated with faecal contamination in water and food. The API 20E is a standardized miniaturized system that was developed by Analytab Products Inc. in the 1970s and still in widespread use for the identification of *Enterobacteriaceae* and some other gram negative rods. Published data suggests that *E. coli* isolates are usually identified with high degree of accuracy (96%) by most rapid commercial identification systems, including the API 20E (Janda & Abbott, 2006). Various studies evaluated the efficacy of API 20E system using a collection of clinically originated isolates with typical and atypical strains, and reported an accurate rate of identification between 93-96% (Washington et al., 1971; Smith et al., 1972; O'Hara et al., 1992). The results reported in this study showed that 95% and 100% of the clinical and the environmental *E. coli* respectively were identified by API 20E system with various level of accuracy (Tables 1 and 2). However, in contrast with other studies, 86% of the clinical *E. coli* and 32% of the environmental isolates were accurately identified (Table 3). This low accuracy is probably due to incubation for 24 h only. O'Hara et al. (1992) noted that after 24 h the accuracy of API 20E identification was 87.7%, however after 48 h incubation, they noted an increase to 96.3% accuracy.

In the current study, 86% of the clinical *E. coli* isolates were identified with high level of discrimination, while only 32% of the environmental isolates were identified with the same degree of discrimination ($P < 0.01$, χ^2) (Table 3). This is probably because most of *E. coli* isolated from clinical specimens (e.g. UTI) are tend to be biochemically typical (Janda & Abbott, 2006), in contrast, environmental *E. coli* isolates may exhibit atypical biochemical characteristic due to physiological changes required for better survival in aquatic habitats, it was particularly found that some proteins are repressed while other are induced under stressful conditions (Roszak & Colwell, 1987; Smith et al., 1994; Jordan et al., 1999). The results of this study may suggest that API 20E may produce more accurate identification with clinical but not environmental *E. coli* isolates.

Fluorogenic substrates that detect the activities of β-D-glucuronidase and β-D-galactosidase, two enzymes specific for *E. coli* and coliforms respectively were proposed by Feng and Hartman (1982). These substrates were employed in culture media for direct detection and/or confirmation of *E. coli* and coliforms in environmental samples, particularly investigating faecal contamination in water and food (Blood & Curtis, 1995; Manafi, 1996). Successful detection and/or confirmation of *E. coli* isolates from various water samples employing fluorogenic substartes is well documented (O'Toole & Chang, 1999; Prats et al., 2008). The results reported in this study using Fluorocult® LMX broth showed 90% and 96% accurate identification of *E. coli* isolates from clinical and environmental sources respectively, with no statistical significant difference ($P > 0.5$, χ^2) (Table 4), these results consistent with results reported elsewhere from investigation on water and food (Suwansonthichai & Rengpipat, 2003; Abulreesh et al., 2004; Nikaeen et al., 2009). Of the 200 *E. coli* isolates examined in this study, 14 isolates were not identified by Fluorocult® LMX broth, four of these isolates were also not identified as *E. coli* by API 20E (Table 5). However, the other 10 isolates were identified as *E. coli* by API 20E, these isolates probably β-D-glucuronidase negative, either they do not have the gene (Chang et al., 1989; Manafi, 1996) or they possess but cannot express the gene (Bej et al., 1991; Feng et al., 1991).

In this study, PCR was used to confirm the accuracy of API 20E and Fluorocult® LMX broth by randomly selecting different isolates that were identified as *E. coli* with various degrees of accuracy, and other isolates that were not identified as *E. coli* by both commercial systems. The primers used (RW01 and DG74) were targeting a fragment of the *E. coli* 16S rRNA gene (Tsai & Olson, 1992). The results showed that PCR further confirm the identification of 70% (n = 14) of the clinical isolates (Table 6) and also 55% (n = 11) of the environmental isolates (Table 7). Few isolates (total 5 out of 40) were only identified by API 20E, but not by Fluorocult® LMX broth and PCR (Tables 6 and 7), this might be incorrectly identification by API 20E as *E. coli*, since three of these five isolates rendered low discrimination (75% and below), incorrect identification of *Enterobacteriaecae* by API 20E has been reported due to aberrant reactions by API and/or atypical strains (Washington et al., 1971; Smith et al., 1972; O'Hara et al., 1992). Also *E. coli* strains in particular are extremely phenotypically diverse and groups of biochemically distinct strains exists (Janda & Abbott, 2006). It can be noted from the results of the PCR that 9 isolates of the 40 examined (4 clinical and 5 environmental isolates) were not identified by PCR despite being identified by API 20E and gave positive reaction with Fluorocult® LMX broth (Tables 6 and 7). Incorrect identification by API 20E is a possible explanation, also it has been reported that there are some genera of the *Enterobacteriaceae* (e.g. *Shigella*, *Klebsiella* and *Citrobacter*) may produce a positive fluorescence response with fluorogenic substrates (Manafi, 1996; APHA, 1998; O'Toole & Chang, 1999). In general the results reported in this study with PCR reaffirm that PCR provide sensitivity in the identification of *E. coli* when compared with biochemical and enzyme-substrate methods (Bej et al., 1991; Feng et al., 1991; Rompré et al., 2002).

In conclusion, identification of *E. coli* isolates of clinical and environmental origins by rapid commercial systems should be interpreted with care, API 20E was found to produce more accurate identification with isolates of clinical but not environmental origin, while Fluorocult LMX broth produced high level of accurate identification of *E. coli* isolates regardless of their origin. Perhaps a combination of two methods might be appropriate. However in routine laboratories this might be laborious, and therefore PCR might be used to further confirm the result of rapid identification systems.

References

Abulreesh, H. H., Paget, T. A., & Goulder, R. (2004). Waterfowl and the bacteriological quality of amenity ponds. *Journal of Water and Health, 2*, 183-189. Retrieved from http://www.iwaponline.com/jwh/002/0183/0020183.pdf

Alkan, U., Elliott, D. J., & Evison, L. M. (1995). Survival of enteric bacteria in relation to simulated solar radiation and other environmental factors in marine waters. *Water Research, 29*, 2071-2081. http://dx.doi.org/10.1016/0043-1354(95)00021-C

APHA. (1998). *Standard Methods for the Examination of Water and Wastewater* (20th ed.). Washington D. C., USA: American Public Health Association.

Ashbolt, N. J., Grabow, W. O. K., & Snozzi, M. (2004). Indicators of microbial water quality. In L. Fewtrell & J. Bartram (Eds.), *Water Quality: Guidelines, Standards and Health: Assessment of Risk and Risk Management for Water-Related Infectious Diseases* (pp. 289-315). London, UK: IWA Publishing.

Barer, M. R., & Harwood, C. R. (1999). Bacterial viability and culturability. *Advances in Microbial Physiology, 41*, 93-137. http://dx.doi.org/10.1016/S0065-2911(08)60166-6

Bej, A. K., McCarty, S. C., & Atlas, R. M. (1991). Detection of coliform bacteria and *Escherichia coli* by multiplex polymerase chain reaction: comparison with defined substrate and plating methods for water quality monitoring. *Applied and Environmental Microbiology, 57*, 2429-2432. Retrieved from http://aem.asm.org/content/57/8/2429.full.pdf+html

Bettelheim, K. A. (2007). The non-O157 shiga-toxigenic (verocytotoxigenic) *Escherichia coli*; under-rated pathogens. *Critical Reviews in Microbiology, 33*, 67-87. http://dx.doi.org/10.1080/10408410601172172

Blood, R. M., & Curtis, G. D. W. (1995). Media 'total' *Enterobacteriaceae*, coliforms and *Escherichia coli*. *International Journal of Food Microbiology, 26*, 93-115. http://dx.doi.org/10.1016/0168-1605(94)00040-D

Chang, G. W., Brill, J., & Lum, R. (1989). Proportion of β-D-glucuronidase-negative *Escherichia coli* in human fecal samples. *Applied and Environmental Microbiology, 55*, 335-339. Retrieved from http://aem.asm.org/content/55/2/335.full.pdf+html

Chen, H. D., & Frankel, G. (2005). Enteropathogenic *Escherichia coli*: unraveling pathogensis. *FEMS Microbiology Reviews, 29*, 83-98. http://dx.doi.org/10.1016/j.femsre.2004.07.002

Davies, C. M., & Evison, L. M. (1991). Sunlight and the survival of enteric bacteria in natural waters. *Journal of Applied Bacteriology, 70*, 265-274. http://dx.doi.org/10.1111/j.1365-2672.1991.tb02935.x

Edberg, S. C., Rice, E. W., Karlin, R. J., & Allen, M. J. (2000). *Escherichia coli*: the best biological drinking water indicator for public health protection. *Journal of Applied Microbiology, 88*, 106S-116S. http://dx.doi.org/10.1111/j.1365-2672.2000.tb05338.x

Feng, P. C. S., & Hartman, P. A. (1982). Fluorogenic assays for immediate confirmation of *Escherichia coli*. *Applied and Environmental Microbiology, 43*, 1320-1329. http://aem.asm.org/content/43/6/1320.full.pdf+html

Feng P., Lum R., & Chang, G. W. (1991). Identification of *uidA* gene sequence in β-D-glucuronidase-negative *Escherichia coli*. *Applied and Environmental Microbiology, 57*, 320-323. Retrieved from http://aem.asm.org/content/57/1/320.full.pdf+html

Janakiraman, A., & Leff, L. G. (1999). Comparison of survival of different species of bacteria in freshwater microcosms. *Journal of Freshwater Ecology, 14*, 233-240. http://dx.doi.org/10.1080/02705060.1999.9663674

Janda, J. M., & Abbott, S. L. (2006). *The Enterobacteria* (2nd ed.). Washington D. C., USA: ASM Press.

Jordan, K. N., Oxford, L., & O'Byrne, C. P. (1999). Survival of low-pH stress by *Escherichia coli* O157:H7: correlation between alterations in the cell envelope and increase acid tolerance. *Applied and Environmental Microbiology, 65*, 3048-3055. Retrieved from http://aem.asm.org/content/65/7/3048.full.pdf+html

Leclerc, H., Mossel, D. A. A., Edberg, S. C., & Struijk, C. B. (2001). Advances in the bacteriology of the coliform group: their suitability as markers of microbial water safety. *Annual Review of Microbiology, 55*, 201-234. http://dx.doi.org/10.1146/annurev.micro.55.1.201

Manafi, M. (1996). Fluorogenic and chromogenic enzyme substrates in culture media and identification tests. *International Journal of Food Microbiology, 31*, 45-58. http://dx.doi.org/10.1016/0168-1605(96)00963-4

Nikaeen, M., Pejhan, A., & Jalali, M. (2009). Rapid monitoring of indicator coliforms in drinking water by an enzymatic assay. *Iranian Journal of Environmental Health Science and Engineering, 6*, 7-10. Retrieved from http://journals.tums.ac.ir/upload_files/pdf/_/12607.pdf

O'Hara, C. M., Rhoden, D. L., & Miller, J. M. (1992). Reevaluation of the API 20E identification system versus conventional biochemicals for identification of members of the family *Enterobacteriaceae*: a new look at an old product. *Journal of Clinical Microbiology, 30*, 123-125. Retrieved from http://jcm.asm.org/content/30/1/123.full.pdf+html

O'Toole, D. K., & Chang, M. M. P. (1999). The use of MUG supplement to detect *Escherichia coli* by the multiple tube method in marine waters in Hong Kong. *Marine Pollution Bulletin, 38*, 921-924. http://dx.doi.org/10.1016/S0025-326X(99)00097-1

Oliver, J. D. (2000). The public health significance of viable but nonculturable bacteria. In R. R.Colwell & D. J. Grimes (Eds.), *Nonculturable Microorganisms in the Environment* (pp. 277-300). Washington D. C., USA: ASM Press. http://dx.doi.org/10.1007/978-1-4757-0271-2_16

Ossmer, R. (1993). *Simultaneous detection of total coliforms and E. coli – Fluorocult® LMX Broth*. 15th

International Syposium in Food Microbiology (August 31-September 3) Bingen, Germany.

Pinto, D., Almeida, V., Almeida Santos, M., & Chambel, L. (2011). Resuscitation of *Escherichia coli* VBNC cells depends on a variety of environmental and chemical stimuli. *Journal of Applied Microbiology, 110*, 1601-1611. http://dx.doi.org/10.1111/j.1365-2672.2011.05016.x

Pommepuy, M., Butin, M., Derrien, A., Gourmelon, M., Colwell, R. R., & Cormier, M. (1996). Retention of enteropathogenicity by viable but nonculturable *Escherichia coli* exposed to seawater and sunlight. *Applied and Environmental Microbiology, 62*, 4621-4626. Retrieved from http://aem.asm.org/content/62/12/4621.full.pdf+html

Prats, J., Garcia-Armisen, T., Larrea, J., & Servais, P. (2008). Comparison of culture-based methods to enumerate *Escherichia coli* in tropical and temperate freshwater. *Letters in Applied Microbiology, 46*, 243-248. http://dx.doi.org/10.1111/j.1472-765X.2007.02292.x

Rompré, A., Servais, P., Baudart, J., de Roubin, M-R., & Laurent, P. (2002). Detection and enumeration of coliforms in drinking water: current methods and emerging approaches. *Journal of Microbiological Methods, 49*, 31-54. http://dx.doi.org/10.1016/S0167-7012(01)00351-7

Roszak, D. B., & Colwell, R. R. (1987). Survival strategies of bacteria in the natural environment. *Microbiological Reviews, 51*, 365-379. Retrieved from http://www.ncbi.nlm.nih.gov/pmc/articles/PMC373117/

Smith, J. J., Howington, J. P., & McFeters, G. A. (1994). Survival, physiological response and recovery of enteric bacteria exposed to a polar marine environment. *Applied and Environmental Microbiology, 60*, 2977-2984. Retrieved from http://aem.asm.org/content/60/8/2977.full.pdf+html

Smith, P. B., Tomfohrde, K. M., Rhoden, D. L., & Balows, A. (1972). API system: a multitube micromethod for identification of *Enterobacteriaceae*. *Applied Microbiology, 24*, 449-452. Retrieved from http://aem.asm.org/content/24/3/449.full.pdf+html

Suwansonthichai, S., & Rengpipat, S. (2003). Enumeration of coliforms and *Escherichia coli* in frozen black tiger shrimp *Penaeus monodon* by conventional and rapid methods. *International Journal of Food, Microbiology, 81*, 113-121. http://dx.doi.org/10.1016/S0168-1605(02)00190-3

Tallon, P., Magajna, B., Lofranco, C., & Leung, K. T. (2005). Microbial indicators of fecal contamination: A current perspective. *Water, Air, and Soil Pollution, 166*, 139-166. http://dx.doi.org/10.1007/s11270-005-7905-4

Tsai, Y. L., & Olson, B. H. (1992). Detection of low numbers of bacterial cells in soils and sediments by polymerase chain reaction. *Applied and Environmental Microbiology, 58*, 754-757. Retrieved from http://aem.asm.org/content/58/2/754.full.pdf+html

Wanjugi, P., & Harwood, V. J. (2013). The influence of predation and competition on the survival of commensal and pathogenic fecal bacteria in aquatic habitats. *Environmental Microbiology, 15*, 517-526. http://dx.doi.org/10.1111/j.1462-2920.2012.02877.x

Washington, J. A., Yu, P. K. W., & Martin, W. J. (1971). Evaluation of accuracy of multitest micromethod system for indentification of *Enterobacteriacae*. *Applied Microbiology, 22*, 267-269. Retrieved from http://aem.asm.org/content/22/3/267.full.pdf+html

Wcisto, R., & Chróst, R. J. (2000). Survival of *Escherichia coli* in freshwater. *Polish Journal of Environmental Studies, 9*, 215-222. Retrieved from http://www.pjoes.com/pdf/9.3/215-222.pdf

Whitman, R. L., Nevers, M. B., Korinek, G. C., & Byappanahalli, M. N. (2004). Solar and temporal effects on *Escherichia coli* concentration at a lake Michigan swimming beach. *Applied and Environmental Microbiology, 70*, 4276-4285. http://dx.doi.org/10.1128/AEM.70.7.4276-4285.2004

Xu, H. S., Roberts, N., Singleton, F. L., Attwell, R. W., Grimes, D. J., & Colwell, R. R. (1982). Survival and viability of nonculturable *Escherichia coli* and *Vibrio cholerae* in the estuarine and marine environment. *Microbial Ecology, 8*, 313-323. http://dx.doi.org/10.1007/BF02010671

7

Why Cyanobacteria Produce Toxins? Evolutionary Game Theory Suggests the Key

Beatriz Baselga-Cervera[1], Camino García-Balboa[1], Eduardo Costas[1] & Victoria López-Rodas[1]

[1] Genetics. Animal Production. Veterinary Faculty. Complutense University, Madrid, Spain

Correspondence: Victoria López-Rodas, Genetics. Animal Production. Veterinary Faculty. Complutense University, Madrid, Spain. E-mail: vlrodas@ucm.es

Abstract

Cyanobacteria are a source of potent toxins among which the microcystin (a hepatotoxic peptide encoded by the *mcy* gene cluster of *Microcystis spp.*) is a frequent cause of poisoning in inland waters worldwide. Although the molecular basis of microcystin production is known, its role is still unknown. It was suggested that microcystin production have a metabolic cost that could be offset by some benefit (e.g. protection from grazing). We check that: i) microcystin-producing and non-producing strains occurs simultaneously in the *Microcystis spp.* blooms, ii) evolutionary forces (mutation, genetic drift) control frequencies of microcystin production and non-production strains, and iii) microcystin producing strains have diminished fitness compared with non-producing strains. We employ evolutionary game theory to explain the maintaining of microcystin-producing genotypes in natural populations of *Microcystis spp.* A two-strategy (to produce or not microcystin), two-players game of cooperators (microcystin-producing genotypes) and cheaters (non-producing genotypes) explains the coexistence of both genotypes in the same bloom. A bloom composed mostly by the microcystin-producing "cooperators" genotype, the "temptation of defection" (increase of non-producing genotypes) is counteracted by kin selection, which enable that natural selection can favour the cooperators. The closest related individuals occur within cyanobacteria blooms, cyanobacteria reproduces asexually providing sets of clones.

Keywords: toxic cyanobacteria, microcystin, kin selection, game theory, cooperators-cheaters

1. Introduction

Cyanobacteria are often competitively superior to other phytoplankton (Dokulil & Teubner, 2000) and recurrently form dense surface blooms in inland waters worldwide (Bartram, Carmichael, Chorus, Jones & Skulberg, 1999), a risk that could increase in frequency as a result of global change (Huertas, Rouco, López-Rodas & Costas 2010; Huertas, Rouco, López-Rodas & Costas, 2011; Rouco, López-Rodas, Flores-Moya & Costas, 2011). Many species of cyanobacteria can produce potent cyanotoxins that threaten humans, livestock and wildlife (Carmichael, 2001), but it has long been known that *Microcystis aeruginosa* Kützing is the most frequent cause of toxicity affecting humans and animals (Sivonen et al., 1990). *M. aeruginosa* produce cyclic heptapeptides called microcystin, which are potent hepatotoxins (Carmichael, 1994; Dawson, 1998). Since microcystins are the most commonly encountered cyanotoxins they are a matter which is of increasing concern. For instance, many people died of acute liver failure as the result of intoxication by microcystin in Brazil (Jochimsen, Carmichael, An, Cardo & Cookson, 1998) and numerous toxic episodes were also reported in China (Chen & Xie, 2005; Zhang, Xie, Liu, Chen & Liang, 2007). Exposure to low concentrations of microcystin through drinking water increases significantly the risk of liver and colorectal cancer (Martínez-Hernandez, López-Rodas & Costas, 2009). Also, microcystin cause catastrophic mass mortalities of fauna even in pristine national parks of wildlife (Alonso-Andicoberry, García-Villada, López-Rodas & Costas, 2002; López-Rodas, Maneiro, Lanzarot, Perdigones & Costas, 2008).

It has long been known that a complex mixture of microcystin-producing and non-producing strains occurs simultaneously during the *M. aeruginosa* blooms (Codd, Bell, & Brooks, 1989; Shirai et al., 1991; Kaebarnick & Neilan, 2001; Vezie et al., 1998). For example, around one-third of strains isolated from lakes and reservoirs in Spain were found to be non-toxic (Alvarez, Basanta, López-Rodas & Costas, 1998; Martin, Carrillo & Costas, 2004; Carrillo et al., 2003).

Over the last years, the molecular basis of how *Microcystis* produce microcystin has been discovered. Microcystin is a peptide synthesised non-ribosomally via a thio-template mechanism (Arment & Carmichael, 1996) by the microcystin synthetase enzyme complex, which is encoded by the microcystin (*mcy*) gene cluster (Kurmayer & Christiansen, 2009). The *mcy* gene cluster of *Microcystis* has been already sequenced (Nishizawa et al., 2000; Tillett et al., 2000). Molecular analysis shows that evolution of the *mcy* genes occurs through asexual reproduction and horizontal gene transfer, with ancestral wild type microcystin-producing genotypes and strains with non-functional genes (Bjørg et al., 2003).

Unfortunately, the molecular basis of the *myc* gene cluster does not explain the biological role of microcystin. The classic approach assumes that microcystin production has a metabolic cost, but generates some advantages that compensate the cost. In this sense, numerous evidences suggest that microcystin production prevent grazing of *Microcystis* by zooplankton, filter feeders and others. Toxin-producing *Microcystis* often have lethal effects upon zooplankton, or significant reduce the growth of these primary consumers (De Mott, Zhang, & Carmichael, 1991; Kinder, 1995; Rohrlack, Henning, & Kohl, 1999; Trubetskova & Haney, 2000). *Daphnia* species actively evade *Microcystis* blooms through vertical migrations (Kinder, 1995). Filter-feeding animals as zebra mussels employ a sophisticated mechanism based on the selective alimentation eliminating living *Microcystis* in pseudo-feces to avoid microcystin producers (Vanderploeg et al., 2001). It is known that animals that consume *Mycrocystis* scum die due to the toxic effect of cyanotoxins (Alonso-Andicoberrt et al., 2002; López-Rodas et al., 2008; Francis, 1878; Galey et al., 1987; Soll & Williams, 1985; Matsunaga et al., 1999). Thus, the protection against grazing will compensate the metabolic cost of microcystin production. Other useful functions for microcystin have been suggested such as binding iron (Utkilen & Gjølme, 1995) or involvement in quorum sensing (Dittmann et al., 2001). Despite this, experiments with cultures of *Microcystis* that had the *myc* gene knockout showed no apparent differences with *m+* functional cultures (Dittmann et al., 2001). Producing microcystin does not seem to be a substantial cost under laboratory conditions. Additionally, other laboratory experiments show that the values of heritability of microcystin production were significantly higher than those found in fundamental physiological characters (López-Rodas et al., 2006). It is generally assumed that only the quantitative traits with little or no selective advantage show high values of heritability (Falconer & Mackay, 1996).

Surely, microcystin production is more complex than which can be explained from the current knowledge of the molecular basis of the myc gene complex, because microcystin production seems to be a complex quantitative trait. In fact, laboratory studies have shown that changes in environmental conditions induce changes in microcystin concentration within a strain by a factor of three to four times, whereas microcystin production between strains grown under identical culture conditions may vary in the three or more orders of magnitude (reviewed by Sivonen & Jones, 1999; Carrillo et al., 2003; Lopez-Rodas et al., 2006).

Consequently, the role played by microcystin production in the biology of *Microcystis* is still far from clear. Einstein said once that you cannot solve a problem with the same mentality it has been posed, but rather you must change the way you approach it. So, we decided to address this controversy in a different perspective. The key question is to analyse why usually microcystin-producing and non-producing strains appear together in the same *Microcystis* bloom and neither of the strains prevail over the other. A population genetics approach could find out what kinds of evolutionary forces are acting (i.e. mutation, natural selection, genetic drift...) on microcystin-producing and non-producing strains and calculate their genotypic frequencies and their evolution (throughout of fitness values, selection coefficients, effective population sizes...).

Consequently, first we performed a classic population genetics experiment to estimate the role played by the evolutionary forces on the frequency of microcystin-producing and non-producing strains. Afterwards, the results obtained were analysed employing evolutionary game theory.

Evolutionary game theory could explain the cases in which producing microcystin has an evolutionary advantage in a population of *Microcystis aeruginosa* and viceversa. Since von Neumann & Morgenstern (1944) and Nash (1950) presented game theory as a mathematical approach to evaluate the results of the different strategic decisions among the complex interactions of self-interested individuals, this discipline has been able to successfully resolve complex problems of population genetics. The application of game theory to evolving populations of life organisms (i.e. evolutionary game theory) is focused on the effect of the frequency with which various competing strategies are found in the population. Evolutionary game theory has been particularly useful to explain complex aspects of biology that had not been explained within the context of Darwinian process by classic population genetics, such as the evolutionary basis of altruistic behaviours (Maynard-Smith, 1982).

The conceptual basis of our approach is based on three steps:

1) Isolate microcystin-producing and non-producing strains from the same population of *Microcystis aeruginosa* (because the evolutionary phenomena occur within the populations).

2) Demonstrate that classic evolutionary forces (mutation, genetic drift and natural selection) can change the frequency of the toxin producing and no toxin producing genotypes (to prove that microcystin production is not only a physiologic phenomenon).

3) Applicate evolutionary game theory to estimate the best strategy (i.e. producing or non-producing toxins) according to the population conditions (i.e. the frequencies of the producing and non-producing cells).

2. Material and Methods

2.1 Isolation of Strains and Culture Conditions

Twenty-one different clonal cultures of *M. aeruginosa,* numbered from one to twenty-one, from the Algal Culture Collection, Veterinary Faculty, Complutense University, Madrid, were used (Figure 1). All the strains were isolated from same area of Guadalquivir basin (SE Spain), in order to compare the population genetics parameters of different strains within the same population, because the evolutionary forces (e.g. natural selection) act within populations. Each strain was isolated as follows: First of all colonies morphologically identified as *M. aeruginosa* based on morphologic characteristics according to Komarek & Anagnostidis (1998) were isolated using micro-pipettes. Only colonies within the middle of the range of cell and colony sizes with clear *M. aeruginosa* morphotypes were isolated. Afterwards, a single vegetative cell was isolated from each colony using a Zeiss-Eppendorf (Carl Zeiss, Hamburg, Germany) micromanipulator–microinjector. Since *M. aeruginosa* cells divide asexually, the clonality of each culture was assumed.

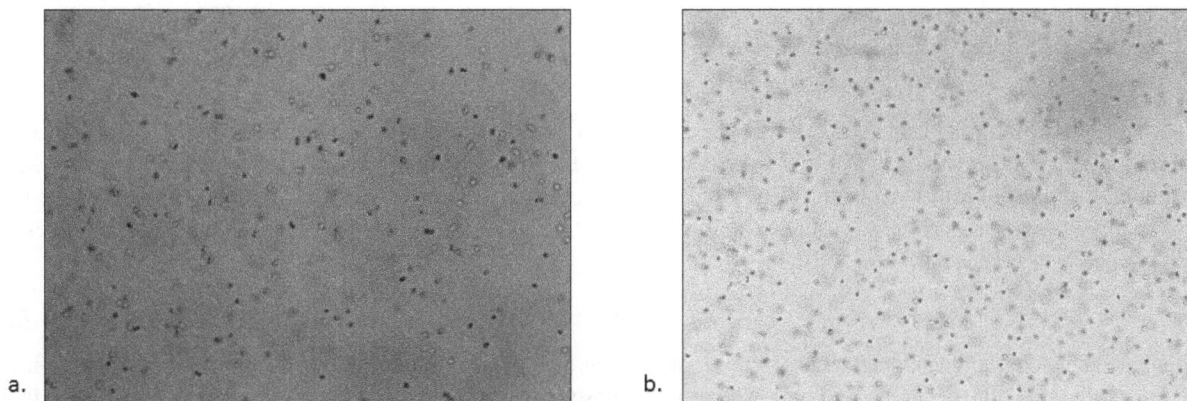

Figure 1. Microphotographs of two different clonal cultures of *M. aeruginosa*

Two microphotographs of two different strains of *M. aeruginosa* from the Algal Culture Collection, Veterinary Faculty, Complutense University, Madrid. a) a non-toxic strain and b) a toxic strain. As can be seeing in the figure *M. aeruginosa* does not make colonies in laboratory cultures.

The strains were grown in 100 mL ventilated cell culture flasks covered with a filter cap (Greiner, Bio-One Inc., Longwood, 155 NJ, USA) with 20 mL of BG-11 medium (Sigma, Aldrich Chemie, Taufkirchen, Germany) at 24°C with a continuous photon irradiance of 80 µmol photons m^{-2} s^{-1} over the waveband 400–700 nm, supplied by daylight fluorescent tubes. Cells were maintained axenically in balanced growth (mid log exponential growth) by serial transfers of an inoculum to fresh medium (Cooper, 1991). Periodic controls (i.e. by observation under fluorescence microscope after acridine orange staining and plating in standard bacterial cultures) assure the absence of detectable bacteria. The position of each culture flask in the incubator chamber was randomly changed once every day. More details on culture of Microcystis are given in Lopez-Rodas et al. (2006) and Rouco et al., (2011). In order to allow full environmental acclimation, the strains were grown under these constant experimental conditions during two months. The maintenance of the strains under the same stable environmental conditions is necessary for disengaging environmental and genetic influences on the phenotypic variability, which ensures that differences found among strains are exclusively due to genetic variability (Brand, 1981; Costas, 1990).

2.2 Microcystin Determination

The microcystin production of the different strains was measured in mid log exponential growth cultures (i.e. acclimated, balanced growth (Cooper, 1991)) of 3×10^6 cell ml^{-1} using a microcystin commercial kit MicroCystest (ZEU-INMUNOTEC, Zaragoza, Spain), based on inhibition of the protein phosphatase 2A (PP2A) by microcystins, and therefore capable of detecting all potentially toxic microcystins with a detection limit of 0.08μg/L and a working range 0.25-2.5 μg/L. This test was approved by Environmental Technology Verification program of the USA Environment Protection Agency (EPA). Extraction of toxins and total microcystin (water dissolved microcystin + microcystin within cells) was measured following the manufacturer's recommendations and expressed as μgrams per litre (μg/L). Two different operators performed triplicate measurements of microcystin. The reliability, reproducibility and precision of microcystin measurements were established according to the British Standards Institute (1979) and Thrusfield (1995). Reliability was determined based on the agreement between three iterations of measurements made by the same observer on the same replicate, whereas reproducibility was determined as the agreement among three sets of observations made by two different observers. The measurements to guarantee reliability, reproducibility and precision were performed on two lots of reagents using (i) microcystin-LR controls (Sigma–Aldrich) at 1.0 and 4 μg/L, and (ii) extracts from Ma2, Ma7 and Ma14 strains containing 3×10^6 cell ml^{-1}. We select these three strains in order to have representation of a non-producing strain (Ma2), a strain with an average production (Ma7) and the highest microcystin-producing strain (Ma14). Precision was determined as the minimum variation in microcystin, which could be detected.

2.3 Fitness Estimation

The Malthusian parameter of fitness (m) was estimated as previously described (Costas, 1991) as the acclimated maximal growth rate from the general equation of growth:

$$N_t = N_0 e^{mt} \tag{1}$$

where N_0 is the initial cell number (after the lag period), and N_t is the cell number after 7 d; therefore, m was computed as Log e $(N_7 / N_0) / 7$. Cells were counted using settling chambers, by two independent observers who counted three replicates per strain. The number of samples counted per replicate was determined using the progressive mean procedure to ensure a counting error of $\leq 5\%$. The Darwinian fitness (i.e. relative fitness) of each genotype was estimated as the ratio:

$$m_i / m_{max} \tag{2}$$

where m_i is the Malthusian parameter of the strain i and m_{max} is the maximum Malthusian parameter. Consequently, the maximum value of fitness will be 1. The growth rates were estimated using triplicates of each genotype. More details on estimation of fitness in *Microcystis aeruginosa* (including heritability and evolution of fitness and microcystin production) are given in López-Rodas et al. (2006) and Rouco et al. (2011). The reliability, reproducibility and precision of microcystin measurements and fitness estimations were established as previously described for *Microcystins*. Ma2, Ma7 and Ma14 strains.

Afterwards, we use linear regression to modelling the relationship between fitness (dependent variable y) and microcystin production (explanatory variable x) using InStat package (GraphPad Software, La Jolla, CA, USA).

2.4 Population Genetics Experiments

We assume the simplest theoretical model of population genetics that might well explain the observed results: i) the functional microcystin-producing genotypes (*m*+) could become non-functional (*m*-) due to occurrence of spontaneous mutation (*m*+ → *m*-); ii) conversely non functional microcystin-producing genotypes (*m*-) can become functional (*m*+) due to reversion (*m*- → *m*+). Within a microcystin-producing population should have a majority of *m*+ genotypes and perhaps a few *m*- genotypes originated by mutation. On the contrary, within a non-producing population should have a majority of *m*- genotypes and possibly a few *m*+ genotypes originated by reversion In a small population the alleles *m*+ and *m*- should be under a mutation–genetic drift balance.

We can essay this model in the laboratory using experimental populations of *Microcystis aeruginosa*. The experimental populations were laboratory cultures microcystin producing (*m*+ strains) and non-producing (*m*-) strains maintained by serial transfers of a small inoculum. In these experimental populations act mutation and genetic drift. The possibility of losing an allele by genetic drift in small populations is high. Consequently, sometimes microcystin-producing cultures could lose their toxicity by random genetic drift, because only *m*- genotypes (originated by mutation *m*+ → *m*-) were transferred in the small inoculum. And just the opposite, sometimes non-producing cultures could became toxic also by random genetic drift, because only *m*+ genotypes (originated by reversion $m-$ → $m+$) were transferred in the small inoculum. To experimentally check this model, two experiments were performed simultaneously:

i) experiments maximizing random genetic drift: Twenty-one clonal cultures were grown until a cell number of about 10^8. Afterwards the cultures were transferred to fresh medium using extremely small inoculum (around 100 cells) to maximize genetic drift. The probability that some allele is lost in such small inoculum is very high. This flash-crash process was repeated periodically during 20 times.

ii) experiments minimizing random genetic drift (controls): The same twenty-one clonal cultures were grown until a cell number of about 10^8 cells. Afterwards the cultures were transferred to fresh medium using large inoculum (around 10^5 cells) to prevent genetic drift. The probability that some allele is lost in such inoculum is extremely low. This serial transfer process was also repeated periodically during 20 times.

In both experiments, the microcystin production was measured at the start of and after the 20 flash-crash (i.e. mutation- genetic drift) cycles in exactly the same conditions.

2.5 Evolutionary Game Theory Model of Microcystin Production

We assume that a bloom is a *Microcystis aeruginosa* population where $m+$ is a microcystin-producing wild type genotype, and $m-$ is a non-producing mutant genotype. Relative frequency of m^+ is p, relative frequency of m^- is q. Fitness of $m+$ genotype is w_1 and fitness of $m-$ genotype is w_2.

We can systematize all possible cases by considering an evolutionary game between two strategies, i) produce microcystin (m^+ genotype) and ii) not produce microcystin (m^- genotype), playing in two different populations, population 1 with predominant m^+ genotypes and ii) population 2 with predominant m^- genotypes.

The payoff matrix of both strategies is:

playing against   (3)

	$m+$ genotype	$m-$ genotype
$m+$ genotype	w_1 (+)	w_1 (-)
$m-$ genotype	w_2 (+)	w_2 (-)

The entries of the matrix denote the fitness for the row player. A $m+$ genotype obtains a payoff w_1 (+) when playing another $m+$ genotype, but payoff w_1 (-) when playing a $m-$ genotype. Likewise, $m-$ genotype obtains a payoff w_2 (+) when playing against $m+$ genotype, or w_2 (-) when playing against $m-$ genotype. This is a case of two-players, two-strategy game, whose evolutionary dynamics has studied in detail (Rapoport & Chammah, 1965; Axelrod, 1984; Taylor & Nowak, 2009).

3. Results and Discussion

3.1 Evolutionary Forces (as Mutation or Genetic Drift) can Affect the Microcystin Production

Measurements of microcystin production were performed with high reliability (98.2%±0.7%), reproducibility (96.8%±1.0%) and precision (0.1 μg/ ml). Despite being grown under identical laboratory conditions, initially 6 strains (Ma2, Ma5, Ma6, Ma10, Ma16, Ma17) were unable to produce microcystin, whereas that the 15 strains produce 1.34 ± 0.23 μg/ml of microcystin LR equivalent (mean ± standard error) in of cultures with 3 x 10^6 cell/ml.

When the 21 clonal strains were subjected to experiments maximizing random genetic drift (the cultures were transferred using inoculum around 100 cells during 20 times), most of the strains maintained similar microcystin production (Table 1). But four clones (Ma3, Ma13, Ma20, Ma21) that initially had produced microcystin became no producing after the experiment (Table1). Inversely, two clones that had not produced microcystin at first (Ma6, Ma16) became able to produce at the end of the experiment (Table 1). In contrast, when the twenty-one clonal strains were maintained by serial transfers using large inoculum (10^5 cells) to prevent genetic drift, toxicity of each strain remained unchanged (Table 1).

Table 1. Population genetics experiments to check the role of mutation, reversion and genetic drift between microcystin-producing and no-producing strains (+ microcystin producing strain; - no-producing strain)

Strain	Initial microcystin production	Final microcystin production after experiments maximizing random genetic drift	Final microcystin production after experiments minimizing random genetic drift (controls)
Ma1	+	+	+
Ma2	-	-	-
Ma3 [a]	+	-	+
Ma4	+	+	+
Ma5	-	-	-
Ma6 [b]	-	+	-
Ma7	+	+	+
Ma8	+	+	+
Ma9	+	+	+
Ma10	-	-	-
Ma11	+	+	+
Ma12	+	+	+
Ma13 [a]	+	-	+
Ma14	+	+	+
Ma15	+	+	+
Ma16 [b]	-	+	-
Ma17	-	-	-
Ma18	+	+	+
Ma19	+	+	+
Ma20 [a]	+	-	+
Ma21 [a]	+	-	+

[a] initially microcystin-producing strains, which became no-producing after the experiments maximizing random genetic drift.

[b] initially no-producing strains, which became producing after the experiments maximizing random genetic drift.

These results suggest that $m-$ genotypes could occur in strains of $m+$ genotype as a result of spontaneous mutation and also could become fixed in populations by genetic drift. Previous work show that genetic drift is an important component in the evolutionary changes affecting to microcystin production, that occur in *Microcystis* populations as response to eutrophication and temperature increase (Rouco et al., 2011) as well as in other harmful algae (Flores-Moya, Costas, & López-Rodas, 2008; Flores et al., 2012).

3.2 Darwinian Fitness of Microcystin-Producing and Non-Producing Strains

Darwinian fitness was measured with high reliability (93.3% ± 1.2%) and reproducibility (90.7% ± 0.9%), whereas precision was of 0.01, high enough to detect inter-strain variability.

Initially, a high inter-strain variation was found for the fitness measured under identical laboratory conditions (Table 2). It should be noted that no microcystin-producing strains has higher fitness than microcystin-producing. Fitness of non-producing strains ranges from 0.8 to 1.0 (mean = 0.90; sd = 0.09), whereas fitness of microcystin producing strains range from 0.4 to 0.8 (mean = 0.65; sd = 0.13). Statistically significant differences (p = 0.0006, t = 4.12, d.f. 0 19 in Student t-test) were observed between producing and non-producing strains. A regression analysis of microcystin production (measured in ppb of microcystin LR-equivalent in cultures with 3 x 10^6 cell/ml) and Darwinian fitness of the 21 different strains of *M. aeruginosa* also show that the genotypes that produce more amount of microcystin have less Darwinian fitness than those non-producers genotypes (statistically significant negative regression, P< 0.001, R^2 = 0.71 was observed; Figure 2).

When these strains were subjected to experiments maximizing random genetic drift most of the strains maintained similar fitness values. But two strains (Ma3 and Ma20) increase in fitness after losing the ability to produce microcystin, whereas the strain Ma20 decrease in fitness after recovering the ability to produce microcystin (Table 2). In contrast, when the twenty-one clonal strains were maintained by serial transfers using large inoculum (10^5 cells) to prevent genetic drift, toxicity of each strain remained unchanged.

Fitness

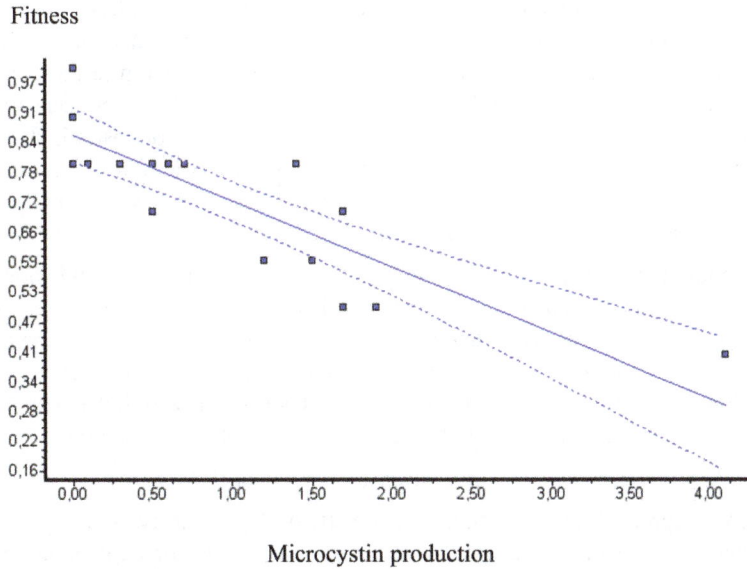

Microcystin production

Figure 2. Linear regression

Linear regression and 95% confidence interval between Microcystin production (in ppb of cultures with 3×10^6 cell ml^{-1}) and Darwinian fitness of 21 different strains of *M. aeruginosa*.

Table 2. Darwinian fitness of 21 different strains of *M. aeruginosa*

Strain	Initial microcystin production	Initial Darwinian fitness	Final Darwinian fitness after experiments maximizing random genetic drift	Final Darwinian Fitness after experiments minimizing random genetic drift (controls)
Ma1	+	0.8	0.7	0.8
Ma2	-	0.8	0.9	0.8
Ma3 [a]	+	0.6*	0.9*	0.6
Ma4	+	0.8	0.7	0.8
Ma5	-	0.9	0.9	0.9
Ma6 [b]	-	0.8*	0.5*	0.8
Ma7	+	0.6	0.6	0.6
Ma8	+	0.8	0.7	0.8
Ma9	+	0.8	0.8	0.8
Ma10	-	1.0	1.0	1.0
Ma11	+	0.8	0.8	0.8
Ma12	+	0.7	0.6	0.7
Ma13 [a]	+	0.7	0.8	0.7
Ma14	+	0.4	0.4	0.4
Ma15	+	0.5	0.5	0.5
Ma16 [b]	-	0.9	0.8	0.9
Ma17	-	1.0	1.0	1.0
Ma18	+	0.7	0.6	0.7
Ma19	+	0.6	0.6	0.6
Ma20 [a]	+	0.5*	0.7*	0.5
Ma21 [a]	+	0.6	0.6	0.6

* statistically significant differences in mean of fitness between initial Darwinian fitness and final Darwinian fitness after experiments maximizing genetic drift, p<0.01, student t test.

[a] initially microcystin-producing strains, which became no producing after the experiments maximizing random genetic drift.

[b] initially no microcystin-producing strains, which became producing after the experiments minimising random genetic drift.

An outstanding fact that emerges from our results is that producing microcystin has a strong cost in fitness. This interesting fact had not been highlighted previously because the studies that characterize the fitness of different strains (genotypes) isolated from the same population are scarce. In the literature there are studies comparing *Microcystis* strains isolates from different populations, forgetting that natural selection acts within populations and not between populations (Lewontin, 1974; Spiess, 1989; Gould, 2002). In contrast, all strains used in our study were isolated as close as possible in space and time, so that they should belong to the same population. A recent study analysing DNA sequencing and genetic variability for physiological and morphological traits show that these strains are very close at genetic level (Lopez-Rodas, 2013).

According to classic population genetics results obtained by fitness valuation suggest that non-producing genotype (*m*-) during a *Microcystis* bloom should be more numerous than microcystin-producing genotype (*m*+) due to it higher fitness. But numerous evidences reveal that the non-toxin producing cells do not prevail massively during a bloom (reviewed in Sivonen & Jones, 1999; data for South Spain populations in Alvarez et al., 1998; Carrillo et al., 2003; Martin et al., 2004; Lopez-Rodas, 2006). Consequently, it is necessary to find an explanation for the natural maintenance of a high frequency of microcystin producing cells in a bloom population.

3.3 The Model of Population Genetics and Game Theory of m+ and m- Genotypes

We propose a theoretical approach of game theory model (two-player and two-strategy model) to explain the abundance of microcystin producing cells. Similar approach was tackle for explaining apparently low fitness behaviours as altruism (Maynard Smith, 1982; Thrusfield, 1995)

Game theory, it has long been employed to determine the best strategy for a genotype interacting with other genotypes to reach the best outcome to himself (Hamilton, 1964; Trivers, 1971; Maynard Smith & Price, 1973). In essence, an evolutionary game is related to reproductive success so that the fitness of a genotype depends on payoff of his strategy interacting in cooperative or non-cooperative strategies with other various strategies in the population (Samuelson, 1997; Hofbauer & Sigmund, 2003; Nowak & Sigmund, 2004; Nowak, 2006).

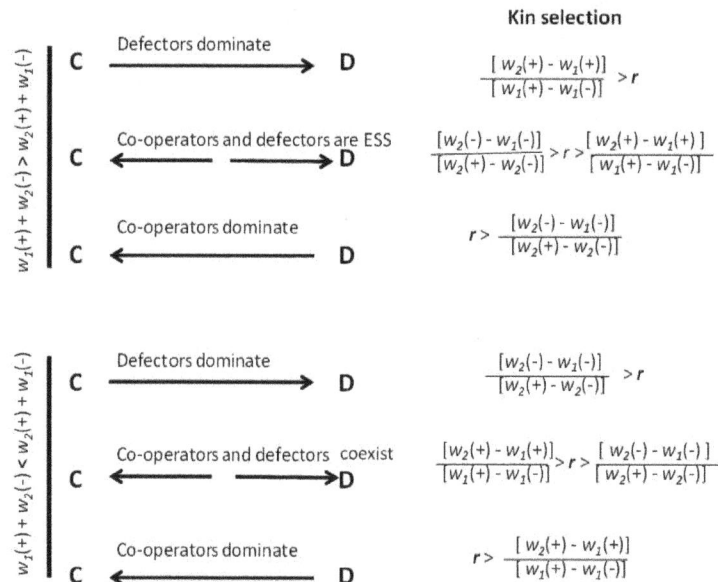

Figure 3. Kin selection can lead to the evolution of cooperation

Being the parameter r the coefficient of genetic relatedness between individuals, for this model we find $w_1(+) + w_2(-) > w_2(+) + w_1(-)$, cooperators and defectors cannot coexist; if $w_1(+) + w_2(-) < w_2(+) + w_1(-)$ cannot be simultaneously a evolutionary stable strategy (ESS).

If $w_1(+) + w_2(-) > w_2(+) + w_1(-)$ then $r_D > r_C$. Three possibilities: i) if $r_D > r_C > r$ then defectors dominate; ii) if $r_D > r > r_C$ then cooperators and defectors are a ESS; iii) if $r > r_D > r_C$ then co-operators dominate.

If $w_1(+) + w_2(-) < w_2(+) + w_1(-)$ then $r_C > r_D$. Three possibilities: i) if $r_C > r_D > r$ then defectors dominate; ii) if $r_C > r > r_D$ then co-operators and defectors are not a ESS; iii) if $r > r_C > r_D$ then cooperators dominate.

Based on the payoff matrix of two players ($m+$ and $m-$) with each of them competing against the other for the same ambient, different strategies can take place as a function of the payoff which leads to different final populations. Evolutionarily stable strategies (i.e. is stable against invasion by a fraction of mutants using the other strategy) can occur depending of payoffs (Figure 2): two strategies of the matrix are Nash equilibrium (the best strategy of a player in order to acquire the maximum possible payoff, knowing the strategy that perform the other player) (Nash, 1950): The first is if $w_1(+) > w_2(+)$ and the second if $w_1(-) < w_2(-)$. In a large mixed population (without small subpopulations), a Nash equilibrium strategy is an evolutionary stable strategy (Maynard Smith, 1982). In practice, the huge blooms of *Microcystis* composed of billions of individuals mixed by the water movements are as close as possible to a theoretical infinite, well-mixed population. Consequently, in such populations 3 different evolutionary stable strategies are possible (Figure 3):

i) If $w_1(+) > w_2(+)$ and $w_1(-) > w_2(-)$, then $m+$ dominates $m-$. In this case is better to produce microcystin (i.e. be a wild type $m+$ genotype) because the expected payoff to produce microcystin ($m+$ genotype) is greater than that of not produce microcystin ($m-$ genotype) in any *Microcystis* population. Consequently, wild type the $m+$ genotypes always prevail in the population.

ii) In the reverse situation $w_1(+) < w_2(+)$ and $w_1(-) < w_2(-)$, then $m-$ dominates $m+$. Obviously, in this case the expected payoff of $m-$ is greater than that of $m+$. Here, $m-$ mutants always prevail in the population.

iii) If both conditions of Nash equilibrium $w_1(+) > w_2(+)$ and $w_1(-) < w_2(-)$ occurs together, then both strategies can be a evolutionary stable strategy. In a population where most genotypes are wild type $m+$ it is the bests to be $m+$. In contrast, in a population where most genotypes are $m-$ mutants it is the bests to be $m-$.

Since in the blooms of *Microcystis* simultaneously coexist microcystin-producing and non-producing genotypes, obviously do not occur Nash equilibrium in natural conditions. But another solution is possible when strategies generate payoff that are different to those that allow Nash equilibrium. This solution is the key to understand why wild type microcystin-producing $m+$ genotypes and non-producing $m-$ genotypes always appear together in the *Microcystis* blooms.

iv) If $w_1(+) < w_2(+)$ and $w_1(-) > w_2(-)$, then there is stable co-existence between both genotypes. In a *Microcystis* bloom where most genotypes are microcystin-producing wild type $m+$, ($p > q$) it is the bests strategy to be a non-toxic $m-$ genotype. Contrarily, in a *Microcystis* population where $p < q$, it is the best strategy to be $m+$ genotype. Thus, equilibrium of frequency-dependent selection is achieved. In this case, there is a stable co-existence between the both genotypes in the *Microcystis* populations. Only this theoretical solution of the payoff matrix presents a stable coexistence of both genotypes (i.e. in a *Microcystis* bloom where most genotypes are $m+$ it is the best strategy to be $m-$ and viceversa). In this case, the frequency dependent selection leads to equilibrium as shown in the *Microcystis aeruginosa* blooms.

At the biological level, there is an explanation to this model of stable coexistence of both genotypes. Although producing toxins is useful to prevent grazing by zooplankton filter feeders and large animals (Alonso-Andicoberry et al., 2002; López-Rodas et al., 2008; De Mott et al., 1991; Kinder, 1995; Rohrlack et al., 1999; Trubetskova & Haney, 2000; Vanderploeg et al., 2001; Frances, 1878; López-Rodas & Costas, 1999), it is also costly in terms of fitness, whereby the toxin-producing $m+$ genotypes showed diminished fitness with respect to non-producing $m-$ genotypes in the absence of grazers (Carrillo et al., 2003; López-Rodas et al., 2006). Our results also show significant negative regression between fitness and toxin production.

Let us assume a *Microcystis* bloom constituted by individuals with the $m+$ ancestral genotype. In this bloom appears a new genotype $m-$ by a rare spontaneous mutation, which does not produce microcystin, so it fitness is slightly above than those of $m+$ ancestral genotypes and is also protected from grazing although he do not produce microcystin. As its fitness is greater than the ancestral $m+$ genotypes, the frequency of $m-$ mutants will increase gradually. In this population it is best to be $m-$, it is tempting to be a $m-$ genotype. But there will come a point where if there are not enough $m+$ genotypes, the bloom would not be protected any more.

This can be interpreted as that the $m+$ genotypes forgo some of their potential for the common good of the population, performing a cooperative strategy. But the temptation for cheating ($m-$ genotype) plays an important role. Under this view, the *Microcystis* blooms comprise two strategies: i) the cooperators ($m+$ genotypes), which would invest in microcystin production for the good of the entire population, and ii) the cheaters ($m-$ genotypes), which would take the advantage conferred by the $m+$ cooperators to acquire the maximum profit from the situation, giving nothing in return.

Consequently, the maintenance of cooperation requires a specific mechanism to enable that natural selection can favour to the cooperators instead of to the cheaters. Kin selection could be the mechanism. Classic view of kin selection is based on the concept that usually the evolutionary game is played among genetic relatives individuals (Hamilton, 1964; Samuelson, 1997; Cavalli-Forza & Felman, 1978). A gene encodes some kind of cooperative behaviour promotes its own survival if it is also present in the beneficiary of the cooperation. Cyanobacteria reproduce asexually. Consequently, cyanobacteria populations are formed by sets of clones (López-Rodas & Costas, 1997) of his act of cooperation becomes a clone of herself. Clones are more likely to benefit from (pseudo)altruistic behaviour than distantly related individuals.

From the seminal works of Maynard-Smith (1982), Nowak (2006) and Taylor & Novak (2009) arises an intriguing idea. Cooperators and cheaters could coexist in a population where the values of the payoff of both strategies ($m+$ cooperators and $m-$ cheaters) meet the following condition:

$$\frac{[w2(+) - w1(+)]}{[w1(+) - w1(-)]} > r > \frac{[w2(-) - w1(-)]}{[w2(+) - w2(-)]} \quad \text{(Hofbauer \& Sigmund, 2003)} \tag{4}$$

$$r_{cooperators} > r_{total\ population} > r_{cheaters}$$

where:

a $m+$ genotype obtains a payoff $w_1(+)$ when playing another $m+$

a $m+$ genotype obtains a payoff $w_1(-)$ when playing a $m-$

a $m-$ genotype obtains a payoff $w_2(+)$ when playing a $m+$

a $m-$ genotype obtains a payoff $w_2(-)$ when playing a $m-$

and r = coefficient of genetic relatedness between individuals (an estimation of the average genetic relatedness between the playing individuals)

Apparently r can be high in the blooms of cyanobacteria, because experimental data show that *Microcystis* blooms are usually constituted by few different genotypes (Alvarez et al., 1998; López-Rodas & Costas, 1997; Martín-Montaño et al., 2000). In addition, the approximation described above determinates that for the coexistence of the $m+$ and $m-$ must be also satisfied that:

$$r_{cooperators} > r_{total\ population} > r_{cheaters} \tag{5}$$

so is said that the coefficient of cooperators should be higher than the total coefficient of genetic relatedness of the population and in turn, the coefficient of cheaters should be lower than the other two. For this to happen, within the population the majority of the population should be the $m+$ genotype, which fit with the experimental evidences (Alvarez et al., 1998; López-Rodas & Costas, 1997; Martín-Montaño et al., 2000). Whereupon it could be said that the strategy perform by cyanobacteria depends in last term on the kinship of the cooperators genotype.

Other causes (such as group selection, graph selection, direct and indirect reciprocity, reviewed by Nowak, 2006; Taylor & Nowak, 2009) could explain microcystin production, but as cyanobacteria are very simple organisms, it seems reasonable to propose a *lex parsimoniae* explanation.

Experimental evidence allows to validate a population genetics model based on a two-strategy (to produce or not produce microcystin), two-players game of cooperators (the wild type $m+$ microcystin-producing genotypes) and cheaters (the $m-$ non-producing mutants), which explain the coexistence of both genotypes in the same population and provides a clear explanation of the role played by microcystin. Our model for explaining toxin production by *Microcystis* is based on the following assumptions: i) mutation from wild type toxin-producing $m+$ genotypes to non-producing $m-$ mutants producers occurs spontaneously; ii) reversion from non-producing $m-$ mutants to toxin-producing $m+$ genotypes also occurs spontaneously; and iii) in the main, toxin-producing $m+$ genotypes have less fitness than non-producing $m-$ mutants.

For the moment, an experimental demonstration of this game theory model of two strategies and two players not easily approachable. Instead, we have provide a list of indirect experimental evidences. In any case, our approach seems to be a "gedanken experiment" type of problem (a thought experiment, a hypothetical scenario carried out to understand a real problem). At present, there is a tendency to interpret the phenomenon of microcystin production under the reductionist perspective of what we know about its molecular basis (specifically on the *myc* gene cluster). These studies have often assumed that *myc* genes operate in a closed system in which the presence or absence of certain gene sequences directly determines microcystin production. However, cyanobacteria blooms exhibit a wide range of variability in microcystin production, which is the result of barely measurable interactions between a particular genotype and its corresponding environment. In such cases, classical models of population genetics have a great power to explain biological phenomenon under an evolutionary level. There is also a tendency to seek new functions to explain the role of microcystin, relegating to a second place its toxic effect, despite that continuously produces countless mass mortalities among the consumers of cyanobacteria. In contrast, our model follow an Ockham´s razor type argument (among competing hypothesis, the one with less assumptions should be selected), which takes into account the environment and operates whatever the genetic basis of toxin production (because the selection unit is the complete genotype). After all, Charles Darwin masterfully explained the mechanisms of evolution nearly 100 years before the development of molecular genetics.

4. Conclusions

1.   Evolutionary forces (i.e. mutation, genetic drift, selection) control the frequency of microcystin-producing and non-producing cells in a process at the population genetics level and not at the physiological level.

2.   Microcystin has useful functions (i.e. grazing prevention) but its production is so costly that the non-producing *m-* genotypes have higher fitness. Consequently, non-producing genotypes should be predominant in a bloom, which contradicts observations from natural *Microcystis* blooms (usually around 60-70% microcystin producing cell, 30-40% non-producing cells)

3.   A two-strategy (to produce microcystin or not), two-players evolutionary game of cooperators (microcystin-producing) and cheaters (non-producing) is the simplest explanation to the simultaneous occurrence of microcystin-producing and non-producing strains. In this game, depending on the payoff matrix could be achieved evolutionary stable strategies (such as Nash equilibrium with domain of producing or non-producing genotypes respectively), but also a stable co-existence between both genotypes if when the microcystin-producing cells dominates being non-producing is the best strategy and vice versa.

4.   The maintenance of microcystin-producing cooperators requires kin selection to enable that natural selection can favour to the cooperators instead of to the non-producing cheaters. Since cyanobacteria reproduce asexually, their populations are formed by sets of clones, which is the closest kinship possible.

Acknowledgements

Supported by CTM2012-34757 grant and CTM 2013-44366-R grant, Spanish government. To Lara de Miguel Fernández for her excellent technical support.

References

Alonso-Andicoberry, C., García-Villada, L., López-Rodas, V., & Costas, E., (2002). Catastrophic mortality of flamingos in a Spanish national park caused by cyanobacteria. *Veterinary Record, 151,* 706-707.

Alvarez, M. J., Basanta, A., López-RodasÓPEZ-RODAS, V., & Costas, E. (1998). Identification of different serotypes during a Microcystis aeruginosa bloom in a SW Spanish reservoir. *Harmful Algae.,* 291-295.

Arment, A. R., & Carmichael, W. W. (1996). Evidence that microcystin is a thio-template product. *Journal of Phycology, 32,* 591-597. http://dx.doi.org/10.1111/j.0022-3646.1996.00591.x

Axelrod, R. (1984). *The Evolution of Cooperation,* Basic Books. http://dx.doi.org/10.1126%2Fscience.7466396.

Bartram, J., Carmichael, W. W., Chorus, I., Jones, G., & Skulberg, O. M. (1999). Chapter 1 Introduction, In, *Toxic Cyanobacteria in Water, A Guide to Their Public Health Consequences Monitoring and Management,* World Health Organization.

Bjørg, M., Boison, G., Skulberg, O. M., Fastner, J., Davies, W., & Gabrielsen, T. M. (2003) Natural Variation in the Microcystin Synthetase Operon mcyABC and Impact on Microcystin Production in Microcystis Strains. *Journal of Bacteriology, 185*(9), 2774-2785. http://dx.doi.org/10.1128/JB.185.9.2774-2785.2003.

Brand, L. E. (1981). Genetic variability in reproduction rates in marine phytoplankton populations. *Evolution, 35*(6), 1117-1127. http://dx.doi.org/10.2307/2408125

Carmichael, W. W. (1994). The Toxins of Cyanobacteria. *Scientific American, 270,* 78-86. http://dx.doi.org/10.1038/scientificamerican0194-78.

Carmichael, W. W. (2001). Health Effects of Toxin-Producing Cyanobacteria, "The CyanoHABs". *Human Ecological Risk Assessment, 7*(5), 1393-1407. http://dx.doi.org/10.1080/20018091095087

Carrillo, E., Ferrero, L. M., Alonso-Andicoberry, C., Basanta, A., Martin, A., López Rodas, V., & Costas, E. (2003). Interstrain variability in toxin production in populations of the cyanobacterium *Microcystis aeruginosa* from water-supply reservoirs of Andalusia and lagoons of Doñana National Park (southern Spain). *Phycologia, 42*(3), 269-274. http://dx.doi.org/10.2216/i0031-8884-42-3-269.1

Cavalli-Forza, L. L., & Feldman, M. W. (1978). The evolution of continuous variation III Joint transmission of genotype phenotype and environment. *Genetics, 90*, 391-425.

Chen, Y., & Xie, J. (2005). Third-Party Product Review and Firm Marketing Strategy. *Marketing Science, 24*(2), 218–240. http://dx.doi.org/10.1287/mksc.1040.0089

Codd, G. A., Bell, S. G., & Brooks, W. P. (1989). Cyanobacterial toxins in water. *Water Science Technology, 21*, 1-13. http://dx.doi.org/10.1080/09670269910001736462

Cooper, S. (1991). *Bacterial growth and division, biochemistry and regulation of prokaryotic and eukaryotic division cycles*. Academic Press.

Costas, E. (1990). Genetic variability in growth rates of marine dinoflagellates. *Genetica, 82*(2), 99-102. http://dx.doi.org/10.1007/BF00124638

Dawson, J. W. (1998). Institutions Investment and Growth, New Cross-Country and Panel Data Evidence. *Economic Inquiry, Western Economic Association International, 36*(4), 603-19.

De Mott, W. R., Zhang, Q., & Carmichael, W. W. (1991). Effects of toxic cyanobacteria and purified toxins on the survival and feeding of a copepod and three species of Daphnia. *Limnology and Oceanography, 36*, 1346-1357. http://dx.doi.org/10.4319/lo.1991.36.7.1346

Dittmann, E., Erhard, M., Kaebernick, M., Scheler, C., Neilan, B. A., & Börner, T. (2001). Altered of a peptide synthetase gene that is responsible for hepatotoxin production in the cyanobacterium *Microcystis aeruginosa* PCC7806. *Molecular Microbiology, 147*, 3113-3119.

Dokulil, M., & Teubner, K. (2000). Cyanobacterial dominance in lakes. *Hydrobiologia, 438*, 1–12. http://dx.doi.org/10.1023/A:1004155810302

Falconer, D. S., & Mackay, T. F. C. (1996). *Introduction to Quantitative Genetics,* Ed 4 Longmans Green, Harlow, Essex, UK.

Flores, A., Rouco, M., García-Sánchez, M. J., García-Balboa, C., Gonzalez, R., & Costas, E. (2012). Effects of adaptation chance and history on the evolution of the toxic dinoflagellate *Alexandrium minutum* under selection of increased temperature and acidification. *Ecology and Evolution, 2*, 1251-1259. http://dx.doi.org/10.1002/ece3.198

Flores-Moya, A., Costas, E., & López-Rodas, V. (2008). Roles of adaptation chance and history in the evolution of the dinoflagellate *Prorocentrum triestinum* under simulated global change conditions. *Naturwissenschaften, 95*, 697-703. http://dx.doi.org/10.1007/s00114-008-0372-1

Francis, G. (1878). Poisonous Australian lake. *Nature, 18*, 11–12. http://dx.doi.org/10.1038/018011d0

Galey, F. D., Beasley, V. R., Carmichael, W. W., Kleppe, G., Hooser, S. B. & Haschek, W. M. (1987). Blue-green algae (*Microcystis aeruginosa*) hepatotoxicosis in dairy cows. *American Journal of Veterinary Research, 48*, 1415-1420.

Gould, J. S. (2002). The Structure of Evolutionary Theory. Harvard University Press 1433pp.

Hamilton, W. D. (1964). The genetical evolution of social behaviour II. *Journal of Theoretical Biology, 7*, 17-52. http://dx.doi.org/10.1016/0022-5193(64)90039-6

Hofbauer, J., & Sigmund, K. (2003). Evolutionary Game Dynamics. *Bulleting of the American Mathematical Society, 40*, 479-519. http://dx.doi.org/10.1090/S0273-0979-03-00988-1

Huertas, I. E., Rouco, M., López-Rodas, V., & Costas, E. (2010). Estimating the capability of different phytoplankton groups to adapt to contamination, herbicides will affect phytoplankton species differently. *New Phytology, 188*, 478-487. http://dx.doi.org/10.1111/j.1469-8137.2010.0.3370.x

Huertas, I. E., Rouco, M., López-Rodas, V., & Costas, E. (2011). Warming will affect phytoplankton differently, evidence through a mechanistic approach. *Proceedings of the Royal Society B, 278*, 3534–3543. http://dx.doi.org/10.1098/rspb.2011.0160

Jochimsen, E. M., Carmichael, W. W., An, J. S., Cardo, D. M., & Cookson, S. T. (1998). Liver failure in death after exposure to microcystin at a hemodialysis center in Brazil. *New England Journal of Medicine, 338*, 873-878. http://dx.doi.org/10.1056/NEJM199807093390222

Kaebarnick, M., & Neilan, B. A. (2001). Ecological and molecular investigations of cyanotoxin production. *FEMS Microbiology Ecoogyl, 35*(1), 1-9. http://dx.doi.org/10.1111/j.1574-6941.2001.tb00782.x

Kinder, K. R. (1995). The effect of Microcystis aeruginosa on Daphnia feeding behavior and vertical distribution, PhD MS Thesis University of New Hampshire, Durham NH, USA.

Komarek, J., & Anagnostidis, K. (1998). *Cyanoprokaryota 1*. In H. Ettl Gärner, H. Heynig, & D. Mollenhauer (Eds.), *Teil: Chroococcales*. Süsswasserflora von Mitteleuropa 19/1, Gustav Fischer, Jena 548p.

Kurmayer, R., & Christiansen, G. (2009). The genetic basis of toxin production in Cyanobacteria. *Freshwater Review, 2*, 31-50. http://dx.doi.org/10.1608/FRJ-2.1.2

Lewontin, R. C. (1974). *The genetic basis of evolutionary change*. Columbia University Press, New York, NY, USA.

López-Rodas, V., & Costas, E. (1997). Characterization of morphospecies and strains of *Microcystis* (Cyanobacteria) from natural populations and laboratory clones using cell probes (lectins and antibodies). *Journal of Phycology, 33*, 446-454. http://dx.doi.org/10.1111/j.0022-3646.1997.00446.x

López-Rodas, V., & Costas, E. (1999). Preference of mice to consume *Microcystis aeruginosa* (toxin producing cyanobacteria), A possible explanation for numerous fatalities of livestock and wildlife. *Research in Veterinary Science, 67*, 107-110. http://dx.doi.org/10.1111/j.0022-3646.1997.00446.x

Lopez-Rodas, V., Costas, E., & Flores-Moya, A. (2013). Phenotypic and genetic diversities are not correlated in strains of the cyanobacterium *Microcystis aeruginosa* isolated in sw Spain. *Acta Botanica Malacitana, 38*, 5-12.

López-Rodas, V., Costas, E., Bañares, E., García-Villada, L., Altamirano, M., & Rico, M. (2006). Analysis of poligenic traits of *Microcystis aeruginosa* (Cyanobacteria) strains by Restricted Maximum Likelihood (REML) procedures, 2 Microcystin net production, photosynthesis and respiration. *Phycologia, 45*(3), 243-248. http://dx.doi.org/10.2216/04-31.1

López-Rodas, V., Maneiro, E., Lanzarot, M. P., Perdigones, N., & Costas, E. (2008). Cyanobacteria cause mass mortality of wildlife in Doñana National Park. *Veterinary Records, 162*, 317-318. http://dx.doi.org/10.1136/vr.162.10.317

Martin, A., Carrillo, E., & Costas, E. (2004). Variabilidad genética para la producción de toxina en poblaciones de *Microcystis aeruginosa* en dos embalses de abastecimiento de Andalucia. Limnetica, 23, 153-158.

Martínez Hernandez, J., Lopez-Rodas, V., & Costas, E. (2009). Microcystins from tap water could be a risk factor for liver and colorectal cancer: a risk intensified by global change. *Medical Hypothesis, 72*, 539–540. http://dx.doi.org/10.1016/j.mehy.2008.11.041

Martín-Montaño, A., Carrillo, E., Costas, E., & Basanta, A. (2000). Identificación de serotipos de Microcystis aeruginosa con distinto grado de toxicidad en un embalse de abastecimiento. *Tecnología del Agua, 199*, 54-59.

Matsunaga, H., Harada, K. I., Senma, M., Ito, Y., Ushida, S., & Kimura, Y. (1999). Possible cause of unnatural mass death of wild birds in a pond in Nishinomiya, Japan, sudden appearance of toxic cyanobacteria. *Natural Toxins, 7*, 81-84. http://dx.doi.org/10.1002/(SICI)1522-7189(199903/04)7:2<81::AID-NT44>3.0.CO;2-O

Maynard Smith, J., & Price, G. R. (1973). The logic of animal conflict. *Nature, 246*, 15-18. http://dx.doi.org/10.1038/246015a0

Maynard Smith, J. (1964). Group selection and kin selection. *Nature, 201*, 145–47.

Maynard Smith, J. (1982). *Evolution and the Theory of Games*. Cambridge University Press. http://dx.doi.org/10.1017/CBO9780511806292

Nash, J. (1950). Equilibrium points in n-person games. *Proceedings of the National Academy of Science USA, 36*(1), 48-49. http://dx.doi.org/10.1073/pnas.36.1.48

Nishizawa, T., Ueda, A., Asayasma, M., Fujii, K., Harada, K. I., Ochi, K., & Shirai, M. (2000). Polyketide synthase gene coupled to the peptide synthetase module involved in the biosynthesis of the cyclic heptapeptide microcystin. *Journal of Biochemistry, 127*, 779-789. http://dx.doi.org/10.1093/oxfordjournals.jbchem.a022670

Nowak, M. A., & Sigmund, K. (2004). Evolutionary dynamics of biological games. *Science, 303*(5659), 793-799. http://dx.doi.org/10.1126/science.1093411

Nowak, M.A. (2006). Five rules for the evolution of cooperation. *Science,* 314 (5805), 1560-1563. http://dx.doi.org/10.1126/cience.1133755

Rapoport, A., & Chammah, A. (1965). *Prisoner's Dilemma*. Ann Arbor, University of Michigan Press.

Rohrlack, T., Henning, M., & Kohl, J. G. (1999). Does the toxic effect of *Microcystis aeruginosa* on *Daphnia galeata* depend on microcystin ingestion rate? *Archiv für Hydrobiologie, 146*, 385-395.

Rouco, M., López-Rodas, V., Flores-Moya, A., & Costas, E. (2011). Evolutionary changes in growth rate and toxin production in the cyanobacterium Microcystis aeruginosa under a scenario of eutrophication and temperature increase. *Microbial Ecology, 62*, 265-273. http://dx.doi.org/10.1002/ece3.198

Samuelson, L. (1997). *Evolutionary games and equilibrium selection*. MIT, Cambridge.

Shirai, K. M., Ohtake, A., Sano, T., Matsumoto, S., Sakamoto, T., & Sato, A. (1991). Toxicity and toxins of natural blooms and isolate strains of *Microcystis spp* (cyanobacteria) and improved procedure for purification of cultures. *Applied and Environmental Microbiology, 57*(4), 1241-1245.

Sivonen, K., & Jones, G. (1999). *Cyanobacterial toxins*. In I. Chorus, & J. Bartram Eds.), *Toxic Cyanobacteria in Water, A Guide to Their Public Health Consequences, Monitoring and Management* (pp. 41–111). London: Spon Press.

Sivonen, K., Carmichael, W. W., Namikoshi, M., Rinehart, K. L., Dahlem, A. M, & Niemelä, S. I. (1990). Isolation and characterization of hepatotoxic microcystin homologues from the filamentous freshwater cyanobacterium Nostoc sp Strain 152 App. *Environmental Microbiology, 56*, 2650-2657.

Soll, M. D., & Williams, M. C. (1985). Mortality of a white rhinoceros (*Ceratotheratium simum*) suspected to be associated with the blue-green alga, *Microcystis aeruginosa. Journal of South African Veterinary Association, 56*, 49-51.

Spiess, E. B. (1989). *Genes in Populations* (2nd ed.). Wiley, New York, NY, USA.

Taylor, C., & Nowak, M. A. (2009). *How to evolve cooperation*. In S. Levin (Ed.), Games, Groups, and the Global Good (pp. 41-56). New York: Springer. http://dx.doi.org/10.1007/978-3-540-85436-4_2

Thrusfield, M. (1995). *Veterinary Epidemiology* (2nd ed.). Blackwell Science.

Tillett, D., Dittmann, E., Erhard, M., vonDöhren, H., Börner, T., & Neilan, B. A. (2000). Structural organization of microcystin biosynthesis in *Microcystis aeruginosa* PCC7806, an integrated peptide-polyketide synthetase system. *Chem Biol, 7*, 753–764. http://dx.doi.org/10.1016/S1074-5521(00)00021-1

Trivers, R. L. (1971). The Evolution of Reciprocal Altruism. *The Quarterly Review of Biology, 46*(1), 35–57. http://dx.doi.org/10.1086/406755

Trubetskova, I., & Haney, J. (2000). The impact of the toxic strain of *Microcystis aeruginosa* on *Daphnia magna Crustacean. Issues, 12*, 457-461.

Utkilen, H., & Gjølme, N. (1995). Iron-stimulated toxin production in *Microcystis aeruginosa. Applied and Environmental Microbiology, 61*, 797-800.

Vanderploeg, H. A., Liebig, J. R., Carmichael, W. W., Agy, M. A., Johengen, T. H., & Fahnenstiel, G. F. (2001). Zebra mussel (Dreissena polymorpha) selective filtration promoted toxic Microcystis blooms in Saginaw Bay (Lake Huron) and Lake Erie, *Can J Fish Aquat Sci, 58*, 1208–1221. http://dx.doi.org 10.1139/cjfas-58-6-1208

Vezie, C., Brient, L., Sivonen, K., Bertru, G., Lefeuvre, J. C., & Salkinoja-Salonen, M. (1998). Variation of microcystin content of cyanobacterial blooms and isolated strains in lake Grand- Lieu (France), *Microbial Ecology, 35*, 126-135. http://dx.doi.org 10.1007/s002489900067

von Neumann, J., & Morgenstern, O. (1944). *Theory of Games and Economic Behavior*. Princeton University Press.

Zhang, D., Xie, P., Liu, Y., Chen, J., & Liang, G. (2007). Bioaccumulation of the hepatotoxic microcystins in various organs of a freshwater snail from a subtropical Chinese Lake, Taihu Lake, with dense toxic *Microcystis blooms. Environmental Toxicology and Chemistry, 26*(1), 171–176. http://dx.doi.org/10.1897/06-222R.1

8

Psoroptes sp. Infestation in Sulawesi Bear Cuscus (*Ailurops ursinus*) in Indonesia

Purwanta[1], Mihrani[1], Sartika Juwita[1], Ahmad Nadif[2], Ali Ma'shum[1] & Muh Arby Hamire[1]

[1] Agriculture Extension College, Gowa, South Sulawesi, Indonesia

[2] Agriculture Quarantine of Pare-Pare, South Sulawesi, Indonesia

Correspondence: Purwanta, Agriculture Extension College, Gowa, South Sulawesi 92171, Indonesia. E-mail: purwantadrhmkes@gmail.com

Abstract

One male and two female sulawesi bear cuscus (*Ailurops ursinus*), weighing 4.4, 5.1 and 4.6 kg was admitted to the Animal Health Center of the Gowa Agriculture Extension College. Upon physical examination auricular skin lesions, and erythematous and pruritic skin lesions, both on the ventral abdomen and on extremities were detected. Microscopic examination of skin scrapings taken from pinnae and hair plucked from the medial extremities region revealed the presence of *Psoroptes sp.* The ventral abdominal and extremitas localization of *Psoroptes sp.* was evaluated as an ectopic infestation. To our knowledge, this is the first report of *Psoroptes sp.* in Sulawesi Bear Cuscus (*Ailurops ursinus*) in Indonesia. The Sulawesi bears cuscus were injected subcutaneously with ivermectin at 0.1 mg/kg of bodyweight, as well as with injected intramuculary a ADE combination to supportive therapy. Three Sulawesi Bears Cuscus became negative for mites after third treatments of ivermectin at seven days interval, and clinical mange did not recur.

Keywords*: Psoroptes sp.*, infestation, Sulawesi bear cuscus (*Ailurops ursinus*)

1. Introduction

Diseases of the sulawesi bear cuscus (*Ailurops ursinus*) rare published. The disease is one threat to the survival of possum if not handled properly and wildlife (Pederson et al., 2007). Sulawesi bear cuscus (*Ailurops ursinus*) is one of the animal species endemic to the island of Sulawesi, which is protected by Indonesian Government Regulation Number. 7/1999. These animals included in the red list of threatened species IUCN 2008 (Salas et al., 2008). Sulawesi bear cuscus is an animal *marsupial* and from family *Phalangeridae*.

Report cases of disease caused *Psoroptes sp.* the possum has not been published, one of a kind *Psoroptes* is *Psoroptes cuniculi* generally attacks on domesticated livestock such as rabbits (Acar et al., 2007; Kyung-Yeon & Oh-Deog, 2010), *Psoroptes* is the causative agent of dermatitis in cattle, sheep, goats, rabbits and turkey (Kurtdede et al., 2007; Lekimme et al., 2008). Type *P. cuniculi* attack on wildlife group *Artiodactyls* reported by Pederson et al. (2007), whereas the incidence of deer in the United States reported by Schmith et al. (1982).

This mite has a host of high specification. *Psoroptes* species that normally live on the host will not infest other host with different species. *Psoroptes* do not dig a tunnel under the skin and only live on the surface, under the scab, and under the accumulated pile of scaly skin, outer ear, *auditory* canal, and obtain food by piercing the skin (Bowman, 1999).

2. Case History and clinical examination

One male and two female sulawesi bear cuscus (*Ailurops ursinus*), weighing 4.4, 5.1 and 4.6 kg was admitted to the Animal Health Center of the Agriculture Extension College, Gowa from Gowa Discovery Park (GDP). Cuscus presented pruritis, alopecic and weight loss. Clinical examination revealed severe bilateral lesions in both pinnae, and erythematous, and alopecic skin lesions with pruritis on the ventral abdomen, perianal and extremities (Figure 1).

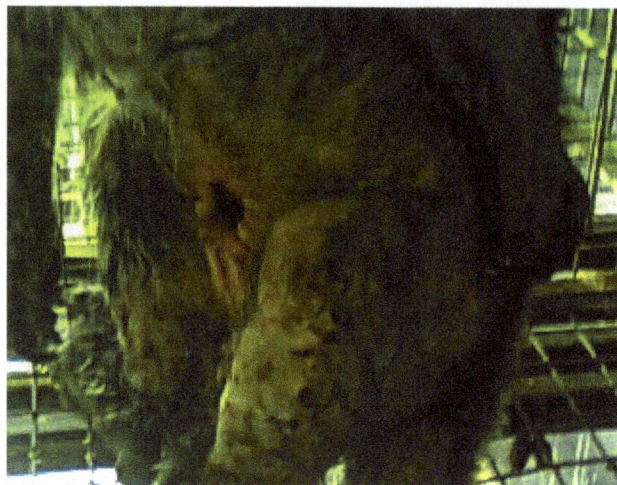

Figure 1. Infection showing spread in extremitas, ventral abdomen and tail

3. Diagnosis

Diagnosis was performed by clinical signs, dermatological and microscopic examination of the skin lesions. Microscopic examination of skin scrapings taken from pinnae and hair plucked from the medial extremities region revealed the presence of Mange. Many mites were detected on microscopic examination of material scraped from the external ventral abdominal and extremitas. Skin scrapings from each region then included in a separate vial containing a solution KOH 10%. Identification is done under a microscope. Results identification diagnosed as *Psoroptes sp.* (Figure 2).

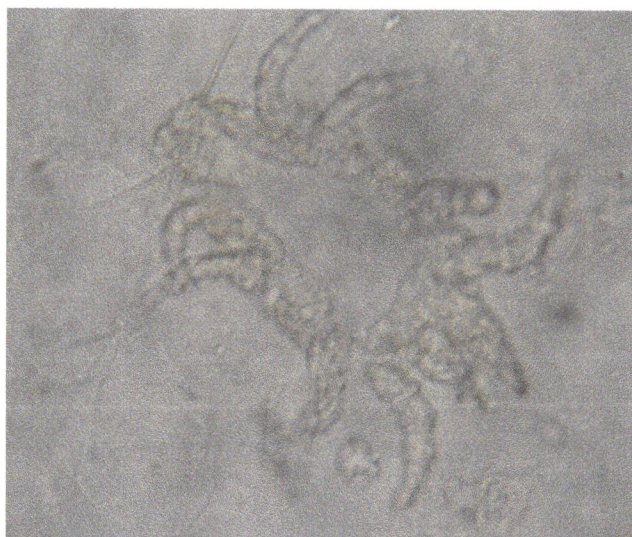

Figure 2. Microscopic examination of skin sample from ventral abdomen of cuscus observes the mite *Psoroptes sp.*

Psoroptes sp. has an oval body shape and length of the section conical front. *Psoroptes* has a long pedicle connected to the carancula. Mites large female with a body length of about 750 μm. Tarsi I and II culminate in karankula while the tarsi III with the same limb size led to long setae (Wall & Shearer, 1997).

4. Treatment

The cuscuses were injected subcutaneously with ivermectin at 0.1 mg/kg of body weight, as well as with injected intramusculary a ADE combination to supportive therapy. Three cuscus became negative for mites after third treatments of ivermectin at seven days interval, and clinical mange did not recur.

5. Discussion

Based on the results of a physical examination on cuscus, there is a dominant lesions include alopecia and purities at the ventral abdominal, extremity, lightly hooked around the eyes and the base of the ear. This is in contrast to the infestation *P.caniculi* on a change in the dominant rabbit ears, however, in many cases *Psoroptes* affected mainly in area of ear, head, neck, abdomen, legs and perianal region of farm rabbits reported severe itching, infections, lesions and swelling. In older or sick animals, or if not treated properly, the parasite can spread to other regions on the body. The mite *P. cuniculi* is a worldwide obligatory ectoparasite, mainly of rabbits, goats, horses, and sheep (Perrucci et al., 2005). Diagnosis of mange is achieeved through observation of clinical signs e.g. itching, pruritis, and wool loss and ultimately through the detection of mites in skin scrapings. Early stages of infestation are often difficult to diagnose and subclinical animals can be a major factor in disease spread (Burgess et al., 2012).

The mites puncture the epidermis of ear, suck lymph and give rise to local inflammatory swelling from which serum exude, coagulates and forms enormous encrustation inside the ear. The mite causes intense pruritus with formation of crusts and scabs, which can completely fill the external ear canal and internal surface of the pinnae in untreated animals (Perrucci et al., 2005). In the rabbit, it is important to mention that in present study, lesions in the experimental infestation with the mite werw detected after 21 day post infestation (Hallal-Calleros et al., 2013) However, this study, Cuscus presented pruritis, alopecic and weight loss. Clinical examination revealed severe bilateral lesions in both pinnae, and erythematous, and alopecic skin lesions with pruritis on the ventral abdomen, perianal and extremities (Figure 1). Many cases *Psoroptes* affected mainly in area of ear, head, neck, abdomen, legs and perianal region of farm rabbits reported severe itching, infections, lesions and swelling (Swarnakar et al., 2014).

According to Jeesup and Boyce (2008), ivermectin administered SC 0.2 mg/kg effective controlling *Psoroptes* infestations in cattle, but this formulation has been ineffective in infested big-horn sheep. Our preliminary studies with both domestic and big-horn sheep have indicated that higher doses of ivermectin (from 0.8 to 1 mg/kg) are requred to achive serum levels comparable to those that are miticidal in cattle. In this case report, the dose used 0.1 mg/kg, with consideration has not been reported on the use of ivermectin cuscus, contraindications arise feared or other side effects adverse. Giving the injection of vitamin ADE as supportive therapy to improve growth, enhance immunity against the disease, especially in young animals, help recovering from illness, and cope with hair loss. Treatment is quite effective in controlling infestations *Psoroptes sp.* after 3 repetitions with 1-week intervals, characterized by microscopic examination of skin scrapings not found *Psoroptes sp.*, healing is also characterized by the growth of the region experiencing hair loss (Figure 3). This is in accordance with Hillyer (1994) recommended the administration of 3 doses.

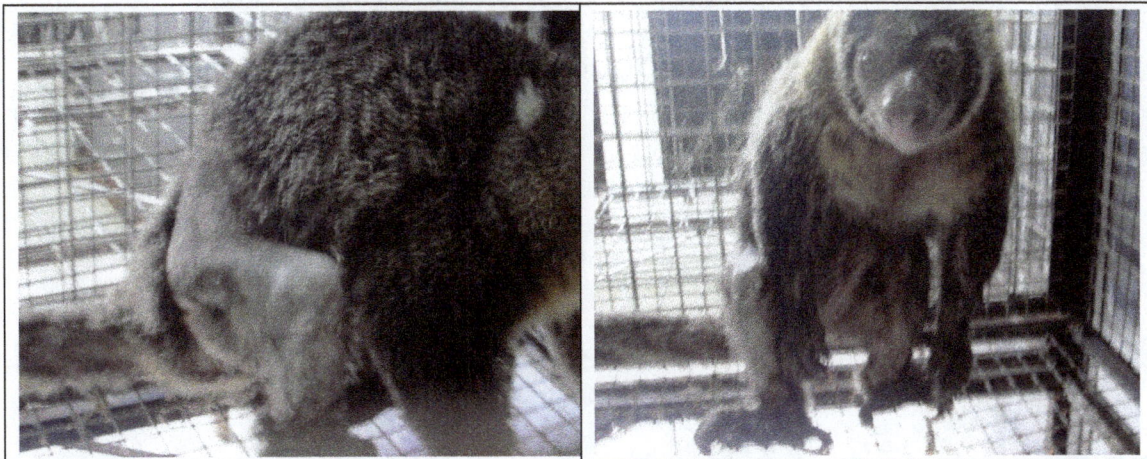

Figure 3. Cuscus after treatment with ivermectin and vitamin ADE injections every week for 3 weeks

Acknowledgement

Authors are grateful to the Operational Manager Sheila Ishak, Gowa Discovery Park (GDP) Makassar South Sulawesi-Indonesia for providing the facilities for the sample collection.

References

Acar, A., Kurtdede, A., Ural, K., Cingi, C. C., Karakurum, M. C., Yaci, B. B., & Sari, B. (2007). An Ectopic Case of *Psoroptes cuniculi* Infestation in a Pet Rabbit. *Turk. J. Vet. Anim. Sci., 31*(6), 423-425.

Bowman, D. D. (1999). *Georgis' Parasitology for Veterinarian* (8nd Ed.). Missouri (US): Saunders.

Burgess, S. T., Innocent, G., Nunn, F., Frew, D., Kenyon, F., Nisbet, A. J., & Huntley, J. F. (2012). The use of a *Psoroptes ovis* serodiagnostic test for the analysis of a natural outbreak of sheep scab. *Parasit Vectors, 5*(7), http://dx.doi.org/10.1186/1756-3305-5-7

Hallal-Calleros, C., Morales-Montor, J., Vázquez-Montiel, J. A., Hoffman, K. L., Nieto-Rodríguez, A., & Flores-Pérez, F. I. (2013). Hormonal and behavioral changes induced by acute and chronic experimental infestation with Psoroptes cuniculi in the domestic rabbit Oryctolagus cuniculus. *Parasites & vectors, 6*(1), 361. http://dx.doi.org/ 10.1186/1756-3305-6-361

Hillyer, E. V. (1994). Pet rabbits. *Vet. Clin. North Am. Small Anim. Pract., 24*, 25-65.

Jessup, D. A., & Boyce, W. M. (2008). *Disease of wild sheep in Zoo and Wild Animal Medicine* (6nd Ed.). In M. E. Fowler, & R. E. Miller (Eds.). WB Saunders Company. Philadelphia, London, Colorado, Montreal, Sydney, Tokyo.

Kurtdede, A., Karaer, Z., Acar, A., Guzel, M., Cingi, C. C., Ural, K., & Ica, A. (2007). Use of selamectin for the treatment of psoroptic and sarcoptic mite infestation in rabbits. *Vet Dermatol., 18*, 18-22. http://dx.doi.org/ 10.1111/j.1365-3164.2007.00563.x

Kyung-Yeon, E., & Oh-Deog, K. (2010). Psoroptic Otocariasis Associated with *Psoroptes cuniculi* in Domestic Rabbits in Korea. *Pak Vet J., 30*(4), 251-252.

Lekimme, M., Focant, C., Farnir, F., Mignon, B., & Losson, B. (2008). Pathogenicity and thermotolerance of entomopathogenic fungi for the control of the scab mite, *Psoroptes ovis. J. Exp. Appl. Acarol., 46*(1-4), 95-104. http://dx.doi.org/10.1007/s10493-008-9171-9

Pedersen, A. B., Jones, K. E., Nunn, C. L., & Altizer, S. (2007). Infectious Diseases and Extinction in Wild Mammals. *Conservation Biology, 21*(5), 1269-1279. http://dx.doi.org/10.1111/j.1523-1739.2007.00776.x.

Perrucci, S., Rossi, G., Fichi, G., & O Brien, D. J. (2005). Relationship between *Psoroptes cuniculi* and the internal bacterium *Serratia marcescens. Exp. Appl. Acarol., 36*, 199-206. http://dx.doi.org/10.1007/s10493-005 -4511-5

Salas, L., Dickman, C., Helgen, K., & Flannery, T. (2008). *Ailurops ursinus*. In IUCN 2013. IUCN Red List of Threatened Species (Ver. 2013.2). Retrieved April 25, 2014 from www.iucnredlist.org

Schmitt, S. M., Cooley, T. M., & Van-Veen, T. W. S. (1982). *Psoroptes cuniculi* in Captive White-Tailed Deer in Michigan. *Journal of Wildlife Disease, 18*(3), 349-351. http://dx.doi.org/10.7589/0090-3558-18.3.349

Swarnakar, G., Sharma, D., Sanger, B., & Roat, K. (2014). Infestation of ear mites *Psoroptes cuniculi* on farm rabbits and its anthropozoonosis in Gudli village of Udaipur District, India. *Int. J. Curr. Microbiol. App. Sci, 3*(3), 651-656.

Wall, R. & Shearer, D. (1997). *Veterinary Entomology: Arthropod Ectoparasites of Veterinary Importance* (1st ed.). London (GB): Chapman and Hall. http://dx.doi.org/10.1007/978-94-011-5852-7

Decline of *Diporeia* in Lake Michigan: Was Disease Associated With Invasive Species the Primary Factor?

Courtney S. Cave[1] & Kevin B. Strychar[1]

[1]Annis Water Resources Institute, Grand Valley State University, Muskegon, Michigan, United States of America

Correspondence: Kevin B. Strychar, Annis Water Resources Institute, Grand Valley State University, Muskegon, Michigan, 49441, USA. E-mail: strychark@gvsu.edu

Abstract

Populations of the freshwater amphipod *Diporeia* spp. have steadily declined in Lake Michigan since the late 1980's. Prior studies have provided inconclusive data on possible reasons for their decline. However, some authors suggest that food competition and/or diseases associated with aquatic invasive species (AIS), such as zebra mussels (*Dreissena polymorpha*), may have caused the collapse of *Diporeia*. In this project, the possibility of pathogens as the cause of the collapse of *Diporeia* has been examined. Linear regression modeling show a significant positive linear association between percent of *Diporeia* exhibiting a pathogenic infection and year (r=0.7202264, p≤0.0124). Chi-square testing for independence was also used to test if there was an association between year and percent infection (X^2 = 50, df = 10, p≤0.0001), implying significant association between year and infection. Hence, the introduction of zebra mussels and the diseases they carry may have been the root cause for the decline of *Diporeia*. Future research is needed to examine other invasive species for similar pathogens, including live studies showing direct causality between zebra mussels and the decline in *Diporeia*.

Keywords: *Diporeia* spp., Lake Michigan, aquatic invasive species, zebra mussels (*Dreissena polymorpha*), disease

1. Introduction

1.1 Background

Diporeia spp. are freshwater amphipods that used to be the most dominant crustaceans in the benthic layer of the Laurentian Great Lakes. High in lipid content, *Diporeia* have previously been considered the primary food source for many bottom feeders in the Great Lakes including whitefish (*Coregonus clupeaformis*), bloater (*Coregonus hoyi*), and slimy sculpin (*Cottus cognatus*) (Nalepa et al., 1998). Since the mid-1980's, however, populations of *Diporeia* began to disappear in Lake Michigan, declining over 95% in the last 15 years in some places (Figure 1). As a consequence, some fish populations also decreased, perhaps resulting from a shift to less nutritional food sources (Nalepa et al., 1998).

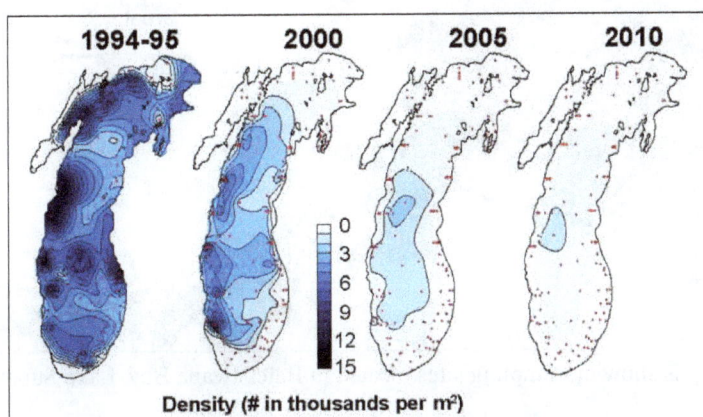

Figure 1. Densities of Lake Michigan *Diporeia* spp. continually decline from 1994 to 2010. Small red dots show NOAA (National Oceanic and Atmospheric Administration) sampling locations (from Nalepa et al., 2014)

The disappearance of *Diporeia* has been postulated to be the result of the invasive Zebra (*Dreissena polymorpha*) and Quagga (*Dreissena rostriformis*) mussels introduced three years prior to the decline of *Diporeia*. Why *Dresseina* spp. had this effect on *Diporeia* is still not completely understood. One hypothesis is that the mussels led to decreased food availability due to food competition (Nalepa, 1989). However, there are some inconsistencies with this hypothesis. *Diporeia* and *Dreissena* coexist in Lake Superior, Lake Cayuga and isolated areas of Lake Michigan. Such observations suggest that the relationship between *Diporeia* and *Dresseina* is more complex than simply competition for food.

The second possibility is that mussels served as a vector for pathogenic organisms infecting *Diporeia*; this hypothesis is still being explored (Fanslow, pers. comm., 2013). In addition, anecdotal evidence indicates that during the early years of decline in *Diporeia*, crustaceans (i.e. shrimp) in many other locations of the United States were experiencing severe population declines, purportedly a consequence of disease. Rickettsia-like infection, i.e. *Haplosporidia* and *Microsporidia*, have all been observed in *Diporeia* tissues (Messick et al., 2004). The origin of these pathogens is not known. However, it is interesting to note that Rickettsia-like infections are found in *Dreissena* located in Greece (Molloy et al., 2001). Similarly, Haplosporidium pathogens were identified as the primary pathogen causing death in *Crassostrea virginica* (Eastern oyster) on the east coast of North America and in fresh water snails (*Physella parkeri*) in Douglas Lake (Michigan; Barrow, 1961). Microsporidia is also a common pathogen in freshwater shrimp (*Gammarus fasciatus*), an amphipod closely related to *Diporeia*. The associated pathologies suggest that any one of these diseases could infect and possibly be the cause of the decline in populations of *Diporeia* in the Great Lakes.

Taxonomically, *Diporeia* spp. belongs to the Phylum Arthropod, Subphylum Crustacea, Class Malacostraca, Order Amphipoda, and Family Pontoporeiidea. In years past, all *Diporeia* were classified as *Pontoporeia hoyi* (anonomous with *P. affinis*), however, taxonomists today believe there may be as many as eight species in the Great Lakes (Cavaletto et al., 1996).

1.2 Objectives of the Study

In order to better understand why *Diporeia* spp. crashed and are still unable to repopulate to their original concentrations, we designed our project to first (1) update the population density of *Diporeia* in Lake Superior's Batchawana Bay. This location has been identified in prior studies as a "safe haven" with high concentrations of healthy *Diporeia* that coexist with a high abundance of zebra mussels. Studies of abundance of *Diporeia* at this location have not been done since 2008. Secondly (2), we wanted to examine preserved (~14 years ago) samples of *Diporeia* tissue for pathogenic infection prior to the introduction of zebra mussels, and immediately there-after, from one general locality (i.e. Lake Michigan). In doing so we would be able to examine what pathogens existed in *Diporeia's* tissues before the presence of zebra mussels in Lake Michigan, and what pathogens exist in their tissues now that the mussels have become established.

2. Materials and Methods

2.1 Field Work

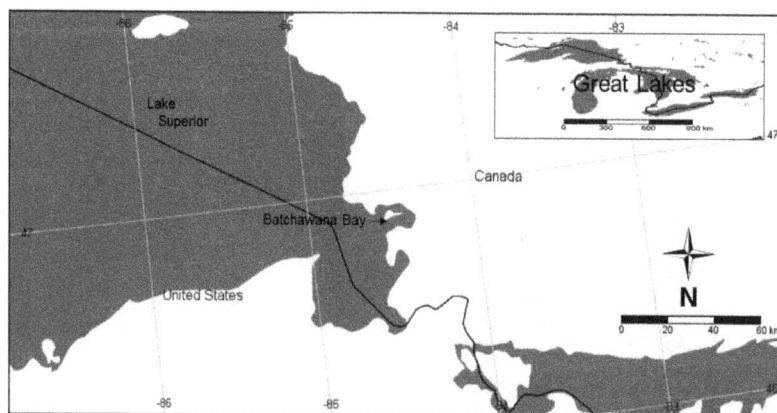

Figure 2. Map is showing sampling sites located in Batchawana Bay, Lake Superior, Canada

In collaboration with the National Oceanic and Atmospheric Association (NOAA) and Great Lakes Environmental Research lab (GLERL), *Diporeia* samples were collected from Lake Superior's Batchawana Bay in Ontario,

Canada (Figure 2). GLERL provided a 7.3 meter research vessel that is equipped with a ponar grab which, when lowered, collects bottom sediment from the benthos. Using numerous site-specific locations known by GLERL after many years of collecting *Diporeia*, the anchored boat at each site allowed technical staff to lower the ponar to the bottom of the lake. Once bottom sediment was collected and retrieved, it was processed through a series of water flushes filtered through a ASTM round all-brass 500 μm sieve until only a mixture of large material plus *Diporeia* remained. Collecting sediment using the ponar was repeated ten times at each location (ten locations) to ensure an accurate abundance of *Diporeia* was collected. After filtering the sediment, *Diporeia* was manually removed with tweezers from the remaining material. The collected *Diporeia* samples were further subdivided into two aliquots. The first aliquot was placed in liquid nitrogen and later transferred to a -80°C freezer for future analyses. The second aliquot was placed into a -20°C freezer and retrieved as needed for this project.

2.2 Labwork: Histology

Diporiea collected from this study in addition to samples provided by NOAA, were prepared for histological studies. Samples of *Diporeia* provided by NOAA were originally collected from Lake Michigan since the late 1980's.

All tissues regardless of their date or site of collection were prepared and analyzed using similar methodologies. All samples were first stained in Rose Bengal dye (Sigma-Aldrich, USA) and then placed in 10% formalin (Sigma-Aldrich, USA) which maintains and preserves the tissue. Samples are then processed for microscopy. The purpose of processing tissue samples is to remove water from the tissue and replace it with a solid medium that will allow for thin sectioning. Individual *Diporeia* are removed from the formalin solution and placed into a histology cassette. Each cassette held ten *Diporeia* (n=10). The *Diporeia* in these cassettes are then processed using a series of increasing graded ethanol (Sigma-Aldrich, USA) solutions to dehydrate the tissue. Once complete, tissue was then placed in xylene (Sigma-Alrich, USA), which is a clearing agent that removes the alcohol from the previous step. Each cassette was placed, in order, in an 80%, 90%, 95%, and two changes of 100% ethanol solutions, followed by two washes of 100% xylene. Each cassette was then incubated for 20 minutes in each respective solution (Bergman, per comm., 2013). Following incubation in graded ethonal, the cassettes were placed in tissue trays sprayed with HistoPrep Mold Releasing Agent (Fisher Scientific, USA). Each sample of *Diporeia* was then placed into a metal embedding tray using tweezers and submerged in liquid paraffin (wax) baths ~30 minutes. Immersion in liquid paraffin for 30 minutes allowed the tissue to become completely infiltrated with wax. Prior spraying of the trays was important as it helped with the removal of the solid wax block after cooling. Blocks not prepared in this manner chipped and fell apart and were too difficult to remove. Once infiltration was complete, each wax block containing a single *Diporeia* sample were left to cool for at least 3 hours at room temperature. Due to the small size of *Diporeia*, the wax outside a 1 cm radius of the organism was heated to 55°C and removed using a vacuum infiltrator and paraffin dispenser (Lipshaw Inc.). Preparring the *Diporeia* sample in this manner assisted in the latter processes of sagitally sectioning the wax block with a sliding microtome (Bausch & Lomb, Rochester, USA).

In order to collect thin-sections of *Diporeia* tissues, each wax block needed to be secured in the sliding microtome. Each collected thin-section (5-8 μm) was transferred to a warm water bath at 36°C to ensure the wax section was free of "wrinkles". The sections were then placed on poly-prep-lysine coated glass slides (Sigma-Aldrich, USA) followed by placement of the slides on a slide warmer (Sigma-Aldrich, USA) at 46°C for 24 hours. This procedure ensured that the thin sections adhered to the slides. After heating the prepared sections for 24 hours, the slides were exposed to a graded series of ethanol and xylene solutions in reverse order as described earlier. The graded series of ethanol and xylene solutions helped to remove the wax from each slide, leaving only the *Diporeia* tissue. Each slide was then incubated in a solution consisting of 65% ethanol and 5% hydrochloric acid for 5 minutes. The purpose of the aforementioned step was to remove any Rose Bengal dye that the tissue may have.

Mayer's Hematoxylin and Eosin Y stains (Sigma-Aldrich, USA) were then used to stain tissues adhering to the glass slides, followed by light and fluorescent microscopy to identify and characterize infected *Diporeia* tissue. The protocol used was a modified version from Lillie (1965), and is as follows:

- Immerse tissue with Mayer's Hematoxylin;
- Incubate for 10 minutes;
- Rinse and run room temperature tap water over sections for 10 minutes;
- Immerse with working Eosin Y Stain;
- Allow to incubate for 30 seconds;
- Rinse with tap water until water runs clear off of a slide;
- Clear, and mount tissue

2.3 Statistical analysis

The statistical analyses and figures in this study only pertain to Lake Michigan *Diporeia* samples. We did not statistically compare pathogens of *Diporeia* tissues collected from Lake Superior as there were too few archived samples to generate any historical trends. Statistically comparing these two localities, one with plenty of replicates collected over time and one with too few (or none), introduces significant error.

However, all Lake Michigan samples were tested using linear regression and chi-square analyses. Linear regression was used, using year as the predictor and percent of pathogens present. A chi-square test for independence was also used to determine if there is association between year and presence of pathogens. All analyses were completed using the R-Statistic (R Core Team, 2013). The pathogen data for *Diporeia* tissue prior to 2005 was collected by Messick et al. (2004).

3. Results

3.1 Field work

The purpose of collecting *Diporeia* from Batchwana Bay (Lake Superior) was to update the status of the population density of *Diporeia* since presence/absence monitoring began in 1977, and to provide researchers with *Diporeia* tissue for future research projects. Based on the average number of *Diporeia* collected in the ponar sample, the observed population was 585 *Diporeia* m^{-2}.

3.2 Lab work: histology

The pathology of *Diporeia* was assessed using both light and fluorescent microscopy in all tissue samples. Tissue was determined to be infected if it had a basophilic body (Figure 3A) and/or a budding structure (Figure 3B). Basophilic bodies (or mass) are structures within the tissue that are abnormal and are darker in color because they absorb more dye (Figure 3A). These masses are involved in the innate immune system of an amphipod's response to pathogens (Martinez, 2007). Basophilic bodies are often found in the legs and along the spine of *Diporeia*. The second prevalent structure found in *Diporeia* tissues were budding structures (Figure 3B). These structures had projections off of the cellular body suggesting possible fungal infection.

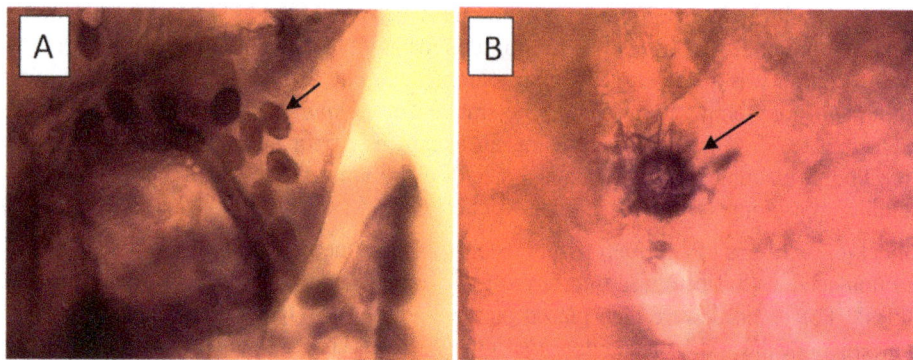

Figure 3. Analysis using fluorescent microscopy. (A) Basophilic bodies (black arrow) observed at 400x. Bodies are observed in proximity to one another. (B) Budding structure (black arrow) observed at 400x. These structures were typically observed scattered throughout the body tissues of *Diporeia* spp.

Examination of Lake Michigan *Diporeia* has been done previously on samples that were collected in 2000 and earlier (Messick, 2004). Examination of these historically relevant samples was to provide more recent insight into the current state of *Diporeia* tissue and disease. For *Diporeia* collected in 2005, it was found that ~18.9% exhibited infection versus 2010 samples, which showed ~29.2% pathogenic infection.

Diporeia that had pathogenic infections are further categorized into those that had (1) basophilic masses or bodies, versus those that had (2) budding structures. In 2005, ~63.6% of *Diporeia* exhibited basophilic bodies or masses, with the remaining being associated with budding structures. In 2010, ~71.4% exhibited basophilic bodies or masses. A linear regression model was used as a predictor to determine if year of infection played a role in the density decline of *Diporeia* populations (Figure 4). The linear regression was significantly positive (r=0.7202264, p≤0.0124). A Chi-square test for independence, (Figure 5), was also used to test if there was an association between year and percent pathogens observed in *Diporeia* tissues. Values obtained were $X^2 = 50$, df = 10, p≤ 0.0001, implying significant association between year and pathogen incidence.

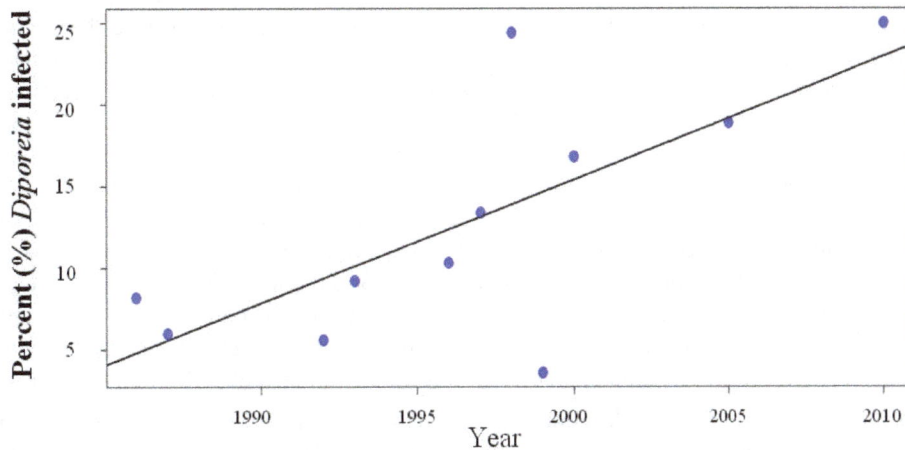

Figure 4. Linear regression analysis using R-statistics. Year was used as a predictor for percent pathogens found in *Diporeia* tissue collected from Lake Michigan (analyses resulted in r = 0.7202264, p≤0.0124)

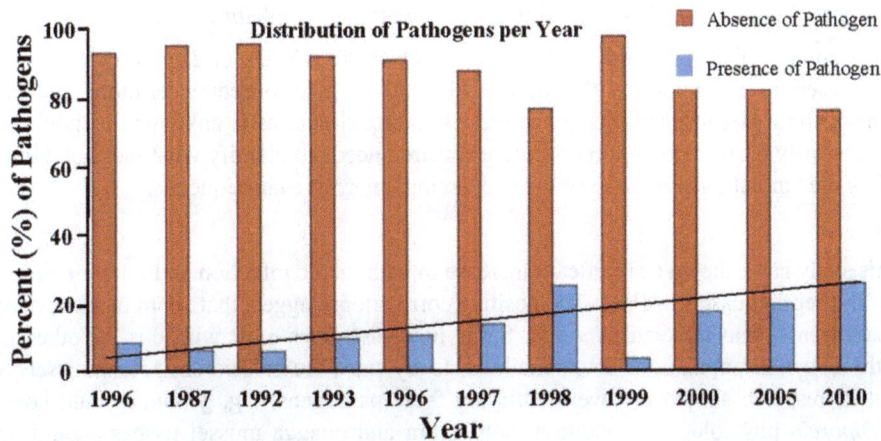

Figure 5. Chi-square analysis for independence using R-statistics for percent pathogens found in *Diporeia* tissue, collected from Lake Michigan (analyses resulted in $X^2 = 50$, df = 10, p≤0.0001)

4. Discussion

The *Diporeia* population in Batchawana Bay, Lake Superior (Canada), does not appear to be affected by the introduction of zebra mussels. In our abundance analyses, we observed an increase in the number of *Diporeia* per square meter. The increase in density of *Diporeia* follows an important trend where the natural population for that locality of *Diporeia* initially fell in 1977 from 675 *Diporeia* m^{-2} to 481 *Diporeia* m^{-2} in 1978; why the population crashed is not known. However, in 2001 the population increased to 529 *Diporeia* m^{-2}, 570 *Diporeia* m^{-2} in 2008, and as we observed, 585 *Diporeia* m^{-2} in 2013. This is particularly important as the density data shows the population of *Diporeia* not only stabilizing, but it continues to exhibit growth. Since zebra mussels are also abundant in this region, we can only speculate that this particular *Diporeia* population may have developed resistance to diseases it shares with the mussels. Alternatively, it is possible that the *Diporeia* living in Lake Superior have evolved/mutated from breeding with other healthier populations and are now, perhaps, a genetic sub-type of the original population. It is also plausible that the population in Lake Superior have developed some type of tolerance to the specific diseases that are causing population crashes elsewhere. Another possibility to consider is food availability as it is also possible that more food exists in Lake Superior compared to other Great Lakes, for instance, Lake Michigan. Nalepa et al. (2006) suggested that the decline of *Diporeia* in Lake Michigan was possibly due to food competition, resulting from the introduction of zebra mussels. We speculate whether more food is available in Lake Superior and as a consequence, there is less competition between *Diporeia* and zebra mussels.

Microscopy analyses of Lake Michigan *Diporeia* samples from 2005 and 2010 compared with prior tissue analyses show an overall increase in the prevalence of pathogens found in *Diporeia* since 1986. This is apparent in both the linear regression model (Figure 4) and the chi-square analysis (Figure 5). The overall increasing trend is consistent with the hypothesis that the invasion of zebra mussels in the Great Lakes has caused *Diporeia's* population crash (Nalepa et al., 2006). Zebra and quagga mussels invaded the Great lakes in the late 1980's (Nalepa et al., 1998); *Diporeia* populations have crashed in most areas since that time. Our data also shows an overall increase in pathogens found in *Diporeia* tissue since the introduction of the zebra mussel. Although it is possible that the increase in tissue pathogens observed in *Diporeia* might be caused by stress associated with increased feeding pressures, it seems more likely that the diseases (not present before the arrival of zebra mussels) are a direct consequence of a foreign invader(s), such as zebra mussels. Hence, an increase in infected *Diporeia* tissue might suggest that competition for food may have been a secondary effect caused by the primary effect, namely disease. The correlation between population decline in *Diporeia*, increased pathogenic infection and disease, and purported increased population of zebra mussels (Nalepa et al. 2006) supports this hypothesis.

Although most areas in the Great Lakes have experienced a decline in *Diporeia* populations since the introduction of zebra mussels, there are some locations that have not been affected by declining populations. Isolated areas of Lake Michigan and Lake Huron still support minimal *Diporeia* populations (Nalepa et al., 1998). Lake Superior's population of *Diporeia* has remained largely unchanged as supported by our sampling of Batchawana Bay. *Diporeia*'s stability in Lake Superior may be attributed to multiple factors as discussed above. One more recent anecdote for their survival is that the greater depths of Lake Superior may have provided a 'safe-haven' compared to other shallower areas that *Diporeia* typically inhabited.

Lastly, it is also possible that the budding structure found in the tissue's of *Diporeia* may not be pathogenic and/or may be present as a commensal (Messick et al., 2004). As a consequence, more studies are needed to confirm any speculation. Because the identity of these budding structures is unknown, it should not be assumed that they are necessarily harmful to *Diporeia*. Future research needs to identify what these budding structures are and whether they are "infecting" *Diporeia* tissue and having a negative consequence.

5. Conclusions

Analyses in this study have shown a significant increase in pathogenic infection and immune-type response since the invasion of the zebra mussels in 1986. The positive correlations suggest that zebra mussels may have acted as a vector for pathogen(s) that infected *Diporeia*. Some inconsistencies exist with this hypothesis, however. For instance, healthy *Diporeia* populations have remained steady since the invasion of zebra mussels in certain areas in Canada. Future research should involve identifying these pathogens (e.g. genomics) and how the infections are affecting *Diporeia* physiology. In addition, both zebra and quagga mussel tissues should be analyzed for similar pathogens that are identified in *Diporeia* tissue.

Acknowledgments

This work could not have been done without support from the R.B. Annis Water Resources Institute Foundation and the D.J Angus-Scientech Undergraduate Student Internships for summers 2013 and 2014 provided by The Annis Water Resources Institute, Grand Valley State University. Additional support was provided by NOAA through David Fanslow whose technical help, in-depth knowledge, and patience made this work possible. We also thank Patrick McEnaney who helped collect samples and Gavin Christie (Division Manager) at the Great Lakes Laboratory for Fisheries and Aquatic Sciences for organizing permits allowing us to collect samples in Canada. We appreciate help from Dr. Sango Otieno at Grand Valley State University, statistical consulting center, who reviewed and analyzed this data, without whom this work would not have seen completion.

References

Barrow, J. H. J. (1961). Observations of a haplosporidian, *Haplosporidium pickfordi* sp. nov. in fresh water snails. *Trans Am Microsc Soc, 80*, 319-32. http://dx.doi.org/10.2307/3223643

Bergman, D. (2013). Personal communication.

Cavaletto, J., Nalepa, T., Dermott, R., Quiggley, W., & Lang, G. (1996). Seasonal variation of lipid composition, weight, and length in juvenile *Diporeia* spp. (Amphipoda) from lakes Michigan and Ontario. *Can J Fish Aquat Sci, 53*, 2044-2051. http://dx.doi.org/10.1139/cjfas-53-9-2044

Fanslow, D. (2013). Personal communication.

Lillie, R.D. (1965). *Histopathologic technique and practical histochemistry*. New York: McGraw-Hill.

Martinez, F. (2007). The Immune System of Shrimp. *Boletines Nicovita*, July-September, 2007. Retrieved from http://www.nicovita.com.pe/cdn/Content/CMS/Archivos/Documentos/DOC_257_2.pdf

Messick, G. A., Overstreet, R. M., Nalepa, T. F., & Tyler, S. (2004). Prevalence of parasites in amphipods *Diporeia* spp. from Lakes Michigan and Huron, USA. *Dis Aquat Org, 59*, 159-170. http://dx.doi.org/10.3354/dao059159

Molloy, D. P, Giamberini, L., Morado, J. F., Fokin, S. I., & Laruelle, F. (2001). Characterization of intracytoplasmic prokaryote infections in *Dreissena* sp. (Bivalvia: Dreissenidae). *Dis Aquat Org, 44*, 203-216. http://dx.doi.org/10.3354/dao044203

Nalepa, T. F., Fanslow, D. L., Foley, A. J. III., Lang, G. A., Eadie, B. J., & Quigley, M. A. (2006). Continued disappearance of the benthic amphipod *Diporeia* spp. in Lake Michigan: is there evidence for food limitation? *Can J Fish Aqua Sci, 63*, 872-890. http://dx.doi.org/10.1139/F05-262

Nalepa, T. F., Hartson, D. J., Fanslow, D. L., Lang, G. A., & Lozano, S. J. (1998). Declines in benthic macroinvertebrate populations in southern Lake Michigan, 1980-1993. *Can J Fish Aqua Sci, 55*, 2402-2413. http://dx.doi.org/10.1139/f98-112

Nalepa, T.F. (1989). Estimates of macroinvertebrate biomass in Lake Michigan. *Can J Fish Aquat Sci, 44*, 515-524. http://dx.doi.org/10.1016/S0380-1330(89)71499-4

Nalepa, T.F., Fanslow, D. L., Lang, G.A., Mabrey, K., & Rowe, M. (2014). Lake-wide benthic surveys in Lake Michigan in 1994-1995, 2000, 2005, and 2010: Abundances of the amphipod *Diporeia* spp. and abundances and biomass of the mussels *Dreissena polymorpha* and *Dreissena rostriformis bugensis*. NOAA Technical Memorandum GLERL-164. NOAA, Great Lakes Environmental Research Laboratory, Ann Arbor, MI, 21 pp. Retrieved from http://www.glerl.noaa.gov/ftp/publications/tech_reports/glerl-164/tm-164.pdf

R Core Team. (2013). R: A language and environment for statistical computing. R Foundation for Statistical Computing, Vienna, Austria. Retrieved from http://www.R-project.org

The Effects of Anabasine and the Alkaloid Extract of *Nicotiana glauca* on Lepidopterous Larvae

Michelle Zammit[1], Claire Shoemake[1], Everaldo Attard[2] & Lilian M. Azzopardi[1]

[1] Department of Pharmacy, Faculty of Medicine and Surgery, University of Malta, Msida, Malta

[2] Division of Rural Sciences and Food Systems, Institute of Earth Systems, University of Malta, Msida, Malta

Correspondence: Michelle Zammit, Department of Pharmacy, Faculty of Medicine and Surgery, University of Malta, Msida, Malta. E-mail: michelle.zammit.01@um.edu.mt

Abstract

For several decades, *Nicotiana glauca* has been known for its content of the pyridine alkaloid, anabasine. The toxicological effects of this metabolite have been extensively studied, as opposed to its potential insecticidal activity. The anabasine content of leaves of *N. glauca*, collected from Malta, was 0.258 ± 0.0042% as determined by High Performance Liquid Chromatography. In the *Pieris rapae* larval bioassay, the median effective concentrations of anabasine and the alkaloid extract were 0.572 and 1.202 mg per larva, respectively. The presence of other interfering metabolites may have resulted in this elevated EC_{50} for the crude extract. Anabasine is quoted to be a very toxic alkaloid not solely to insects, but also to other animals, and its use in minute concentrations in insect traps may well prove it to be an effective natural insecticide.

Keywords: *Nicotiana glauca*, anabasine, *Pieris rapae*, insecticide

1. Introduction

Extracts of plants have been used as insecticides by humans since before the time of the ancient Romans, a practice that continues to present day with many of the species of plants known to have insecticidal properties (Balandrin, Klocke, Wurtele, & Bollinger, 1985; Isman, 2006). Plant products may be more rapidly degraded in the environment than synthetic compounds and some may have increased specificity that may favour insects beneficial to the plant. However, sublethal effects may still be experienced by these natural enemies or beneficial insects (Desneux, Decourtye, & Delpeuche, 2007).

The *Nicotiana* genus is well known for its insecticidal effects (Puripattanavong, Songkram, Lomlim, & Amnuaikit, 2013). *Nicotiana glauca* belongs to the plant family *Solanaceae* and is an evergreen, bluish green, erect, slender, sparsely-branched perennial, soft-woody shrub to small tree; usually up to 4 m tall or more which reproduces only by seeds (Halvorson & Guertin, 2003). *N. glauca,* Tree Tobacco or Indian Tobacco, was native to South America but is now naturalized in California, Australia, the Mediterranean, and Africa (Ollerton et al., 2012).

In Malta, the Tobacco Tree was introduced as an ornamental, but is now extensively naturalised especially on rubble (Weber & Kendzior, 2006). As its Maltese name implies (Tabakk tas-swar) it is also commonly seen growing on Maltese bastion walls. The stems are slender and loosely branching and break easily. The leaves are evergreen and alternate, and covered with a thin waxy coating on both surfaces and with a whitish powder which rubs off easily. When crushed the leaf gives off an odour that is always described as "unpleasant". The tubular flowers are yellow, about 5cm long, and are borne on large leafless branches at the ends of the stems throughout the year. The five lobes are very short, so there is little spread. The seedpods are brown, contain many seeds, 1-1.25 cm long, egg-shaped or oblong, on curving stalks from which they hang downward. The seeds are dark brown and kidney-shaped, about 0.3 cm long, with a rough surface (Haslam, Sell, & Wosley, 1977; Halvorson & Guertin, 2003).

Like other *Nicotiana* species, *N. glauca* is known for the presence of pyridine alkaloids (Andersson, Wennström & Gry, 2003). The alkaloids nicotine and anabasine (Figure 1) have been widely used as pesticides. Nicotine is the predominant alkaloid in *Nicotiana tabacum* (da Silva et al., 2013) and *Nicotiana rustica* (Lisko, Stanfill, Duncan & Watson, 2013) whereas anabasine is the predominant alkaloid in *Nicotiana glauca* (Lisko et al., 2013;

Slyn'ko, Tatarova, Shakirov, & Shul'ts, 2013). Nicotine is a powerful insecticide towards aphids (Puripattanavong et al, 2013) and larvae of lepidopterous pests (Shao, Dong, & Zhang, 2007). Both nicotine and anabasine are well known to exert their insecticidal effect by interacting with nicotinic acetylcholine receptors (Shao et al., 2007; Glennon & Dukat, 2000).

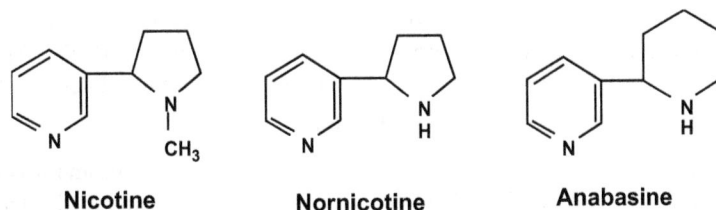

Nicotine   **Nornicotine**   **Anabasine**

Figure 1. Anabasine (first named Neonicotine) and other pyridine alkaloids of Nicotiana species, Nicotine and Nornicotine

Anabasine is the major alkaloid in *N. glauca* (Hawley, 1977; Saitoh, Noma, & Kawashima, 1985; Abd-El-Khalek, Abd-El-Nabey, & Abdel-Gaber, 2012) and all alkaloidal effects of the plant are attributed to this compound (Keeler, Balls, & Panter, 1981; Lisko et al., 2013). In *N. glauca*, anabasine makes up 1.2% of the plant material in fruits and 1.1% of the plant material in leaves. The roots, flowers and stems contain lower quantities of anabasine (Khafagy & Metwally, 1968). Anabasine [3-(2-Piperidinyl)pyridine, 2-(3-pyridyl)piperidine], neonicotine, $C_{10}H_{14}N_2$, molecular weight = 162.23 (Harborne & Baxter, 1995) was first synthesised in 1930 by Smith. It contains a pyridine ring attached at a β-position to the α-position of a pyrrolidine or piperidine ring, chemically very closely related to nicotine and the compound was therefore first named neonicotine (Ginsburg & Schmitt, 1935). Anabasine is a colourless viscous liquid which turns brown upon aging in contact with air. It has a boiling point of 280 °C to 281 °C and is less volatile with steam than is nicotine. Anabasine freezes at 9 °C. The compound is very stable, appreciably basic in reaction, forms salts in typical alkaloidal fashion and is soluble in water in all proportions and in most organic solvents (Haag, 1933).

This study is aimed at isolating and detecting anabasine in local *Nicotiana glauca* collected in Malta and testing of anabasine for pesticidal properties on cabbage white caterpillars (*Pieris rapae*) so as to determine the concentration of anabasine in the local plant and its effectiveness against local larvae, Data regarding anabasine content in the Maltese plant was not available in literature.

2. Method

2.1 Chemical Investigations

2.1.1 The Extraction and Separation Procedure

The leaves of approximately the same size of *N. glauca* were collected from Iklin in Malta, in July whilst the plants were flowering. The leaves were identified. The extraction and separation procedures for anabasine followed were those modified from standard procedures for alkaloidal extraction (Trease & Evans, 1978) and separation (Cunniff, 1995). Briefly, the leaves were dried at 35 °C for 48 hours and 10 g of the resulting powder was then shaken for 4 hours at 200 rpm with 200ml of 0.5% sodium hydroxide. The volume of the filtrate was reduced to 70 ml at 35 °C. 20 ml of chloroform were added and after five additions, the combined chloroform solution was then added to five portions of 20 ml of 0.05 M hydrochloric acid. The combined acidified aqueous extract was neutralised with ammonia solution to a pH of 7.

2.1.2 Chemical Test Analysis

The solutions prepared after the extraction and separation procedure were tested with the Dragendorff reagent (Potassium bismuth iodide) (Wagner & Bladt, 1996). UV analysis, using a Pharmacia LKB Ultrospec III UV/Vis Spectrophotometer was carried out on both the acidified and the neutralised extracts after the determination of the wavelengths at which to test. Anabasine standard (Sigma-Aldrich Chemie GmbH Germany) was used for all the analyses as control.

The extract was analysed by a Shimadzu LC-10A HPLC (Shimadzu, Kyoto, Japan) using a C18 MicroBondapak column, 250 × 4.6 mm, 10 mm, with a mobile phase of 40% methanol containing 0.2% phosphoric acid buffered to pH 7.25 with triethylamine, a flow rate of 0.5 ml/min and a run time of 20 minutes (Saunders & Blume, 1981).

The anabasine standard was injected at the same concentration using different volumes to obtain different values for the calibration of absorbance with anabasine concentration. The same procedure was repeated in duplicate for the *N. glauca* extract. Loss of anabasine and total anabasine content was calculated.

2.2 Pharmacological Investigations

2.2.1 Rearing of *Pieris rapae*

P. rapae larvae were obtained from fields and grown in cages until they formed pupae. After a few days the pupae developed into butterflies and these were grown in special cages and mated. The females were then left to lay eggs which hatched into new larvae which were then used in the experiment (Dickson, 1992).

2.2.2 Topical Bioassay With Anabasine Standard and *N. glauca* Extract

The 48-hr topical bioassay method used to test the *Pieris rapae* larvae with anabasine was one modified from an established method (Hsin & Coats, 1987; Wright et al., 2000). Larvae were examined after 48 hours; the criteria used to judge paralysis was placing the larvae on a cabbage leaf on the bench surface for 10 mins with constant, light prodding.

In the reference method, 1 µL of insecticide was applied to each organism. Due to practical reasons dilutions were adjusted to 200 µL.

Briefly, one larva was placed in each of thirty five wells of six-well plates (Nunc AS, Copenhagen, Denmark). Six anabasine dilutions were prepared containing the anabasine standard and 0.2 ml of each dilution was pipetted in five of the wells containing the larvae, with final concentrations ranging between 0.025 mg to 5 mg per larva. 0.2 ml of distilled water was pipetted in another five wells. The well plates were left in a well-lit, well-ventilated room at 25 °C for 24 h. The solution was then dried from the wells and the caterpillars left inside the wells in the same room for a further 24 h. The larvae were then slowly lifted out of the wells one after the other and placed on a cabbage leaf on the bench surface for 10min and prodded. Their behaviour was noted.

The same procedure was followed for the *Nicotiana glauca* extracts. In this case, one larva was placed in each of ninety wells. Eight dilutions of *N. glauca* neutralised extract from 2.1.1 were prepared and 0.2 ml of each dilution was pipetted in ten of the wells containing the larvae with final concentrations ranging between 0.7 mg to 2.8 mg per larva. 0.2 ml of distilled water was pipetted in another ten wells.

The range of concentrations (0.7 mg to 2.8 mg) for the *N. glauca* extract was selected such that the range of activity included dilutions at which the alkaloids were not active and dilutions which exhibited 100% activity. For both standard and extract, the median effective concentration (EC_{50}) was calculated using LdP LineR software Version 1.0 (Ehabsoft, 2014) based on Finney Probit Analysis (1971).

2.2.3 Statistical Analysis

One-Way Analysis of Covariance was used to determine if there is any statistical significant difference between the activity of anabasine standard and the anabasine extracted from *N. glauca*. This was carried out using the program BMDP statistical software (Engelman, 1990). Following transformation and normalization of the data, the two trends were compared using non-linear regression (Akaike's method) and the respective EC_{50} values were compared by using GraphPad Prism version 5.00 for Windows, GraphPad Software, San Diego California USA, www.graphpad.com. P values less than 0.05, were considered as significant.

3. Results

In the chemical investigations, some of the *Nicotiana glauca* extracts, alongside the anabasine, tested positive with the Dragendorff's reagent, which was used as the alkaloid identification test. In particular, the acidified and the final neutralised solutions tested positive. The presence of anabasine in the extracts was further confirmed by the spectrophotometric analysis giving peaks at 231 and 257 nm (Table 1). HPLC analysis revealed that the average retention time of anabasine in the sample was 10.368 ± 0.03 min and the average concentration of anabasine in the sample was 258.633 ± 4.217 ppm after adding the calculated 0.73% loss (Figure 2). The similarity index of the anabasine in the sample to the standard was 0.999 (v = 4). A 0.73% loss was the value obtained in all cases of the different standard concentrations showing a high degree of precision.

The bioassay on *Pieris rapae* larvae showed that the standard anabasine exhibited an EC_{50} of 0.572 mg/larva (or 0.286%) while the *N. glauca* final extract exhibited an activity of 1.202 mg/larva (or 0.601%).

Table 1. UV absorbance at two wavelengths for crude *N. glauca* extracts and pure anabasine

Solution	Absorbance	
	231 nm	257 nm
Anabasine standard	3.793	3.661
Acidified extract	0.769	2.295
Neutralised extract	2.417	2.153

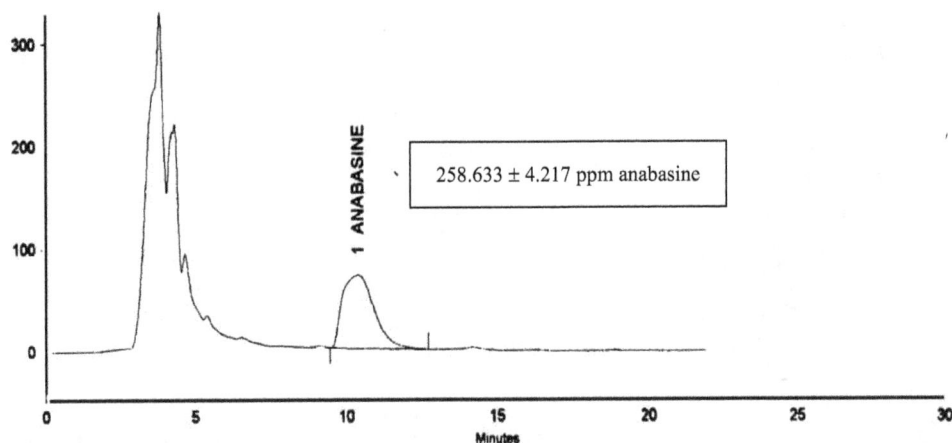

Figure 2. HPLC spectrum of *N. glauca* extract. The average retention time of anabasine in the sample was 10.368 ± 0.03 min and the average concentration of anabasine in the sample was 258.633 ± 4.217 ppm after adding the calculated 0.73% loss

Table 2. EC_{50} of standard and extract. This difference in value is not statistically significant

	EC_{50}^{a} (mg/larva)	lower limit[b]	upper limit[b]	EC_{50} (%)	ANCOVA[c]	EC_{50}Comparison[d]
Standard	0.572	0.466	0.702	0.286	p=0.1554	p<0.001
Extract	1.202	1.091	1.324	0.601		

[a]LDPline; [b]95% Confidence Intervals; [c]ANCOVA for slopes; [d]Non-linear regression analysis.

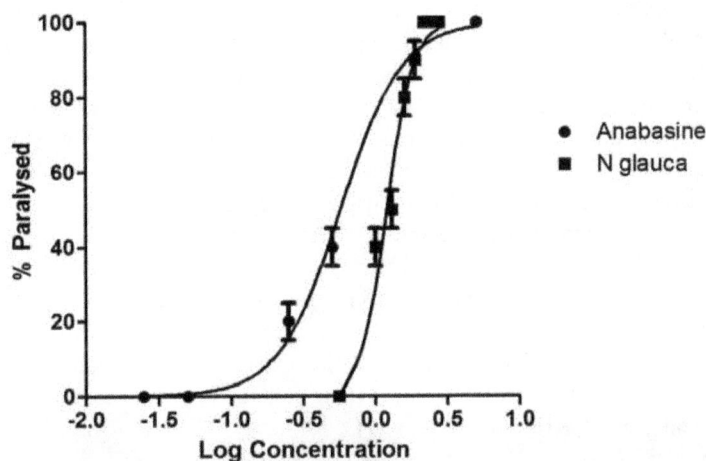

Figure 3. Percentage paralysed larvae with the anabasine and *N. glauca* extract. The bioassay on *Pieris rapae* larvae showed that the standard anabasine exhibited an EC_{50} of 0.572 mg/larva (or 0.286%) while the *N. glauca* final extract exhibited an activity of 1.202 mg/larva (or 0.601%)

4. Discussion

4.1 Chemical Investigations

Dragendorff's reagent (Potassium bismuth iodide) combines with alkaloids to give a characteristic orange brown precipitate. The alkaline and organic solutions did not result in a precipitate as opposed to the acidified and the final neutralised solution. In the latter two cases, the alkaloids must have been extracted from the plant by the sodium hydroxide, into chloroform and then into the hydrochloric acid showing that all the alkaloids had been extracted during this process. Anabasine standard gives two prominent peaks at 231 nm and 257 nm when analysed spectrophotometrically. Both the *N. glauca* acidified and neutralised extracts gave absorbances at both these wavelengths, thus confirming the presence of anabasine and/or related pyridine alkaloids (Shkurina,et al, 1960). The content of nicotinic alkaloids in various plants has been investigated using various methods. The wild form of *N. glauca* found in Egypt was found to contain 1.2% and 1.1% anabasine in the fruits and leaves, respectively (Khafagy & Metwally, 1968). More recently, anabasine content was determined by gas chromatography. This constituent made up 0.2% of the Israeli *N. glauca* (Mizrachi, Levy, & Goren, 2000). In another study, it was found that the young plant in Arizona contained 0.233 ± 0.0061 % anabasine (Keeler et al., 1981) which concords very closely with the results obtained in this study (0.258 ± 0.0042%). In another HPLC determination, the anabasine content of *N. glauca* plants in California, was 0.143% (Plumlee, Holstege, Blanchard, Fiser, & Galey, 1993).

The alkaloidal content in plants changes with factors such as environmental conditions, season and age of the plant. In fact, the results obtained in the this study (0.258 ± 0.0042%) reflect the alkaloidal content of the plant at the time of collection of the leaves (Cordell, 1994) and compares well with the content reported in studies performed on plants collected in other counries.

4.2 Pharmacological Investigations

Control larvae and larvae that were not affected by the applied solutions described in 2.2.2, moved off from the cabbage leaf they were placed upon, and when placed again on the leaf, they continued feeding. However, those that were affected by the treatments, were unable to move the appendages. These larvae were considered as paralysed. These can be distinguished from dead larvae as in such cases the larvae would be limp, discoloured and brown in colour. Only paralysis was observed with the alkaloidal concentrations used in this study.

The water content in controls (200 μL) did not have any physical effects on the larvae, such as the respiratory system (Fraenkel & Herford, 1938).

The bioassay with the standard anabasine determined the effect of anabasine on *Pieris rapae* larvae (0.572 mg/larva or 0.286%). The higher EC_{50} in the extract (1.202 mg/larva or 0.601%) demonstrates that alongside anabasine and other possible nicotinic alkaloids, interfering substances that reduced the response of the caterpillars to anabasine may be present. This occurred in spite of the fractionation and partial purification of the alkaloids in the extract. The rate of activity, as determined through the Analysis of Covariance, did not show any significant difference between the two treatments (p = 0.1554). To a certain extent this explains a similar activity between the two treatments and hence it can be partially concluded that pyridine alkaloids are provoking the effect for the *N. glauca* extract as is evident in the anabasine treatment. On the other hand, from the EC_{50} comparative analysis, anabasine showed a higher potency compared to the *N. glauca* extract. This was observed in other studies that clearly show that the single constituents are in general more potent than the crude extract (Keeler, Crowe, & Lambert, 1984; Ganfon et al., 2012; Rizvi et al., 2010). Although it is expected that the extract with a known concentration of anabasine could be compared to the pure anabasine standard, the presence of interfering substances will definitely reduce the potency of the *N. glauca* extract. It is known that alkaloids interact with triterpenoids and derivatives (Mali & Borges, 2003), amino acid decarboxylase enzymes (Beeker, Smith & Pennington, 1992) amongst other metabolites. Besides the pyridine alkaloids, *N. glauca* stores such metabolites (Bagni, Creus, & Pistocchi, 1986; Skliar, Curino, Milanesi, Benassati, & Boland, 2000). Some studies argue that crude extracts may have a better activity than pure single metabolites (Keung, Lazo, Kunze, & Vallee, 1996). This most probably would be possible if the crude extract contains solely or in a relatively high concentration the class of compounds represented by the pure single metabolites. In spite of all this, the alkaloidal effects of the *N. glauca* extract may be attributed to anabasine, as already noted in other phytochemical studies and bioassays (Keeler et al., 1981).

5. Conclusion

Attention has recently been directed to use of botanical insecticides vs. the use of synthetic insecticides especially in developing countries where synthetic insecticides are not affordable (Isman, 2006), where there is resistance to conventional synthetic insecticides as well as increased concern for the environment. Research into botanicals containing active insecticidal phytochemicals is becoming increasingly important to help address these issues.

It has already been reported in literature that *N. glauca* is toxic to humans and other mammals (Haag, 1933; Keeler et al., 1981; Panter, Keeler, James, & Bunch, 1992; Castorena, Garriott, Barnhardt, & Shaw, 1987; Plumlee et al., 1993; Mellick, Makowski, Mellick, & Borger, 1999). Nicotiana alkaloids have been also detected in animal products such as milk (Panter & James, 1990).

Although the potential toxicity of this extract may not justify the use of *N. glauca* extracts as a direct insecticide, with suitable chemoreception metabolites (van Loon, 1990), this extract can be confined in pheromone traps, hence reducing the incidence of mammalian or beneficial insect toxicity. In conclusion, this study has determined the presence and quantity of the alkaloid anabasine in local *N. glauca* extract which is comparable to that reported in earlier documented studies in other countries and its potential effectiveness against the larvae of local *Pieris rapae*.

References

Abd-El-Khalek, D. E., Abd-El-Nabey, B. A., & Abdel-Gaber, A. M. (2012). Evaluation of Nicotiana Leaves Extract as Corrosion Inhibitor for Steel in Acidic and Neutral Chloride Solutions. *Portugaliae Electrochimica. Acta., 30*(4), 247-259. http://dx.doi.org/10.4152/pea.201204247

Andersson, C., Wennström, P., & Gryo, J. (2003). *Nicotine alkaloids in solanaceous food plants.* TemaNord, Copenhagen, Denmark.

Bagni, N., Creus, J., & Pistocchi, R. (1986). Distribution of Cadaverine and Lysine Decarboxylase Activity in *Nicotiana glauca* Plants. *Journal of Plant Physiology, 125*(1), 9-15. http://dx.doi.org/10.1016/S0176-1617(86)80238-3

Ballandrin, M. F., Klocke, J. A., Wurtele, E. S., & Bollinger, W. H. (1985). Natural plant chemicals: Sources of industrial and medicinal materials. *Science, 228*, 1154-60. http://dx.doi.org/10.1126/science.3890182

Beeker, K., Smith, C., & Pennington, S. (1992). Effect of cocaine, ethanol or nicotine on ornithine decarboxylase activity in early chick embryo brain. *Developmental Brain Research, 69*(1), 51-57. http://dx.doi.org/10.1016/0165-3806(92)90121-C

Castorena, J. L., Garriott, J. C., Barnhardt, F. E., & Shaw, R. F. (1987). A fatal poisoning from *Nicotiana glauca. Clinical Toxicology, 25*(5), 429-435. http://dx.doi.org/10.1016/0378-8741(88)90058-X

Cordell, G. A. (Ed.). (1994). The Alkaloids. *Chemistry and Pharmacology* (Vol. 44). Access Online via Elsevier.

Cunniff, P. (Ed.). (1995). *Official methods of analysis of AOAC International* (16th ed.,Vol. 1). Virginia: AOAC international.

da Silva, F. R., Erdtmann, B., Dalpiaz, T., Nunes, E., Ferraz, A., Martins, T. L., ... da Silva, J. (2013). Genotoxicity of *Nicotianatabacum* leaves on *Helix aspersa. Genetics and Molecular Biology, 36*(2), 269-275. http://dx.doi.org/10.1590/S1415-4757201300500002

Desneux, N., Decourtye, A., & Delpeuch, J. M. (2007). The Sublethal effects of Pesticides on Beneficial Arthropods. *Annual Revie of Entomology, 52*, 81-106. http://dx.doi.org/10.1146/annurev.ento.52.110405.091440

Dickson, R. (1992). *A lepidoperist's handbook* (2nd ed., pp. 32-57). Middlesex: The Amateur Entomologist's Society. (The Amateur Entomologist, Vol. 13).

Ehabsoft. (2014). http://www.ehabsoft.com/ldpline

Engelman, L. (1990). One-way analysis of covariance. In W. J. Dixon, M. B. Brown, L. Engelman, & R. I. Jennrich (Eds.), *BMDP Statistical Software Manual* (Vol. 2). Berkley: University of California Press.

Finney, D. J. (1971). *Probit analysis* (3rd ed.). New York, Ny: Cambridge University Press.

Fraenkel, G., & Herford, G. V. B. (1938). The respiration of insects through the skin. *Journal of Experimental Biology, 15*(2), 266-280.

Ganfon, H., Bero, J., Tchinda, A. T., Gbaguidi, F., Gbenou, J., Moudachirou, M., ... Quetin-Leclercq, J. (2012). Antiparasitic activities of two sesquiterpenic lactones isolated from *Acanthospermumhispidum* DC. *Journal of Ethnopharmacology, 141*(1), 411-417. http://dx.doi.org/10.1016/j.jep.2012.03.002

Ginsburg, J. M., & Schmitt, J. B. (1935). Comparative toxicity of anabasine and nicotine sulphate to insects. *Journal of Agricultural Research, 51*, 349-354.

Glennon, R. A., & Dukat, M. (2000). Central nicotinic receptor ligands and pharmacophores. *Pharmacochemistry Library, 31*, 103-114. http://dx.doi.org/10.1016/S0165-7208(00)80006-9

Haag, H. B. (1933). A contribution to the pharmacology of anabasine. *The Journal of Pharmacology and Exeperimental Therapeutics, 48*, 95-104.

Halvorson, W. L., & Guertin, P. (2003). Factsheet for *Nicotiana glauca Graham.* Retrieved from http://sdrsnet.srnr.arizona.edu/data/sdrs/ww/docs/nicoglau.pdf

Harborne, J. B., & Baxter, H. (Eds.). (1995). *Phytochemical dictionary: A handbook of bioactive compounds from plants.* United Kingdom: Taylor & Francis. http://dx.doi.org/10.1016/0031-9422(95)90167-1

Haslam, S. M., Sell, P. D., & Wolseley, P. A. (1977) A Flora of the Maltese Islands. Malta University Press, Malta.

Hawley, G. G. (1977). *The condensed chemical dictionary.* Van Nostrand Reinhold, London.

Hsin, C. Y., & Coats, J. R. (1987). Bendiocarb metabolism in adults and larvae of the southern corn rootworm, *Diabrotica undecimpunctata howardi. Journal of Pesticide Science, 12*, 405-413. http://dx.doi.org/10.1584/jpestics.12.405

Isman, M. B. (2006). Botanical Insecticides, Deterrents, and Repellents in Modern Agriculture and an Increasingly Regulated World. *Annual Review of Entomology, 51*, 45-66. http://dx.doi.org/10.1146/annurev.ento.51.110104.151146.

Keeler, R. F., Balls, L. D., & Panter, K. (1981). Teratogenic effects of *Nicotiana glauca* and concentration of anabasine, the suspect teratogen in plant parts. *The Cornell Veterinarian, 71*, 47-53.

Keeler, R. F., Crowe, M. W., & Lambert, E. A. (1984). Teratogenicity in swine of the tobacco alkaloid anabasine isolated from *Nicotiana glauca. Teratology, 30*(1), 61-69. http://dx.doi.org/10.1002/tera.1420300109

Keung, W. M., Lazo, O., Kunze, L., & Vallee, B. L. (1996). Potentiation of the bioavailability of daidzin by an extract of *Radix puerariae. Proceedings of the National Academy of Sciences, 93*(9), 4284-4288. http://dx.doi.org/10.1073/pnas.93.9.4284

Khafagy, S. M., & Metwally, A. M. (1968). Phytochemical study of *Nicotiana glauca* R. Grah. grown in Egypt. *Journal of Pharmaceutical Sciences, U.A.R., 9*, 83-97.

Lisko, J. G., Stanfill, S. B., Duncan, B. W., & Watson, C. H. (2013). Application of GC-MS/MS for the Analysis of Tobacco Alkaloids in Cigarette Filler and Various Tobacco Species. *Analytical Chemistry, 85*(6), 3380-3384. http://dx.doi.org/10.1021/ac400077e

Mali, S., & Borges, R. M. (2003). Phenolics, fibre, alkaloids, saponins, and cyanogenic glycosides in a seasonal cloud forest in India. *Biochemical Systematics and Ecology, 31*(11), 1221-1246. http://dx.doi.org/10.1016/S0305-1978(03)00079-6

Mellick, L. B., Makowski, T., Mellick, G. A., & Borger, R. (1999). Neuromuscular Blockade After Ingestion of Tree Tobacco (Nicotiana glauca). *Annals of Emergency Medicine, 34*(1), 101-104. http://dx.doi.org/10.1016/S0196-0644(99)70280-5

Mizrachi, N., Levy, S., & Goren, Z. (2000). Fatal poisoning from *Nicotiana glauca* leaves: identification of anabasine by gas-chromatography/mass spectrometry. *Journal of Forensic Sciences, 45*(3), 736-741.

Ollerton, J., Watts, S., Connerty, S., Lock, J., Parker, L., Wilson, I., ... Stout, J. C. (2012). Pollination ecology of the invasive tree tobacco *Nicotiana glauca*: Comparisons across native and non-native ranges. *Journal of Pollination Ecology, 9*(12), 85-95.

Panter, K. E., & James, L. F. (1990). Natural plant toxicants in milk: a review. *Journal of Animal Science, 68*(3), 892-904.

Panter, K. E., Keeler, R. F., James, L. F., & Bunch, T. D. (1992). Impact of plant toxins on fetal and neonatal development: A review. *Journal of Range Management,* 52-57. http://dx.doi.org/10.2307/4002525

Plumlee, K. H., Holstege, D. M., Blanchard, P. C., Fiser, K. M., & Galey, F. D. (1993). *Nicotiana glauca* toxicosis of cattle. *Journal of Veterinary Diagnostic Investigation, 5*(3), 498-499. http://dx.doi.org/10.1177/104063879300500340

Puripattanavong, J., Songkram, C., Lomlim, L., & Amnuaikit, T. (2013). Development of Concentrated Emulsion containing *Nicotiana tabacum* Extract for Use as Pesticide. *Journal of Applied Pharmaceutical Science, 3*(11), 016-021.

Rizvi, W., Rizvi, M., Kumar, R., Kumar, A., Shukla, I., & Parveen, M. (2010). Antibacterial Activity of *Ficuslyrata* - An In vitro Study. *The Internet Journal of Pharmacology, 8*(2), 7.

Saitoh, F., Noma, M., & Kawashima, N. (1985). The alkaloid content of sixty *Nicotiana* species. *Phytochemistry, 24*(3), 477-480. http://dx.doi.org/10.1016/S0031-9422(00)80751-7

Saunders, J. A., & Blume, D. E. (1981). Quantitation of major tobacco alkaloids by high performance liquid chromatography. *Journal of Chromatography, 205*, 147-154. http://dx.doi.org/10.1016/S0021-9673(00)81822-1.

Shao, Y. M., Dong, K., & Zhang, C. X. (2007). The nicotinic acetylcholine receptor gene family of the silkworm, Bombyxmori. *BMC genomics, 8*(1), 324. http://dx.doi.org/10.1186/1471-2164-8-324

Shkurina, T. N., Alashev, F. D., Zvorykina, V. K., & Gol'dfarb, Y. L. (1960). Ultraviolet absorption spectra of some pyridine and nicotine derivatives. *Bulletin of the Academy of Sciences of the USSR, Division of chemical science, 9*(6), 1041-1045. http://dx.doi.org/10.1007/BF00903985

Skliar, M., Curino, A., Milanesi, L., Benassati, S., & Boland, R. (2000). *Nicotiana glauca*: another plant species containing vitamin D3 metabolites. *Plant Science, 156*(2), 193-199. http://dx.doi.org/10.1016/S0168-9452(00)00254-5

Slyn'ko, N. M., Tatarova, L. E., Shakirov, M. M., & Shul'ts, E. E. (2013). Synthesis of N-aryloxyalkylanabasine derivatives. *Chemistry of Natural Compounds, 49*(2), 294-301. http://dx.doi.org/10.1007/s10600-013-0585-1

Trease G. E., & Evans, W. C. (1978). *Trease and Evans' pharmacognosy* (11th ed.). London: BalliereTindall.

van Loon, J. J. (1990). Chemoreception of phenolic acids and flavonoids in larvae of two species of *Pieris*. *Journal of Comparative Physiology A, 166*(6), 889-899.

Wagner, H., & Bladt, S. (1996). *Plant drug analysis: A thin layer chromatography atlas* (2nd ed.). New York: Springer-Verlag Berlin Heidelberg. http://dx.doi.org/10.1007/978-3-642-00574-9

Weber, H. C., & Kendzior, B. (2006). *Flora of the Maltese Islands: A field guide*. Weikersheim: Margraf Publishers.

Wright, R. J., Scharf, M. E., Meinke, L. J., Zhou, X., Siegfried, B. D., & Chandler, L. D. (2000). Larval susceptibility of an insecticide-resistant western corn rootworm (Coleoptera: Chrysomelidae) population to soil insecticides: laboratory bioassays, assays of detoxification enzymes, and field performance. *Journal of Economic Entomology, 93*(1), 7-13. http://dx.doi.org/10.1603/0022-0493-93.1.7

Breeding Biology of Blackheaded Wagtail *Motacilla feldegg* Michahhelles, 1830 (Passeriformes, Motacillidae, Motacillinae) in South of Russia

E. A. Artemieva[1] & I. V. Muraviev[1]

[1] Ulyanovsk State Pedagogical University of I. N. Ulyanov, Russia

Correspondence: E. A. Artemieva, Ulyanovsk State Pedagogical University of I. N. Ulyanov, the Centenary of V.I. Lenin's Birth sq., 4, Ulyanovsk, 432700, Russia. E-mail: hart5590@gmail.com; pliska58@mail.ru

Abstract

Species-specific features of blackheaded wagtail *Motacilla feldegg* Michahelles, 1830 (Passeriformes, Motacillidae, Motacillinae) breeding biology were identified in south of Russia. A tendency to current species range shift is traced. Critical estimation of literary information about some peculiarities of reproduction and ecology of black-headed wagtail is carrying out on boundary XIX-XXI centuries, estimation of contemporary quantity, limited factories and regularities of species distribution on research territory of European part of Russia are gived. Distribution and quantity *M. feldegg* are irregular in this region and determine by presence of nesting biotopes and potential forage reserve. General character of distribution of this species estimates as a local and not numerous that gives foundation to include *M. feldegg* to some region Red Data Books of Russia and neighboring countries.

Keywords: population, species, nests, layings, nestlings, nesting biotope, birds, "yellow" wagtails, Russia

1. Introduction

Blackheaded wagtail *Motacilla feldegg* Michahelles, 1830 (Passeriformes, Motacillidae, Motacillinae), that is referred to the group of "yellow" polytypic complex of wagtails *Motacilla flava sensu lato*, just to its west forms, is regarded as a separate species in this study (Red'kin, 2001). This bird is treated like a strictly protected species within Europe according to the Appendix II of the Convention on the Conservation of European Wildlife and Natural Habitats (Genovesi & Shine, 2004). At present, data on possible nesting of the blackheaded wagtail in many habitats (regions) of its range are not available. Moreover, limiting factors of this species populations decline and impairment are unknown.

The objective the work is to study (a) nesting phenology; (b) nidiology; (c) oology; (d) nestlings of *M. feldegg* under environment conditions of southern Russia.

Information on possible nesting black-headed wagtail in a number of regions in the area at the present time are sketchy, based on the individual (sporadic) cases, the overall picture of the nature of the distribution of populations of the species in the area is missing. Unknown causes of downsizing and deterioration of populations, the exact data on the number and distribution of species, some aspects of biology. *M. feldegg* narrow local view is extremely demanding on the nesting habitats, which leads to extremely dispersed breeding distribution of this species in the space area. Complex investigations and geographical distribution of populations of the black-headed wagtail *M. feldegg* space area in the south of Russia (Figure 1).

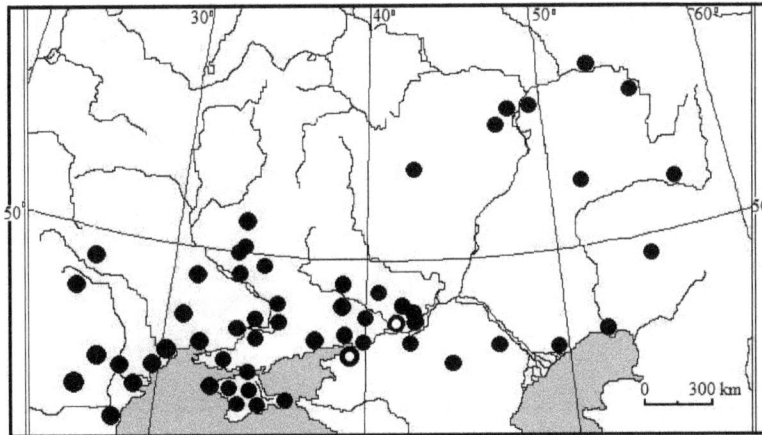

Figure 1. Location of revealed *Motacilla feldegg* in the south of Russia and neighboring countries in 2000-2013 and according to the literature (unshaded poissons shown find nests)

Shown to reduce the boundaries of the form as a reflection of its vulnerability to the combined effect of environmental factors. Identified potential habitat populations *M. feldegg* in southern Russia. Identified limiting factors and patterns of modern species distribution within the European part of Russia and adjacent territories.

In most previous studies on the biology and ecology of *M. feldegg* were delineated boundaries of the historic and current range and abundance of the species identified, but almost nothing was known about the reliability of the findings of nests of this species in the study area, so this study is necessary and urgent. Despite the findings of adult in nature, discover the nest is not simply due to violation of conservation of habitats and nesting behavior of birds breeding complex in modern conditions altered human habitat. Therefore, the existence and condition of populations of *M. feldegg* raises serious concerns in southern Russia. Presented in this study finds nests made for the first time in last 50 years.

2. Materials and Methods

Field studies were carried out during seasons 2000-2010 in the south of Russia and neighboring countries and 2011-2013 in the Rostov oblast of Russia: in the Aksai district, in the vicinity of the Bolshoe Mishkino station, the Don-Aksai water-meadow, from the 4th to 11th of June, 2011 and from the 1th to 10th of May, 2013; and in the Azov district, in the vicinity of the settlement of Kagalnik, maritime meadows of bottomland and mouth of the Kagalnik river, from the 13th to 18th of June, 2012. Nidicolous material (nests): n = 3. Oological material (layings , eggs): $n_1 = 3$, $n_2 = 13$. The material on nestlings: n = 4. The work was done using the following methods: mapping of nesting settlements and meetings, survey of plots according to traditional techniques, ringing, the study of the diet of adults and nestlings. The diets were examined by imposing cervical ligature chicks (were analyzed food samples chicks), as well as by analysis of stomach contents of dead birds.

3. Results

The range of *M. feldegg* extends by a broad band in steppe and desert zones of the Northern Palaearctic: from southern Europe (Balkan Peninsula) to the river Volga delta and further to the east up to the south-east of Kazakhstan (Gladkov, 1954; Dolgushin et al., 1970; Abdusalyamov, 1973; Stepanian, 1990; Gavris', 2003). The species occurs on the territory of Russia in the Rostov oblast, Stavropol' and Krasnodar kray, the Northern Caucasus, comes up over the steppes to the Orenburg oblast, the Krasnoyarsk kray and the Irkutsk oblast. It is found outside of Russia in the southern Ukraine, Crimea and Moldova (Figure 2).

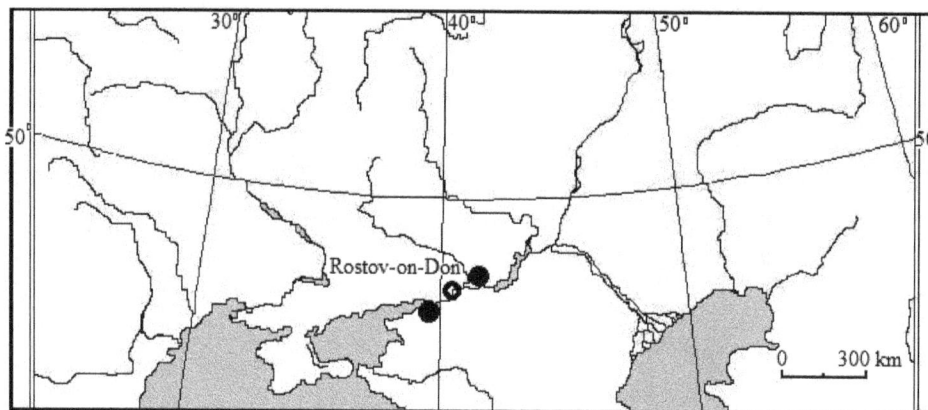

Figure 2. Location of revealed *Motacilla feldegg* nests (nesting sites) in the Rostov oblast in 2011-2013

Two subspecies forms, *Motacilla feldegg* f. *feldegg* Michahelles (1830) and *Motacilla feldegg* f. *melanogrisea* Homeyer (1878) are represented on the territory of Russia and CIS. The blackheaded wagtail subspecies *M. feldegg* f. *melanogrisea* inhabits the delta of the Volga river, the Caspian Sea coast, the Orenburg oblast, the Krasnoyarsk kray and the Irkutsk oblast in the Volga Region (eastern part of the range); the nominative subspecies *M. feldegg* f. *feldegg* lives in the Rostov oblast, on the Black Sea coast, in Crimea and in the Caucasus (the western part of the range) (Zarudny, 1897; Stepanian, 1990; Koblik et al., 2006). It hybridizes with *Motacilla flava* f. *flava* L., 1758 on the southern boundaries of the distribution (Bakhtadze, 1987; Gavris', 2003).

Perennial dense nesting settlements of *M. feldegg* f. *feldegg* are observed in the Rostov oblast: in lowlands of the Don delta, floodplains of it tributaries (as the Aksai river) and interfluve of the Don and Manych rivers, up to the Manych-Gudilo lake. This area could be considered a real refugium of the species where it reaches enough high abundance and could be a background species; *M. feldegg* f. *feldegg* is registered sporadically and is quite rare on other territories within the range.

Joint nesting settlements of *M. feldegg* f. *feldegg* numbered up to 45 breeding pairs in the Rostov oblast, Aksai district, in the Don-Aksai water-meadow, floodplain of the Aksai river (4-10.06.2011). *M. feldegg* f. *feldegg* density was on average 8.51 specimens/ha at M = 8.51 ± 0.21 within this area of floodplain of the Aksai river. Nesting sites of *M. feldegg* f. *feldegg* are sufficiently dispersed in the Don-Aksai water-meadow and floodplain of the Aksai river.

Four nesting pairs and their first broods were registered over the study area of 0.75 ha in vicinity of the settlement of Kagalnik, the Rostov oblast. From 6 to 7 males and 4 females of *M. feldegg* f. *feldegg* were observed and recorded during two counts on 15-16.06.2012.

According to long-term survey, population density of *M. feldegg* nesting settlements in the Kagalnik bottomland on the maritime meadows of the Azov Sea might exceed 160 specimens/km[2].

4. Discussion

4.1 Nesting Habitats

In the Rostov oblast of Russia blackheaded wagtail *M. feldegg* f. *feldegg* prefers to nest in halophytic mesophilic habitats of the river lowland floodplains. The nesting sites are usually removed at 100-300 m from the forage resources (shores of water reservoirs, reed beds, agrocoenoses). The nests are built in a sparse grass cover on the ground, forming small clustered settlements of 3-5, 8-9 breeding pairs or dense colonies on relatively small nesting sites (up to 100-500 m[2]).

The nesting settlement of *M. feldegg* f. *feldegg*, that we discovered and studied in the Aksai district, in the vicinity of the Bolshoe Mishkino station, the Don-Aksai water-meadow, floodplain of the river Aksai on 4-10.06.2011, occupied the following plant formations: bottomland steppe-meadows, halophytic suffrutescent meadow-steppe with echinoid licorice, halophytic dry grass meadows with spurge, sedge-grass meadows. Adult birds feed in meadow parcels near water reservoirs, wetlands around eriks (creeks, channels). Males jointly patrol the nesting sites from possible occurrence of birds of pray (kestrel, red-footed falcon, harriers, etc.) (Muraviev & Artemieva, 2012).

Inhabiting the Don-Aksai water-meadow (floodplain of the Aksai river, the Rostov oblast) the blackheaded wagtail mostly prefers plant associations that are attributed to echinoid licorice – meadow brome, meadow brome – Seguier's spurge, Austrian wormwood – meadow brome, downy brome – common wormwood, Kentucky bluegrass – Austrian wormwood. Echinoid licorice (*Glycyrrhiza echinata* L., Fabaceae) is an important part of the *M. feldegg* nesting microbiotopes due to formation of typical microlandscapes in halophytic suffrutescent steppe and bottomland dry meadows. Birds tend to build nests at the base of the plant *G. echinata*, in the thick turf of grasses.

Large nesting colony of diffuse type was found near Kagalnik, in the mouth of the river Kagalnik, the vicinity of Azov from the Rostov oblast on 15-16.06.2012. This nesting settlement occupies the maritime halophytic mesophilic water meadows along the Azov Sea coast and the delta of the Don.

Nesting biotope of *M. feldegg* f. *feldegg* in the Kagalnik river bottomland and delta of the Don, in the vicinity of Azov, is a maritime halophytic mesophilic water meadow with marsh mallow (*Althaea officinalis* L.), Siberian statice (*Limonium gmelinii* (Willd.) O. Kuntze), high goniolimon (*Goniolimon elatum* (Fisch.) Boiss.), Tatar seakale (*Crambe tataria* Sebeók), tufted hair grass (*Deschampsia caespitosa* (L.) P. Beauv.) on meadow parcels, couch grass (*Elytrigia repens* (L.) Nevski), saltpeter wormwood (*Artemisia nitrosa* Weber) on alkaline lands, curly dock (*Rumex crispus* L.), etc. Nesting microstation (plant association) of *M. feldegg* f. *feldegg* composed of Siberian statice (*Limonium gmelinii*) high goniolimon (*Goniolimon elatum*), tufted hair grass (*Deschampsia caespitosa*) on meadow parcels, saltpeter wormwood (*Artemisia nitrosa*) on alkaline lands. Shrub forming sparse thickets in the Kagalnik bottomland and on the coast of the Azov Sea is wolf-willow (*Elaeagnus commutata* L.). Marsh mallow is a key plant species of the nesting microstation, which is a dominant of the characteristic dwarf subshrub halophytic community. Unlike the yellow wagtail that is common in dry meadows and agrocoenoses in the Rostov oblast, the blackheaded wagtail nests in lowland areas with relief depressions, preferring grasslands of floodplains and on the coast of the Azov Sea.

Nesting communities of the birds from the Don-Aksai water-meadow and the Kagalnik bottomland composed of the next indicator species for the *M. feldegg* f. *feldegg* nesting settlements (N. = nesting bird): great cormorant (*Phalacrocorax carbo* L.) inhabits overflow lands, N.; great egret (*Egretta alba* L.) inhabits overflow lands, N.; little egret (small white heron, *Egretta garzetta* L.) inhabits overflow lands, N.; purple heron (*Ardea purpurea* L.) inhabits overflow lands, N.; glossy ibis (*Plegadis falcinellus* L.) inhabits overflow lands, N.; tawny eagle (*Aquila rapax* Temminck) inhabits steppe, N.; red-footed falcon (*Falco vespertinus* L.) inhabits floodplain, N.; common pheasant (*Phasianus colchicus* L.) inhabits meadows, N.; lesser grey shrike (*Lanius minor* J. F. Gmelin) inhabits thickets in bottomland, N.; stonechats (*Saxicola torquata* f. *torquata* L., *S. t.* f. *maura* Pallas) inhabits meadows with echinoid licorice and marsh mallow, N.; corn bunting (*Emberiza calandra* L.) inhabits meadows with echinoid licorice and marsh mallow, N.

Blackheaded wagtail specializes feeding mainly on representatives of Acridoidea, Sphaerininae, Histeridae (*Saprinus*), Chrysomelidae, Psylloidea, *Noctuidae*, Curculionidae, Ichneumonidae, Arachnida (Aranei), species of *Messor* and *Musca* (Gladkov, 1954; Dolgushin et al., 1970, Gavris', 2003). In food samples chicks and when opening the stomachs of dead birds were found next insect species and spiders in the diet of these birds *Ligus pratensis* L. (7), *Helophorus griseus* (Herbst) (4), *Sphaeridium scarabaeoides* L. (1), *Phyllobius oblongus* L. (1), *Pachybrachis tesselatus* (G. A. Olivier) (2), *Philidrus* sp. (1), *Stenus* sp. (1) and various Arachnida (4), Ephydridae, Cicadellidae (2), Muscidae (2), small Diptera (20) and pupae of Diptera (3), larvae of Aradidae (9) (Gudina, 2009). As part of the stomach contents from 09.06.2011 individuals *M. feldegg* detected: the male – Mollusca (Gastropoda: Planorbidae: *Planorbis spitorbis*; Lymneidae – *Galba glabra*) – 19.1%; Diptera (Chironomidae: *Chironomus* sp.) – 14.8%; Homoptera (Aphirophoridae: *Lepyronia coleoptrata* L.; *Philaenus spumarius* L.) – 13.7%; Odonata (Coenagrionidae) – 11.3%; Aranea – 11.6%; Hemiptera (Rhopalidae: *Corizus hyosciami* L.; Pentatomidae: *Aelia acuminata* L.) – 10.9%; Coleoptera (Carabidae: *Amara* sp.) – 10.2%; Hymenoptera (Formicidae) – 4.3%; Neuroptera (Chrysopidae: *Chrysopa* sp.) – 4.1%. Accordingly, the female: Lepidoptera – 32.2%; Orthoptera – 27.8%; Diptera – 15.4%; Homoptera – 14.6%; Hemiptera (Miridae) – 6.3%; Aranea – 3.7%. Adult birds readily collected herbaceous plants chironomid (*Chironomus* sp.) after their mass breeding along the shoreline of Taganrog Bay. Diet chicks includes the following groups of invertebrates (n=4): Aranea – 1.8%; Orthoptera – 28.2%; Homoptera – 11.4%; Hemiptera – 2.0%; Coleoptera – 2.3%; Lepidoptera – 2.7%; Diptera – 24.7%; aquatic invertebrates – 2.9%.

The differences in forage preferences for *M. feldegg* f. *feldegg* males and females were registered. Males mainly collect Mollusca (Gastropoda: Planorbidae – *Planorbis spirorbis* L.; Lymneidae – *Galba glabra* (O. F. Mueller)) – 19.1%; Diptera (Chironomidae: *Chironomus* sp.) – 14.8%; Homoptera (Aphirophoridae: *Lepyronia coleoptrata* L.; *Philaenus spumarius* L.) – 13.7%; Odonata (Coenagrionidae) – 11.3%; Aranea – 11.6%; Hemiptera

(Rhopalidae: *Corizus hyosciami* L.; Pentatomidae: *Aelia acuminata* L.) – 10.9%; Coleoptera (Carabidae: *Amara* sp.) – 10.2%; Hymenoptera (Formicidae) – 4.3%; Neuroptera (Chrysopidae: *Chrysopa* sp.) – 4.1%. Females forages by chasing various Lepidoptera – 32.2%; Orthoptera – 27.8%; Diptera – 15.4%; Homoptera – 14.6%; Hemiptera (Miridae) – 6.3%; Aranea – 3.7%. Adults of *M. feldegg* f. *feldegg* readily utilize as a food supply the mosquito chironomids of *Chironomus* genus during mass breeding, collecting them in the maritime meadows of the Azov Sea. Birds use warmed up by the sun shallow waters, ground roads, pathways as to collect insects attracted by the warmth of the soil and water surface, catching up them in flight.

4.2 Nesting Phenology (a)

M. feldegg f. *feldegg* start to arrive s in early April, nidification is registered at the end of April and in May, juvenile birds are already flying in early June. Counts on the routes (the number of specimens per km^2) during breeding period revealed gradual increase in density of birds along with arrival of wagtails and occupying nesting grounds, it grows from 3 to 174 sp/km^2 in some years.

In the Rostov oblast in late July – early August the broods of blackheaded wagtail migrate to the coast of the Taganrog Bay in the Black Sea along floodplains of the tributaries of the Don and Manych rivers. The first broods of *M. feldegg* f. *feldegg* from the Kagalnik bottomland on the Azov Sea coast appear in mid-June, their number increses by migrating broods from other nesting sites in floodplains of the tributaries and delta of the Don river. Terms of this species nesting period are rather extended. Presence of the first brooms could be simultaneously combined with the second laying of the same female. The second breeding cycle occurs when the weather conditions are favourable in early spring and the forage resources are available.

4.3 Nidiology (b)

M. feldegg builds nests on the ground, under the cover of low shrub or grass stand, sometimes low over the ground in the bush or on clean sand, but under the effuse tuft of grass (Gladkov, 1954; Dolgushin et al., 1970; Abdusalyamov, 1973; Gavris', 2003; Gudina, 2009; Muraviev & Artemieva, 2012).

We found two nests with complete layings in the floodplain of the river Aksai, the Don-Aksai water-meadow, the Rostov oblast on 06.06.2011 and 09.06.2011. The nests were in a joint nesting settlement of the model type and located on the site of the halophytic suffrutescent meadow-steppe with echinoid licorice. Registered nests were located at the base of the echinoid licorice low shrub, deep into the turf of dry plants of Volga fescue (*Festuca valesiaca* Schleich.). They were deeply embedded and hidden in the dry turf grasses. Entrance to the nest was carefully masked by hanging down and twisted stems of grasses. In the structure of the nests were incorporated small dry stems of Volga fescue, and in the trays were identified horsehair and pet wool. Small "niche" or a pit directly adjoined each of the nests and served for the males from each nesting pair to stay overnight (Figures 3-5).

Figure 3. Nest of *Motacilla feldegg* with complete layings and chicks (the Don-Aksai water-meadow, the Rostov oblast on 06.06.2011)

Figure 4. Nest of *Motacilla feldegg* with complete layings (the Don-Aksai water-meadow, the Rostov oblast on 09.06.2011)

Figure 5. Nest of *Motacilla feldegg* (Kagalnik river, the Rostov oblast on 16.06.2012)

The nest of *M. feldegg* f. *feldegg* was found with five strongly incubated eggs (stage 7) on 16.06.2012. The eggs were clearly related to the second laying of the season. This nest located on the meadow bleakness parcels with such low grasses as Siberian statice (*Limonium gmelinii*) and tufted hair grass (*Deschampsia caespitosa*), besides there were low shrubs of marsh mallow (*Althaea officinalis*). In the structure of the nest wall were incorporated small dry stems of grasses, the trays was evident and with a few small feathers of gulls (Table 1).

Table 1. Parameters (Size) of the *Motacilla feldegg* nests (in *mm*, n = 3) observed in the Rostov oblast

Nest number	Date of observation	Nest diameter (D)	Tray diameter (d)	Nest height (H)	Tray height (h)
1.	06.06.2011	85	60	-	55
2.	09.06.2011	90	55	55	38
3.	16.06.2012	80	60	-	40
Lim		80-90	55-60	55	38-55

4.4 Oology (c)

Full laying of M. *feldegg* includes six or, at least, five eggs (Gladkov, 1954; Dolgushin et al., 1970; Gavris', 2003).

There were found two eggs and three newly hatched nestlings (chicks) in the nest No. 1 on 06.06.2011.

Coloration of eggshells is light yellowish-brown, ornamentation is in the form of dark, dense mottles that become thicker to the infundibular end. The laying of five eggs was found in the nest No. 2 on 09.06.2011. Coloration of eggs grayish-olive, brownish ornamentation is not clearly defined. The female from this nest with a laying had been already ringed and tagged during study of this species nesting in the Don-Aksai water-meadow (04-10.06.2011). One of the ringed females of blackheaded wagtail laid an egg when it was released from ornithological net (08.06.2011). Coloration of this egg differs visually from oological descriptions prepared on the basis of previously revealed two layings in the nests. Main background coloration of the egg is khaki-greenish-gray, ornamentation is not clearly defined on the eggshell. Size of the egg is 18.4 × 14.4 mm (Table 2).

Table 2. Parameters of the *Motacilla feldegg* eggs (in *mm*, n = 12) observed in the Rostov oblast

Number of laying	Date of observation	Eggs length	Eggs diameter	Lim	M
1.	06.06.2011	18.9	14.1	18.9-19.1 × 14.1	19.0 × 14.1
2.	06.06.2011	19.1	14.1	-	-
3.	09.06.2011	18.9	14.3	18.9-19.2 × 13.9-14.4	19.04 × 14.4
4.	09.06.2011	19.1	13.9	-	-
5.	09.06.2011	18.9	14.0	-	-
6.	09.06.2011	19.1	14.4	-	-
7.	09.06.2011	19.2	14.1	-	-
8.	16.06.2012	19.1	14.9	18.0-19.1 × 13.0-14.9	18.5 × 13.8
9.	16.06.2012	19.0	14.0	-	-
10.	16.06.2012	18.1	14.1	-	-
11.	16.06.2012	18,0	13,1	-	-
12.	16.06.2012	18.2	13.0	-	-
	In total:	18.8	14.0	18.0-19.2 × 13.0-14.9	18.8 × 14.0

The laying of five eggs, that had been strongly incubated (stage 7) and had been the second laying of the season, was found in the nest of *M. feldegg* f. *feldegg* in the Kagalnik river bottomland of the Rostov oblast, in maritime meadows of the Azov Sea on 16.06.2012. Eggshell is slightly shiny. The background coloration of eggshells is milky yellowish-grey, ornamentation is in the form of dense mottles and strokes, yellowish-brown, becomes darker and thicker to the infundibular end (Figures 6, 7).

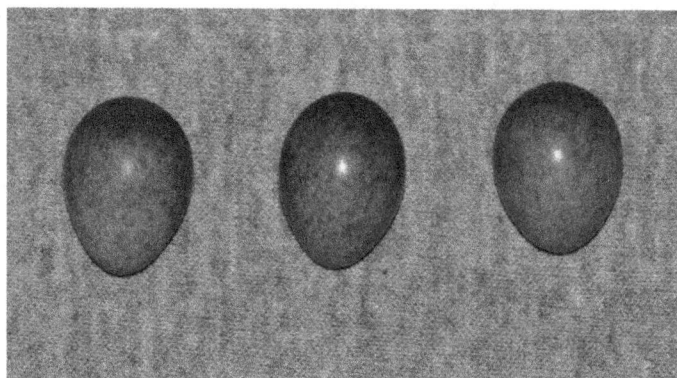

Figure 6. Eggs of *Motacilla feldegg* (Kagalnik river, the Rostov oblast on 16.06.2012)

Figure 7. Polymorphism of eggs coloration of *Motacilla feldegg* (the Don-Aksai water-meadow, the Rostov oblast on 09.06.2011)

For *M. feldegg* f. *feldegg* an assumption about existence of polymorphism was done after visual assessment of eggs coloration. The differences in coloration and ornamentation of eggs of this bird in the Don-Aksai water-meadow indicate possible existence in groups of blackheaded wagtails, at least, two environmental nesting (biological) races. Similar phenomenon was previously registered by the biochemical analysis for yellow-backed and citrine wagtails in the Penza oblast (Titov et al., 1997).

4.5 Nestlings (d)

Female *M. feldegg* f. *feldegg* incubates laying for 12 days and the chicks are in the nest up to 14-15 days in Ukraine, according to data of Gavris' (2003).

There were noted two eggs and three newly hatched chicks in the nest of *M. feldegg* f. *feldegg* studied in the area of floodplain of the river Aksai, in the Don-Aksai water-meadow, in the Rostov oblast on 06.06.2011. One new nestling (chick) appeared the next day, on 07.06.2011. There were registered totally four chicks in the same nest when we examined it on 09.06.2011.

For the first time in the Rostov oblast it was recorded bigamy phenomenon for blackheaded wagtail. There were done observations of active courtship display and coupling of males with other females of the same species, when they already had at that time own nesting territories with nests (layings and chicks). It was also indicated courtship display and coupling of males with two different females of their species during watching period (June 2012) on behavior of *M. feldegg* f. *feldegg* at the time of nesting in the Kagalnik river bottomland and maritime meadows of the Azov Sea. One of those females had a nest with second laying and even so was finishing feed of flying chicks from the first brood. Different females of the same bigamy family had distinctive morphological characters (female No. 1 was grey-headed, female No. 2 was dark-headed). Females of the same male were presented by different ecological (biological) races (morphs). Incubation was carried out by females only in all registered layings.

5. Conclusions

Trends and patterns in the choice of the *M. feldegg* nesting conditions within studied territory of the Rostov oblast are conditioned by mesophilic communities, halophytes and topography, which define the nature of plant associations and food supply in the nesting biotopes. Blackheaded wagtail is very sensitive to the choice of nesting habitats and disappears when they are strongly waterlogging, or under the presence of steep shores and hilly terrain (mineral salts washout). Therefore, this species can be used as an indicator one for floodplain, mesophilic, halophytic, meadows and steppe coeonoses in river valleys, saline clay steppes on a gently sloping shores (Muraviev & Artemieva, 2012).

Basic trends in the modern boundaries shift of the *M. feldegg* distribution in the Rostov oblast of Russia is the sharp reduction of the range boundaries as a whole caused by human activities, and forced concentration of the species in lowland areas with relief depressions along floodplains of the rivers Don and its tributaries, in the delta of the Don and the coast of the Azov Sea. Saving the groups of species in this area became possible thanks to the preservation of halophilic herbaceous and shrubby grasslands and halophytic suffrutescent meadow-steppe with echinoid licorice in the floodplains of the tributaries of the river Don (the river Aksai), maritime halophytic mesophilic water meadow with marsh mallow in the delta of the Don and the Azov Sea coast, and the preservation in those biotopes the key facilities of fodder base. Following the widespread reduction in population size and boundaries of the range of *M. feldegg*, it was changed the choice of food and nesting biotopes in the south of the European part of Russia. The blackheaded wagtail uses the shores of ponds and eriks, farmlands (corn fields), the treatment industrial plants of large settlements (the Rostov-on-Don city). Reduction of distribution boundaries and abundance of the species under the human economical activity, its accumulation in floodplains reflects vulnerability and relatively low tolerance of the species to the combined effect of environmental factors.

Identified species-specific features of the nesting biology and ecology of *M. feldegg* in the Rostov oblast include definite degree of moisture, salinity and topography of the nesting biotopes for development of specific plant associations, two types of nested structures (open and covered) that depends on micro-relief; marked limits of the nests and eggs parameters; certain composition of the building material for the construction of nests, characteristic set of prey insects for feeding adults and chicks; bigamy structure of families and the associated with third polymorphism of eggs, presence of the second cycle of reproduction.

Acknowledgements

The authors express their sincere gratitude to Professor V. M. Loskot and Professor V. A. Paevsky (Laboratory of ornitology of Zoologichesky Institute of Russian Academy of Sciences), Associate Professor A. V. Maslennikov and L. A. Maslennikova (Department of Botany of the Ulyanovsk State Pedagogical University), Professor S. V. Titov and D. G. Smirnov (Department of Zoology and Ecology, Penza State University), A. V. Zabashta (ornithologist of Rostov-on-Don Airport).

References

Abdusalyamov, I. A. (1973). *The fauna of the Tajik SSR. - T. XIX, Part 2. Birds* (pp. 343-345). Dushanbe: The Academy of Sciences of the Tajik SSR. Institute of Zoology and Parasitology of E. N. Pawlowski.

Bakhtadze, G. B. (1987). *Distribution of gray- (M. flava L.) and black- (M. feldegg Mick) yellow wagtails in the south of the European part of the USSR // Proceedings of the North Caucasus Research Centre of higher education*. Science Series. Rostov-on-Don, 11. Dep. in VINITI 18.08.87, No. 1921.

Council of Europe Publishing. (2004). European Strategy on Invasive Alien Species: Convention on the Conservation of European wildlife and habitats (Bern Convention) By Piero Genovesi, Clare Shine. *Nature and environment, 137*.

Dolgushin I. A., Korelov, M. I., Kuzmina, M. A., Kovshar, A. F., Borodikhin, I. V., & Rodionov, E. F. (1970). *Birds of Kazakhstan. - T. 3*. Alma- Ata: Nauka Publishing House of the Kazakh SSR, pp. 341-347.

Gavris' G. G. (2003). Pliska chornogolova Motacilla feldegg. *Ptakha Ukrainy pid receptionists Bernskoï konventsiï. - Delhi* (pp. 218-220).

Gladkov, N. A. (1954). *Birds of the Soviet Union. - T. 5*. Moscow: Soviet Science, pp. 594-690.

Gudina, A. N. (2009). *Rare and little-studied birds in eastern Ukraine. Passeriformes. - T. 3. -* Zaporozhye: Dnipro metallurgist, p. 182.

Koblik, E. A., Redkin, Y. A., & Arkhipov, V. Y. (2006). *List of birds of the Russian Federation* (pp. 146-148). Moscow: KMK Scientific Press Ltd.

Muraviev, I. V., & Artemieva, E. A. (2012). By the breeding biology and ecology of Blackheaded wagtail Motacilla feldegg Michahelles, 1830 (Passeriformes, Motacillidae, Motacillinae). *Scientific. guided. Belgorod State. Univ. Eats. Science., 9*(128), 113-122.

Red'kin, Y. A. (2001). Taxonomic and evolutionary relationship forms complexes of young birds on the example of the genus Motacilla L., 1785 (taxonomic revision of the subgenus Budytes): Author. dis. Candidate. biol. Science. - M., p. 19.

Stepanian, L. S. (1990). *Synopsis of the ornithological fauna of the USSR* (p. 366). Moscow: Nauka.

Titov, S. V., Muraviev, I. V., & Logunova, I. Y. (1997). On studying pigmentation of bird eggshell. *ZOOLOGICHESKY ZHURNAL, 76*(10), 1185-1192.

Zarudny, N. A. (1897). Additions to the "Ornithological fauna of the Orenburg region". Materials to the knowledge of the fauna and flora of the Russian Empire. *Dep. Zool., 3*, 171-312.

Evaluating the Viability of Lactic Acid Bacteria and Nutritional Quality of *Hibiscus Sabdariffa* Stored Under Natural Condition

Eguono Esther Anomohanran[1]

[1] Department of Microbiology, Delta State University, Abraka, Nigeria

Correspondence: Eguono Esther Anomohanran, Department of Microbiology, Delta State University, Abraka, Nigeria. E-mail: a_eguono@yahoo.com

Abstract

The viability of lactic acid bacteria and nutritional quality of *Hibiscus sabdariffa* were investigated under ambient temperature to evaluate the health implication associated with the consumption of the beverage obtained from the plant. This was carried out by employing the pour plate method and the gravimetric technique of estimating the indigestible fibre in a beverage. The result obtained from the study showed that the viable bacteria counts increase steadily for the first six days of storage and decreased afterwards. The study also showed that isolates of *Bacillus spp. Streptococcus spp.* and *Staphylococcus spp.* were identified at the initial phase of the storage while isolates of *Aspergillus spp., Geotrichum spp and Lactobacillus* were identified in the later phase of the storage period. The study further showed that the pH, carbohydrate, protein and vitamin C content increased gradually throughout the period of storage. The risk associated with the beverage could be reduced if at all stages from harvesting to selling, good hygienic conditions are imbibed. Hence the *Hibiscus sabdariffa* beverage could serve as a convenient substitute to carbonated drinks, which may not be affordable by a lot of people.

Keywords: *Hibiscus sabdariffa*, Roselle, Zobo, lactic acid bacteria, red calyces, beverage

1. Introduction

Hibiscus sabdariffa is a dicotyledonous and autogamous plant of the Malvaceae family. It is native to Africa and widely grown in tropical and subtropical regions of both hemispheres and many areas of India and parts of Asia, America and Australia (Sie et al., 2011). *Hibiscus sabdariffa* also known as Roselle is also found in many countries of the world which include Malaysia, Indonesia, Mexico and Nigeria. The high content of anthocyanins in the calyx of the plant makes it possible for the production of drinks and tea. The calyces are used for making fresh Roselle wine, jelly, gelatin, beverages, cakes, tea, marmalade, ices, ice-cream, butter, sauces and other desserts.

The *Hibiscuss sabderiffa* plant is also considered to have antihypertensive properties hence it has been used in the medical field as a diuretic, sedative and treatment of cardiac, nerve diseases, cancer, and liver disorders (Sie et al., 2011; Sengupta & Banik, 2011; FIIRO, 2014). Study has also shown that the extract from this flower has demonstrated hypocholesterolemic properties (Sindi et al., 2014). It is also agreed that the concentrated *Hibiscus sabdariffa* beverages lower blood pressure in patients with hypertension and diabetes compared with black tea (Mckay et al., 2010).

The plant is used as a beverage that helps to lower the body temperature while the water extracts of hibiscus flowers are said to have a relaxing effect on the uterus and lower the blood pressure. Clinical evaluation has shown that the plant reduces cholesterol by a factor of 8-14% within a month (Lin et al., 2007). The solution obtained from the dried Roselle calyces is known to contain a lot of chemical constituents which include alkaloids, L-ascorbic acid, citric acid, anthocyanin quercetin and anisaldehyde.

The plant Roselle calyces are also rich in minerals such as calcium, magnesium, iron, potassium and sodium (Peter et al., 2014). The red calyces of the plant are increasingly exported and used as food colourings. The red calyx of the *Hibiscus sabdariffa* is the major component possessing the sour taste obtained from the beverage.

In Nigeria, the beverage obtained from the plant is produced from the red calyces and its consumption has gained wide popularity all over the country. However, the low shelf life of the drink produced by the traditional method has placed serious limitations on its general acceptance and consumption in Nigeria. Another factor which has limited its general acceptance is the growing concern that the crude extract from many of our natural plants are unsafe for consumption (Sulaiman et al., 2014). This is the reason why this study was carried out to evaluate the microbiological and nutritional properties of the crude extract from *Hibiscus sabdariffa*. This study will provide information on the risk associated with the storage of the beverage and how to overcome it.

2. Materials and Methods

2.1 Collection of Plant Material

Dried reddish petals of *Hibiscus sabdariffa* plant were obtained from the open market and transferred to the Microbiology Laboratory, Delta State University, Abraka, Nigeria in a polyethylene bag. The plant was prepared in accordance with the method explained by Sulaiman et al. (2014). In accordance with this method, 100 g of the dried flower was mixed with 1000 cm^3 of water and boiled for 30 minutes to obtain the plant beverage. The crude extract was allowed to cool and then filtered using a clean sieve.

2.2 Determination of Microbial Counts

The viable bacteria counts were evaluated by pour plate method using 1 ml of serially diluted beverage samples in Nutrient agar and incubated at room temperature for 48 hours (Cheesbrough, 2002). Duplicate plates were made for each dilution. The yeast and mould counts were determined using Potato Dextrose Agar incorporated with antibiotics and incubated at room temperature for 72 hours (Cheesbrough, 2002). Pure isolates were obtained by picking discrete colonies from the growth media plate. The colonies were then subculture by streaking onto fresh growth media plates until pure isolates were obtained. The isolates were then transferred onto agar slants of the same medium as stock culture. They were stored in a refrigerator at 0-4°C from where they were taken for identification test.

2.3 pH Test

The pH test was carried out by using the HI99131N pH meter manufactured by Hanna Instrument. A phosphate buffer solution of pH 7.02 was used to standardize the meter at room temperature before the readings were taken.

2.4 Determination of Carbohydrate, Vitamin C and Protein Content

The test for carbohydrate, protein and vitamin C was carried out by employing the gravimetric method as adopted by the Association of Official Analytical Chemists (AOAC, 1990). This method gives an estimate of the indigestible fibre in the beverage. The insoluble residue is collected by filtration, dried, weighed and ashed to correct for mineral contamination of the fibre residue.

2.4.1 Carbohydrate Test

The carbohydrate test was carried out by measuring 1 ml of the beverage sample into a test tube followed by the addition of 0.9 ml of water and 5 ml of concentrated sulphuric acid. These were mixed thoroughly and allowed to stand for a period of 30 minutes. The result was read on a spectrophotometer at a wavelength of 485 nm and compared with a calibration curve to calculate the carbohydrate content (AOAC, 1990).

2.4.2 Test for Vitamin C

Vitamin C was determined by measuring 2 ml of the test sample into 100 ml volumetric flask. 50 ml of metaphosphoric acid was added to the flask content and was stirred for 10 minutes. The mixture was filtered using a Whatman filter paper. 20 ml aliquot of the filtrate was transferred into a flask and titrated against 0.01 N iodine, using starch solution as indicator to blue black end point. This was used to determine the vitamin C content of the sample (AOAC, 1990).

2.4.3 Test for Protein Content

This was carried out by collecting a known weight of the sample and digested with 20 ml of tetraoxosulphate VI acid. The digested sample was added to 8 ml of distilled water and 2 ml of Nesseler reagent. This was left for 20 minutes at room temperature for colour development and then read at 520 nm in a spectrophotometer. The result obtained was multiplied by a factor to convert to protein content (AOAC, 1990).

3. Results and Discussion

The total viable bacteria and fungi counts of the *Hibiscus sabdariffa* beverage obtained during processing and storage under ambient temperature are presented as shown in Table 1. The result shows that the total variable count decreases sharply at first and then increases gradually up to the sixth day before decreasing again all through the storage period. There were no fungi detected at the beginning of the storage period, indicating that the heat applied during the processing has eliminated all the associated fungi. However, as the storage progresses, fungi were detected and the viable count followed the trend observed with the bacteria count. It is observed from Table 1 that ecological successions amongst the microorganisms exist. The period from 0-4 days were dominated by *Bacillus* sp., *Staphylococcus sp.* and *Streptococcus sp.* while *Lactobacillus sp.* and fungi dominated the late phase (Tables 2, 3, 4).

Table 1. Microbiological quality changes in *Hibiscus sabdariffa* beverage during storage at 28°C

Period (days)	Bacterial Count (CFU/ml)	Fungal (CFU/ml)
0	1.4×10^6	ND
2	2.5×10^2	1.1×10^1
4	10.2×10^4	2.0×10^1
6	11.5×10^6	3.1×10^4
8	8.0×10^5	2.5×10^2
10	6.5×10^4	1.5×10^2

ND – Not detected.

Table 2. Characteristics used for the identification of bacteria isolates from *Hibiscus sabdariffa* beverage

Characterization and Tests	Isolate A	Isolate B	Isolate C	Isolate D
CULTURAL: SURFACE	Smooth	Smooth	Smooth	Smooth
FORM	Circular	Spreading	Circular	Circular
ELEVATION	Convex	Slightly Convex	Convex	Convex
EDGE	Entire	Entire	Entire	Entire
SIZE	1-2mm	2-4mm	2-0mm	1-0mm
CHROMOGENESIS	Yellow	White	Creamy	Creamy
OPACITY	Opaque	Slightly Opaque	Opaque	Opaque
MORPHOLOGICAL:				
GRAM STAIN	+	+	+	+
CELL SHAPE	Cocci	Rods	Rods	Rods
SPORE STAIN	-	+	+	-
ARRAGEMNENT	Single/Clusters	Chains	Single	Single/Chains-
MOTILITY	-	+	+	-
BIOCHEMICAL:				
CATALASE	+	+	+	-
OXIDASE	-	-	+	-
COAGULASE	-	-	-	-
INDOLE	-	-	-	-
METHYL RED	-	-	-	-
VOGES PROSKAUER	+	+	-	-
CITRATE UTILIZATION	+	+	+	+
HYDROGEN SULPHIDE	-	-	+	-
PHENYLALANIWE	-	-	-	-
SUGAR UTILIZATION	-		-	-
GLUCOSE	AG	A	AG	AG
LACTOSE	A	-	-	A
MANNITOL	A	A	-	A
PROBABLE ISOLATE	*Staphylococcus sp.*	*Bacillus sp*	*Lactobacillus sp.*	*Streptococcus sp.*

Key: + = Positive; - = Negative; A = Acid production; AG = Acid and gas Production.

Table 3. Characterization of yeasts isolated from *Hibiscus sabdariffa* beverage

Attribute	Isolate D (Result)	Isolate E (Result)
Appearance of culture in medium	White firm flat like mass on potatoes Dextrose Agar	Brown convex with irregular edge dull and opaque in potato dextrose Agar (PDA)
Hyphae Nature	Septate	No hyphae
Conidiophores	Neither conidiospore or sporangiophore	Neither conidiophore or Sporangiophore
Types of asexual spores	Arthrospores	Ascospores
Characteristic of asexual spore head	Hyphae breaks into arthrospore at the tip	No spore head
Microscopic appearance of spore head	Oral aerial spores and cylindrical submerged ones	Spherical, globase and occurs in pairs or singles
No ascospores in asci	No ascospores	1-Bascospores
Type of budding	No budding	Multilateral budding
Biochemical	NT	+
Assimilation of carbon compounds		
Glucose	NT	+
Sucrose	NT	+
Mallose	NT	+
Lactose	NT	+
Mannitol	NT	-
	NT	-
Formation of Pseudomycellium		Absent
Foot cell	-	Absent
Rhiziod	-	Absent
Probable Identity	***Geotrichum sp.***	***Saccharomyces sp.***

Key: NT – Not tested; - =Negative; + = Positive.

Table 4. Characteristics of mould species isolated from *Hibiscus sabdariffa* beverage.

Attribute	Isolate (Result) A	Isolate (Result) C	Isolate (Result) D
Appearance of culture in medium	Freely branched greenish mycelium in Potato Dextrose Agar (PDA)	Colony profusely branched which appear brownish black	White colony grow on Potato Dextrose Agar (PDA)
Hyphae	Septate	Septate	Non-septate
Nature of conidiophores	Perpendicular septate that branches toward the top	Arises from foot cells organised septate terminating in a swollen vesicle	Upright sporangiophore connected by stolons
Types of asexual spores	Conida	Conidia	Sporangiophore
Characteristic of asexual head spore	Sterigmata arising from metule rear conidia in chains	Vesicles bearing in chains	Dark pear shaped sporangium on hemispherical columella
Microscopic appearance of sexual spore	Round light conidia in chains	Brownish black conidia in chains	Small globose sporangiosphore
Production and type of spore	Ascospores	Ascospores	Zygospores
Special features stolons	Absent	Absent	Present
Rhizoids	Absent	Absent	Present
Foot cells	Absent	Absent	Present
Probable	***Penicillum sp.***	***Aspergillus sp.***	***Rhizopus sp.***

Table 5. pH, and titratable acidity content of *Hibiscus sabdariffa* beverage during storage

Period (days)	pH values	Titratable acidity (%)
Day 2	2.98	0.042
Day4	2.78	0.045
Day6	2.67	0.048
Day8	2.62	0.058
Day10	2.61	0.067

Table 6. Proximate composition of *Hibiscus sabdariffa* beverage during storage

Period (days)	Carbohydrate (mg/g)	Protein (mg/g)	Vitamin C (mg)	Total soluble solid (%)
Day 0	0.03	0.29	0.18	0.01
Day 2	0.06	0.34	0.31	0.01
Day4	0.08	0.39	0.49	0.03
Day6	0.09	0.44	0.56	0.04
Day8	0.06	0.37	0.54	0.03
Day10	0.04	0.35	0.48	0.02

The effect of processing and storage on the properties of the beverage is presented as shown in Table 5. Table 5 shows that the pH decreased all through the storage period, indicating that the acidity increased throughout the period of storage. This result agrees with the finding of Olayemi (2011) which shows that the pH of the plant beverage is acidic with values of between 2.53 and 2.67. The result of the proximate composition of the beverage from the plant during storage is shown in Table 6. The total carbohydrate content, protein content, vitamin C content and total soluble solid increased gradually from the second day up to the sixth day and thereafter decreased gradually all through the remaining period of storage.

The results of the investigation as shown in Table 2, 3 and 4 indicate that *Hibiscus sabdariffa* beverage contained a reasonable percentage of bacteria and fungi population. The large number of bacteria and fungi may be due to the fact that the beverage is susceptible to bacteria and fungi contamination during the processing and storage stages. The lactic bacterial isolated from the beverage are *Lactobacillus* sp., *Bacillus* sp., *Staphylococcus* sp. and *Streptococcus* sp. (Table 2) while *Aspergillus sp*, *Penicillium sp*, *Rhizopus sp*, *Geotrichum sp* and *Saccharomyces sp* were the fungi isolates (Table 4). This is in agreement with some earlier works carried out by Akinyosoye and Akinyele (2000). This result also agrees with earlier work carried out by Braide et al. (2012) where they identified the presence of some microorganism associated with the plant.

The presence of lactic acid bacteria is expected since they are mostly found in the fermentative mash. The presence of these lactic bacteria prevents the survival of other pathogenic microorganism since their ability to produce lactic acid reduces the pH of the food medium. The presence of these organisms is of advantage to the overall health quality of the beverage.

The presence of *Streptococcus sp.* and *Staphylococcus sp* may have been enumerated from the beverage as a result of the handlers since they are associated with hand, hair or nasal cavities. This could have been discharged into the preparation stage through sneezing or coughing. The handler could also have been a carrier of the organism and so easily distributes the organism. It is observed from this study that the presence of *Baccilus sp.* in the beverage could have resulted from the fact that it is a spore former and as such the spores were easily distributed and able to withstand high temperature and pH to fully germinate.

The fungal isolates found in the plant beverage could be traced to the time when the petals were either being harvested or stored. They may have produced spores, which were attached to the petals and overcome adverse condition during the preparation and finally germinated in the finished product. This may be responsible for the high count recorded in the late phase of the storage period.

The carbohydrate contents in the beverage increased throughout the storage period (Table 6). This increase can be associated with the activities of the various organisms. Similarly, the increases in the protein content may be due to the release of bound protein associated with the microbial activities and the content of the microorganism themselves. The result therefore shows that the conversion of carbohydrate and protein to other products such as alcohol, acids and other metabolites may have contributed to the low pH and high titratable acidity recorded.

4. Conclusion

The general method for producing the beverage of the plant will always contain an unusually large population of fermentative beneficial organisms and very likely, some pathogenic microorganisms. The isolation of some of this pathogen such as *Staphylococcus sp.*, *Streptococcus sp.* and *Aspergillus sp.* could be indicative of health risk, although this is not to cause extreme worries because their populations have been inhibited to an extent by the acid produced by the lactic acid bacteria. The risk associated with the beverage could be reduced if at all stages from harvesting to selling, good hygienic conditions are imbibed. If this is done, the rate of contamination will be reduced tremendously and hence *Hibiscus sabdariffa* drink could serve as a convenient substitute to carbonated drinks which may not be affordable by a lot of people.

References

Akinyosoye, F. A., & Akinyele, B. J. (2000). *Microorganisms associated with "zoborodo", a Nigerian beverage.* Nigerian Society of Microbiology Book of Abstract.

AOAC. (1990). Official methods of food analysis (15th ed.). Association of Official Analytical Chemists, Washington DC., USA.

Braide, W., Oranusi, S., & Peter-Ikechukwu, A. I. (2012). Perspectives in the hurdle techniques in the preservation of a non alcoholic beverage, zobo. *African Journal of Food Science and Technology, 3*(2), 46-52.

Cheesbrough, M. (2002). Medical laboratories manual for tropical countries (1st ed.). Butterworth Kent, London,.

FIIRO. (2011). Industrial profile on zobo drink production and preservation, FIIRO Publication 2014. Retrieved from http://www.manufacturingtoday.com.ng

Lin, T., Lin, H., & Chen, C. (2007). *Hibiscus sabdariffa* extract reduces seru cholesterol in men and women. *Nutritional Research, 27*, 140-145.

McKay, D. L., Chen, C. O., Saltzman, E., & Blumberg, J. B. (2010). *Hibiscus Sabdariffa* L. Tea (Tisane) Lowers Blood Pressure in Prehypertensive and Mildly Hypertensive Adults. *The Journal of Nutrition.* http://dx.doi.org/10.3945/jn.109.115097

Olayemi, F., Adedayo, R., Muhummad, R., & Bamishaiye, E. (2011). The nutritional quality of three varieties of Zobo (*Hibiscus sabdariffa*) subjected to the same preparation condition. *American Journal of Food Technology, 6*, 705-708. http://dx.doi.org/10.3923/ajft.2011.705.708

Peter E., Mashoto, K. O., Rumisha, S. F., Malebo, H. M., Shija, A., & Oriyo, N. (2014). Iron and Ascorbic Acid content in *Hibiscus sabdariffa* Calyces in Tanzania: Modelling and Optimization of Extraction Conditions. *International Journal of Food Science and Nutrition Engineering, 4*(2), 27-35.

Sengupta, R., & Banik, J. K. (2011). Evaluation of *Hibiscus sabdariffa* leaf mucilage as a suspending agent. *International Journal of Pharmacy and Pharmaceutical Sciences, 3*(5), 184-187.

Sie, R. S., Charles, G., Diallo, H. A., Kone, D., Toueix, Y., Dje, Y., & Branchard, M. (2011). Breeding of *Hibiscus sabdariffa L.*: Evaluation of resistance to Fusarium oxysporum Schlecht. Emend. Snyd. and Hans in two varieties. *Agriculture and Biology Journal of North America, 2*(1), 125-133.

Sulaiman, F. A., Kazeem, M. O., Waheed, A. M., Temowo, S. O., Azeez, I. O., Zubair, F. I., ... Adeyemi, O. S. (2014). Antimicrobial and toxic potential of aqueous extracts of *Allium sativum*, *Hibiscus sabdariffa* and *Zingiber officinale* in Wistar rats. *Journal of Taibah University for Science, 8*(4), 315-322. http:dx.doi.org/10.1016/j.jtusci.2014.05.004

13

Polymorphism and Damage of Aphids
(*Homoptera: Aphidoidea*)

Alla Vereschagina[1] & Elena Gandrabur[1]

[1] Russian Academy of Sciences, All Russian Institute for Plant Protection (VIZR), Saint-Petersburg, Pushkin, Russia

Correspondence: Elena Gandrabur, All Russian Institute for Plant Protection, Podbelski road, 3, Saint-Petersburg, Puskin 196608, Russia. E-mail: helenagandrabur@gmail.com

Abstract

The paper presents an analysis of original and published data on the ways of interaction of aphids with their host plants, the abiotic environment, other aphids and various other organisms, in relation to the specific traits of their biology, epigenesis, and migrations. Special attention is paid to the factors determining the colonial mode of life and the factors controlling the formation of alate morphs during the vegetation period. It is shown that damage of aphids is determined by their mode of feeding on plant sap, gregariousness, short- and long-range dispersal, and intraspecific variation. The factors controlling polymorphism in aphids are classified for the first time and a diagram showing their action is proposed. The importance of alate morph development in aphids as a factor and an indicator of their damage is demonstrated.

Keywords: aphids, polymorphism, crowding, migrations, life cycle, damage

1. Introduction

Aphids are a large group of insects (Insecta: Homoptera) including about 4700 species, of which about 1000 occur in Europe. Despite the general diversity of the biocenotic roles of aphids, the attention of experts is mainly focused on the "cereal" aphid species since they can cause considerable yield loss due to their abundance, the specific mode of feeding on plant sap, ecological polymorphism, and the ability to transmit viral diseases of plants even at very low abundance. The alate aphids not only disperse viruses but also provide links between the aphid populations of various regions. The resulting interpopulation gene exchange creates the basis for selection of aphid clones with different life cycles, host preferences, and other properties. Among the ways of aphid control the first place is still held by the chemical method, but this method should be avoided due to environment pollution and the resulting health risks. Increasingly greater attention is paid to more environment-friendly means of plant protection, such as biological and agrotechnical methods, selection of resistant cultivars, development of transgenic genotypes, etc. The existing methods of short-term and long-term prediction of the aphid population dynamics are not always efficient. Due to their intrapopulation variability, aphids quite rapidly get adapted to insecticides, resistant cultivars, and climatic factors. The use of entomophages is hindered by the sheltered life of many aphid species among leaves or in galls and by their escape behavior (dropping off the plant), whereas the immune system of aphids is sometimes capable of protecting them from parasites. Aphids may possess some other defense mechanisms as well. The outbreaks of aphids may be unpredictable because of the gaps in the knowledge of their biology and interactions with the host plant. These gaps may be partly filled by considering some key aspects of the colonial mode of life and formation of alate morphs responsible for dispersal and host plant selection in the different periods of the aphid life cycle. Studies carried out on aphids as model objects may also help to understand the general biological effects of crowding on animal development and the mechanisms of biological invasions.

This communication presents a brief review of the original material and data published by various authors on the problem of aphid polymorphism, which appeared in the course of historical development of their life cycles and is controlled by the gregarious mode of life, interactions with other taxa, and climatic and anthropogenic factors, in relation to the kinds of damage inflicted to plants. Special attention is devoted to the gregarious mode of life and formation of alate morphs during the aphid life cycle.

2. The Life Cycles and Morphs of Aphids

The damage to cultivars caused by aphids is always determined by the specific traits of their trophic relations with the host plant (including the way of feeding on the plant sap) and their life cycles. Only the representatives of the families *Adelgidae* and *Phylloxeridae* feed on the plant parenchyma while all the remaining species of aphids consume phloem sap (Vereschagin et al., 1985). The established life cycles of aphids reflect their trophic relations. Development of holocyclic aphids usually includes alternation of one sexual and 10–20 parthenogenetic generations and is accompanied by polymorphism, whereas in heteroecious species it also involves transition from the primary (winter, arboreal) host plant to the secondary (summer, herbaceous) one. Host alternation is known in only 10% of aphid species, whereas the majority is autoecious and monophagous. However, most of the aphids damaging cultural plants are either heteroecious species or autoecious but polyphagous ones (Dixon, 1977).

The sexual generation may be "lost" in the aphid populations developing under constant climatic conditions, for example, in the southern parts of their ranges or in greenhouses; this results in a reduced, or anholocyclic, development cycle. Some aphid species, in particular *Eriosoma lanigerum*, *Rhopalosiphum maidis*, and *Aphis gossypii*, have lost the sexual generation due to disappearance of their primary host plants. During the vegetative period the aphid population includes several adult morphs somewhat differing in morphology, behavior, and demographic parameters (Shaposhnikov, 1986; Powel & Hardie, 2001; Vereschagina, 2008; Webster, 2012). The complete life cycle of aphids usually includes the following morphs: parthenogenetic fundatrices and alate emigrants in spring, alate and apterous viviparous females (viviparae) in summer, and alate remigrants (sexuparae, gynoparae, androparae) as well as males and oviparous sexual females in autumn. The emigrants fly from the primary hosts to the secondary ones at the beginning of summer, whereas the remigrants return from the secondary hosts to the primary ones in autumn. On the primary host, the sexuparae produce the larvae of males and sexual females (gynoparae, only of females; androparae, only of males). Males of some heteroecious species can be produced on the secondary hosts. In a few species remigration takes place in the middle of summer rather than in autumn, resulting in a shortened life cycle (Popova, 1967). Some unusual morphs can be found in a number of aphid taxa, for example, pseudofundatrices in the family *Adelgidae*, "soldier" larvae incapable of further development and reproduction, in some species of the families *Pemphigidae* and *Hormaphididae*, and flying larvae in the genus *Periphyllus* Hoev (Shaposhnikov, 1986). Due to polymorphism, certain functions are partially distributed between groups of specialized individuals, or morphs. Each morph has its own specific role and appears in specific time during the life cycle. The apterous and alate females are settled and migrating individuals that differ in their morphology, epigenesis, metabolism, behavior, function in the life cycle, and damage to plants.

Damage of aphids can be conditionally subdivided into three categories. The first kind of damage is determined by consumption of considerable amounts of plant sap by aphid colonies, and also the chemical action on the plant. The second kind is related to the possibility of rapid intrapopulation variation. The third kind of damage is related to the ability of aphids to transmit viral diseases of plants. All the three categories of damage are related in one way or another to polymorphism of aphids.

3. The Gregarious Mode of Life and the Influence on the Host Plant

Such traits as parthenogenesis, viviparity, polyvoltinism, and formation of "telescoping generations" (with embryos developing inside the larva before it is born) facilitate mass reproduction and aggregation of aphids on host plants. Such aggregation is especially frequently observed in spring and early summer, when plants are more favorable for aphid feeding.

Different authors interpret the phenomenon of aphid aggregation in different ways (Way & Banks, 1967; Way & Cammell, 1970; Bonner & Ford, 1972; Kidd, 1976; Loxdal et al., 1993, etc.). Since aphids are adapted to feeding on assimilates contained in plant sap, they are attracted to the plant's growth zones where they aggregate and reproduce in great numbers, mostly due to rapid development and high fecundity of apterous parthenogenetic females (Sloggett & Weisser, 2004). There is an interesting hypothesis according to which the first step in the evolution of aptery was related to the possible fecundity gains in the brachypterous forms due to autolysis of the wing muscles. The species producing only alate morphs (the family *Drepanosiphidae*) and occurring on arboreal hosts have a lower fecundity as compared to the species which also have apterous morphs (Dixon, 1985).

According to our data, the realized fecundity in the bird cherry-oat aphid *Rhopalosiphum padi* varied between the morphs. During feeding on bird cherry, the mean fecundity was 49–187 larvae in the fundatrices; 39–103 larvae in the apterous fundatrigenous females; 15–55 larvae in the emigrants. During feeding on favorable secondary hosts, the fecundity was 30–93 larvae in the apterous viviparae and 15–52 larvae in the alate viviparae. Thus, apterous morphs had higher fecundity than alate ones, both on the primary and the secondary host (Vereschagina, 2007).

The colonies always include more apterous viviparae than alate ones, except for the periods of spring emigration from the primary to the secondary host, return migration in autumn, and the flight peaks in summer.

Due to their colonial mode of life, aphids can "precondition" their trophic substrate for intake by injecting large quantities of salivary gland secretion, and sometimes even gut content, containing digestive enzymes and a powerful set of glycolytic and antioxidant enzymes (Auclair, 1963; Vereschagina, 1982; Vereschagina & Gandrabur, 1988; Urbanska et al., 1998, 2004; Miles, 1999; Lukasik et al., 2004). As the result of such preconditioning of plant sap, the colonial larvae of *Megoura viciae* developed faster and more uniformly than isolated ones (Bonner & Ford, 1972). Two species of cereal aphids, *Rh. padi* and *Rh. maidis*, revealed lower mortality rates in case of gregarious, rather than solitary distribution on different genotypes of corn (Vereschagina, 2001). A positive correlation was observed between the colony size in two biotypes of *Diuraphis noxia* and the success of its reproduction and offspring development on three cultivars of *Triticum aestivum* (Michaud et al., 2006). In general, the trophic conditions for aphids were shown to be more favorable within the colony than in the case of solitary feeding, as long as the density remained below a certain limit (Way & Cammell, 1970). The specific microclimate in the colonies may be also important for aphid development.

An increase of the colony size and density beyond a certain limit leads to the local deficiency of food and triggers the intraspecific mechanism of abundance regulation, namely migration of apterous females and formation of alate individuals (Way & Banks, 1967; An et al., 2012). The load on the host plant is thus reduced to the optimum level.

According to other authors (Kidd, 1976; Loxdale et al., 1993; Ban et al., 2008), aggregation of the larvae and adults of the lime aphid *Eucallipterus tiliae* is determined by attractiveness of the aphids themselves, rather than by certain properties of the feeding sites. In their opinion, the "group effect" reflects the contact communication of aphids via tactile stimuli perceived by the receptors on their legs, antennae, and bodies. When produced by insects of other species or even by plant fibers, such mechanical stimuli cause a "pseudo-crowding" effect (Lees, 1967). Our experiments confirmed that the "group effect" was mediated by trichoid sensilla located on the legs of aphids. When feeding on the VIR-44 corn variety, the colonial apterous viviparae of *Rh. padi* with amputated tibiae produced on average 0.3% of alate offspring in 10 days, as compared to 44% in aphids with intact tibiae. The ability of contact communication allows the aphids to employ the "dropping" escape response when disturbed by entomophages. In some solitary aphid species the leading role in communication may belong to visual perception (Bonnemaison, 1967; Lees, 1966). However, these behavioral traits do not contradict the role of attractiveness of the plant growth zones, but demonstrate one more positive aspect of the colonial mode of life in aphids.

Thus, the reasons for aggregation of aphids are parthenogenesis, high rates of development, attractiveness of certain feeding sites (the plant growth zones), and the presence of an aggregation attractant; the consequences of aggregation are better trophic conditions, high abundance, timely dispersal, and defense against aphidophages by means of contact communication.

In turn, such a general influence of aphids on the plant induces a variety of defensive and pathological responses involving transport and transformations of primary and secondary metabolites. At low abundance, some species of aphids can even stimulate growth and photosynthesis in their host plants, due to general metabolic enhancement as a manifestation of the initial immune response. In particular, a colony of *Brevicorinae brassicae* feeding on the mustard plant induced a 70–80-fold increase of sucrose inflow to the feeding site (Way & Cammell, 1970). According to the data of Ehrhardt (1968), a phloem sap feeding aphid can imbibe the amount of liquid equal to 130% of its body mass. Considering that one plant can be inhabited by 1000 and more aphids, the loss of sap due to their feeding may be as great as 0.5 ml per hour.

Consumption of plant sap by aphids is accompanied by release of great amounts of excrements (honeydew). The amount and composition of honeydew, as well as of food, vary depending on the aphid species, the host plant, and the external conditions and comprises from 2.0 to 133.0% of the insect's body mass (Auclair, 1963). Since aphid excrements are rich in sugars, they are colonized by abundant epiphytic microflora (soot-dew, etc.) which reduces the photosynthetic abilities of the host plant by blocking part of the light, causes premature aging of the leaves, and hinders their transpiration and respiration.

Damage by aphids decreases the total solids content as well as the sugar, protein, and vitamin C content in the leaves of many cultivars. The damaged plants show defects of growth and fruit development and often produce seeds of poor quality (Goszczynski & Cichocka, 1998).

The agents affecting the trophic substrate are mostly carried by the salivary gland secretions which contain digestive hydrolases and various compounds toxic to plants (Miles, 1999). The interaction of the salivary components of aphids with the defensive secondary metabolites of plants was best studied by the example of cereal aphids. The group of secondary metabolites responsible for aphid resistance in wheat includes hydroxamic acids, with DIMBOA (2.4-dihydroxy-7-methoxy-1.4-benzoxazine-3-one) as the main component. As the oral stylets of the aphid damage the plant tissues and reach the phloem, the damaged cell walls and cytoplasmic vacuoles release β-glucosidase which splits phenolic and flavonoid glucosides to aglucons. Aglucons, in particular DIMBOA, are more toxic to plants and phytophages than their β-glucoside forms. Synthesis of phenolic compounds in the plant tissue is limited, whereas in the intact cells these compounds are protected from enzymatic oxidation by their isolated position and the presence of antioxidants. However, depletion of the antioxidant reserves in the damaged cells may lead to necrosis (El Naghy & Shaw, 1966; Miles, 1998). The aphid saliva was found to contain β-glucosidase, which seems strange since aglucons produced by this enzyme are toxic to aphids; apparently these insects possess a certain mechanism of aglucon detoxification. In addition, it is known that polyphenol oxidase and peroxidase present in the aphid saliva and gut content can rapidly detoxify a great number of phenolic compounds contained in plant sap (Peng & Miles, 1988; Miles, 1999; Urbanska et al., 2004).

Besides secondary metabolites, the defensive mechanisms of plants may be associated with the structure and function of the sieve tube elements of the phloem. Mature sieve tubes contain great quantities of P-proteins, which seal the pores of the sieve plates in case of damage or at the onset of winter. In case of prolonged damaging influence, the plates of young sieve tube elements are sealed with callose, an easily hydrolysable 1,3-β-glucopolysaccharide (Medvedev, 2004). Callose is highly hygroscopic and protects the cells from drying. In young cells it also serves as a source of carbohydrates since such cells do not possess amyloplasts. Large accumulations of callose have been found in the phloem cells damaged by aphids and around their stylets. Deposits of P-proteins and callose in the pores of the sieve plates block the flow of phloem sap into the aphid's food canal, forcing the insect to change the feeding place. However, aphids may possess some mechanism preventing the formation of seals of P-proteins and callose, since they do not always leave the blocked site. At any rate, during mass feeding on the phloem, when sieve tube sealing appears more probable, the crowded aphids showed a higher level of intestinal proteolytic activity as compared to the solitary females. The sheath of saliva surrounding the aphid stylets protects the plant from mechanical damage and may thus prevent the release of β-glucosidase and other defensive compounds (Evert et al., 1968; Vereschagina, 2012).

Drying, necroses, and many kinds of deformities of plants caused by aphids impair their trophic conditions. However, some kinds of the plants' response to damage facilitate the development of aphids. For example, gall formation protects aphids from adverse external conditions and creates a favorable feeding zone for them, "prolonging" the plant growth phase. Galls are formed as the result of hypertrophic and hyperplastic development of plant tissue in response to prolonged action of aphids. At the same time, the plant benefits from the feeding of aphids being restricted to the gall zone (Miles, 1999). The gall can be considered as an active biochemical filter that modifies the sap according to the aphids' requirements. It should be noted that modifications occur both in primary and in secondary metabolites. The main components of the aphid saliva participating in gall formation may be β-indoleacetic acid and amino acids. It is assumed that β-indoleacetic acid is supplied to the salivary glands via the intestine or is synthesized by oxidative deamination of tryptophan by the saliva oxidases (Miles, 1972). The greatest number of gall-forming species (37) were found in the subfamily *Pemphiginae*. The representatives of the families *Aphididae*, *Hormaphididae*, *Drepanosiphidae*, and some species of *Pemphiginae* produce false galls in the form of rolled, bent, and otherwise deformed leaves (Chakrabarti, 2004).

Thus, the plant's response to aphids affecting its metabolism by extraintestinal chemical influence, mechanical damage, and sap consumption has a complex biochemical nature and manifests itself in disruption of many physiological processes. The compounds injected by aphids into the plant tissues usually produce local effects (drying, necrosis, deformities); however, if transported by the vessels, these compounds or products of their metabolism can disrupt the functions of individual organs or the whole plant.

The physiological and biochemical adaptations of different aphid morphs to feeding on primary and secondary hosts are poorly studied. Still, some detoxifying, antioxidant and extracellular enzymes in the intestine of *Rh. padi* were shown to have a very low activity in fundatrices, a higher activity in the fundatrigenous generation, and the maximum activity in all the morphs during spring emigration (Lukasik et al., 2004). Fundatrices of *Rh. padi* died when placed on aging bird cherry leaves although the oviparae could successfully feed on them (Leather & Dixon, 1981). We also demonstrated an increase in the proteolytic activity in the intestines of the apterous viviparae of *Rh. padi* and *Rh. maidis* feeding on aphid-resistant corn cultivars (Vereschagina & Gandrabur, 1988).

It may be assumed that responses of different aphid morphs to the composition of plant sap and its changes during the plant ontogenesis may vary depending on the biological specificity of these morphs.

These and other intricate relations between the aphids and their host plants, allowing the two groups to coexist in nature, were established in the process of evolution. Cultural plants, produced by artificial selection, can change these relations and the ability of aphids to inflict damage.

The physiologically active secondary metabolites of plants, and also insecticides and bacterial toxins are the most widespread and typical inductors of microevolutionary processes in the populations of aphids and other phytophages.

4. Intrapopulation Variability in Aphids and Their Damage

The conditions of most modern agrobiocenoses are conducive to faster development of adaptations in insect populations. These processes lead to loss of sensitivity of phytophages to the regularly used chemical agents and the cultivars employing certain resistance mechanisms. Adaptations are known to appear at the greatest rate in the most variable and ecologically flexible insect species.

Aphids provide a perfect object for studying microevolution in insects. Intrapopulation variability in aphids may sometimes lead to the appearance of morphs with different trophic, temperature, and other preferences, and change their status as pests. The second category of damage to plants inflicted by aphids is related to this feature. The intrapopulation changes in the aphid population structure result from preservation and propagation of new morphs, enriching the gene pool. A special role in this process is played by the behavioral and developmental traits of apterous and alate morphs.

The range of phenetic diversity of aphids is enlarged due to polymorphism which is typical of this group. The ontogenetic morphs vary in their time of appearance and sensitivity to trophic and living conditions; therefore they are affected by different selective factors of the environment. There is a mass of evidence concerning the differences in development and behavior of the alate and apterous morphs, in particular their ability to produce males and amphigonic females, demographic parameters, interactions with the host plant, ability to produce alate viviparae in small and large colonies, morphology, and color. Such features as the rapid growth of aphid colonies, the ability of emigrants to settle on differently suitable plants, and the restricted ability of the first apterous generations of the emigrants' progeny to change the host plant when necessary, are all conducive to microevolutionary processes. The alate and apterous morphs also differ in the composition of honeydew, mycetocyte symbionts (Campbell & Dreyer, 1985), and activity of digestive hydrolases (Vereschagina & Gandrabur, 1988). These and other parameters are the sources of variation and the basis for selection, both within the clones and among them (Shaposhnikov, 1981). Aphids are characterized by holocentric chromosomes and the thelytokous mode of reproduction, determining a higher level of intraspecific variation and rapid fixation of the accumulated changes (Blackman, 1981).

Different karyotypes of *Rh. maidis* are known to damage different host plants (Brown & Blackman, 1988); different forms resistant to insecticides and host plant toxins are also known in *Myzus persicae, Sitobion avenae, Acyrthosiphom pisum* (Sandström, 1995; Figueroa et al., 2004; Eleftherianos et al., 2004), and many other species. Some authors have considered the probable increase of intrapopulation changes and range expansion of aggressive aphid forms in connection with global warming.

A demand has recently arisen for control of appearance and expansion of especially harmful forms of aphids. For this purpose, changes in the genetic structure of aphid populations are studied using molecular markers and the possibility of genetic mapping is discussed (Llewellin et al., 2003).

Thus, during the life cycle of aphids, the clones best adapted to particular living and feeding conditions undergo selection based on the biotic parameters of the ontogenetic morphs comprising them. In view of this, the abundance of alate migrants is especially important for dispersal of adapted clones.

5. Alate Morphs in Aphids and Their Role in Interactions With Plants and Other Groups of Organisms

5.1 Specific Traits of Damage of Alate Aphid Morphs

The third category of damage caused to plants by aphids depends not only on their abundance but also on the specific traits of formation and behavior of alate morphs capable of migrating to short and long distances. Alate migrants and viviparae quickly colonize the adjacent territories and can also be carried far away by wind, providing links between the populations in different geographic zones. Alate aphids spread the genotypes resistant to cultivars and insecticides onto new territories (Loxdale et al., 1993), which may result in sudden pest outbreaks. Due to parthenogenesis, a single migrating aphid can potentially found a new population in a previously uninhabited region, or change the clonal composition of the existing local population. If males and sexual females are apterous, the composition of trophic forms can vary even between the fields positioned nor very far apart; this was demonstrated by the example of *Acyrthosiphon pisum* clones (Sandström, 1995).

Aphids hold the first place among insects in their ability to transmit viral diseases. Of them, 300 species are presently known to transmit about 200 species of viruses (Dyakonov, 2000). Damage of aphids as virus vectors is manifested even at low abundance, since one alate individual flying from plant to plant can infest many of them. The activity of virus transmission and aphid dispersal is closely connected with the formation of alate morphs and the factors regulating it.

5.2 Rhythmicity of Wing Induction in Aphids

The rhythmicity of wing induction in aphids is an ambiguous phenomenon. On the one hand, it has an endogenous nature and does not depend on the density, since alate morphs appear in certain periods of the year. The endogenous factors are the holocyclic or anholocyclic life cycle and the epigenetic factors determining the development of a population. In anholocyclic populations the appearance of first flying adults is mostly determined by the quality of the host plant and the crowding of the colony. Emigrants and remigrants appear in the ontogenesis of heteroecious holocyclic populations according to the innate program of host change (Glinwood & Petterson, 2000). The so-called fundatrix factor determines the obligatory time gap since the beginning of production of larvae by the fundatrix after which alate gynoparae can start to appear (Lees, 1966). Both alate and apterous viviparae also seem to possess endogenous rhythms of production of offspring belonging to particular morphs. In our long-term research, most clones (96.9%) of *Rh. padi* emigrants transferred from bird cherry to wheat shoots of the Leningradka variety produced no more than 10% of alate offspring or no alate individuals at all. It follows from these data that the first summer generations of *Rh. padi* developing on secondary hosts (the descendants of the emigrants) are usually apterous (Vereschagina, 2008). Therefore, the main spring vectors of viral diseases are alate emigrants which leave the bird cherry in great numbers and visit a wide range of plants in search of suitable hosts.

The first larvae produced by alate viviparae are also always apterous. This was demonstrated by us in two species of cereal aphids: *Rh. padi* and *Rh. maidis* (Vereschagina & Shaposhnikov, 1997). Subsequently the fraction of alate adults in the offspring of alate viviparae increases but their quantity and behavior depend on a number of factors. First, even under the same trophic conditions alate individuals start to appear in the offspring of different clones at different times since the onset of reproduction; second, their number varies without any definite correlation with abundance or colony density, apparently as the result of combination of exogenous and endogenous rhythms; third, alate adults may be not only migratory but also settled (Vereschagina & Vereschagin, 2013). The offspring of apterous viviparae always includes alate individuals, their number and behavior being similar to the previous variant.

5.3 Factors Controlling Polymorphism in Virginoparae

5.3.1 Crowding

The principal factor inducing development of the alates in the summer aphid generations is crowding, or the "group effect" (Sutherland, 1969; Vereschagina & Gandrabur, 1988; An et al., 2012). The group effect modifies the epigenesis of aphids by an endocrine mechanism. In the opinion of most researchers, formation of alate morphs in aphids is controlled by the endocrine system of the maternal generation. A decrease in the activity of *corpus allatum* reduces secretion of juvenile hormone, thus facilitating wing development (Hardie, 1981).

The embryos remain in the neutral state until they reach a certain stage. The phase of development sensitive to the group effect is species-specific and may be either pre- or postnatal (Ankersmit & Dijkman, 1983). The crowding effect may sometimes be produced by keeping the aphids together for only several minutes, though usually it takes several hours to develop. Both the age and the density of individuals are significant (Bonnemaison, 1967). At the same time, contrary to what was observed by some authors (Delisle et al., 1983) working with other aphid species, we did not reveal any correlation between the age of apterous viviparae of *Rh. padi* and *Rh. maidis* and the fraction of alate individuals in their offspring (Vereschagina & Gandrabur, 1988).

The development of alate or apterous viviparae or sexuparae is not solely controlled by the crowding effect. According to some authors, the less important factors are the photoperiod, the host plant, and the temperature (An et al., 2012). A shorter photoperiod has little effect on wing induction in the summer viviparae of both holocyclic and anholocyclic clones, but it is very important for production of autumn remigrants in the period when both the photoperiod and the temperature change (Bonnemaison, 1967; Tsitsipis & Mittler, 1977). It has been suggested that the effect of environmental factors may be greater at a certain colony density, which was demonstrated by the example of *Schizaphis graminum* (An et al., 2012).

5.3.2 The Host Plant

There is no consensus as to the role of the trophic factor in wing induction in aphids. Some authors believe that the high percentage of alate adults in aphid colonies does not depend on the biochemical changes induced by their feeding and salivary secretion (Forrest, 1974). By contrast, other authors have shown that the host plant may be the prevalent factor of wing dimorphism in various aphid species, due to its attractiveness, trophic value, and the presence of substances impairing feeding and affecting embryonic development (Harrewijn, 1978). Experiments with artificial substrates revealed a correlation between the number of alate aphids and concentration of amino acids and sucrose in the diet (Mittler & Kleinjan, 1970). Most studies confirm the influence of the host plant on wing induction in aphids. It is known that apterous females of *Sitobion avenae* produce fewer alate offspring on winter wheat in the flowering phase when the trophic conditions are favorable, than during the less favorable tillering or ripening phases (Ankersmit & Dijkman, 1983). Our research showed that feeding on the resistant corn variety A-619 differently increased the number of alate individuals in the offspring of corn and bird cherry-oat aphids (Vereschagina & Gandrabur, 1988). The number of alate individuals may reflect different trophic specialization of conspecific clones. For example, noticeable changes in the alate morph production were observed when clones of *Acyrthosiphum pisum* adapted to feeding on pea, clover, or alfalfa were transferred onto other species of host plants (Sandström, 1995). We also demonstrated the effect of the winter wheat cultivar on abundance and composition of offspring in different clones of apterous viviparae of *Rh. padi.* In the latter case, the number of alate adults and nymphs in the summer offspring of apterous viviparae varied by several times depending on the diet optimality.

Application of fertilizers changes the trophic conditions of the plants and, correspondingly, affects the phloem sap composition, growth rate, and other physiological parameters. Application of plant growth bioregulators together with fertilizers enhances these changes, which are bound to affect the development of aphids. Nitrogen fertilizers particularly affect the quality of plants and the abundance of aphids and entomophages. Although the response of aphids to fertilizers is species-specific, enhancement of plant growth usually increases the size and density of aphid colonies (Dudnik, 1985; Garratt et al., 2010; Stafford et al., 2012) and, correspondingly, the fraction of alate morphs in the offspring and the migration intensity.

5.3.3 Consort Associations

Wing induction and migration in aphids are also affected by such factors as infestation with viruses and parasites and the presence of ants and other predators (Bonnemaison, 1967; Gutierres et al., 1980; Gildow, 1983; Sudd, 1987).

For example, predators visiting aphid colonies disturb the aphids, which may increase the number of alate females in their offspring, but lowering the colony density may have the opposite effect. It is known that attendance by various species of ants (*Formicinae*) reduces the number of alate morphs in the aphid population (Sudd, 1987). However, the various kinds of relations between ants and aphids, ranging from mutualism to predation, can modify this effect since, on the one hand, ants reduce the number of other disturbing visitors, and on the other hand, they can also consume and disturb aphids (Novgorodova, 2005; Stadler & Dixon, 2005). When attacked by predators, aphids synthesize and release the alarm pheromone (E)-β-farnesene. This volatile compound can act as a communication agent causing the "pseudo-crowding" effect and increasing the number of alate morphs in the offspring of crowded aphids but not of solitary ones (Kunert et al., 2005). Aphids infected with viruses become more sensitive to the alarm pheromone and leave the plant quickly (Ban et al., 2008). As distinct from the effects of parasites and predators, infestation with entomopathogenic fungi and viruses was shown to induce alate morph formation in *Acyrthosiphon pisum* and *Dysaphis plantaginea* independently of physical contacts between aphids; this effect was not related to that of pseudo-crowding (Ryabov et al., 2009; Hatano et al., 2012).

The various factors and their most probable effects on aphid polymorphism are shown in Figure 1. The system shown in the figure optimizes the levels of abundance and dispersal of aphids in relation to their mode of life, epigenesis, and behavior as well as the abiotic, biotic, and anthropogenic environmental factors.

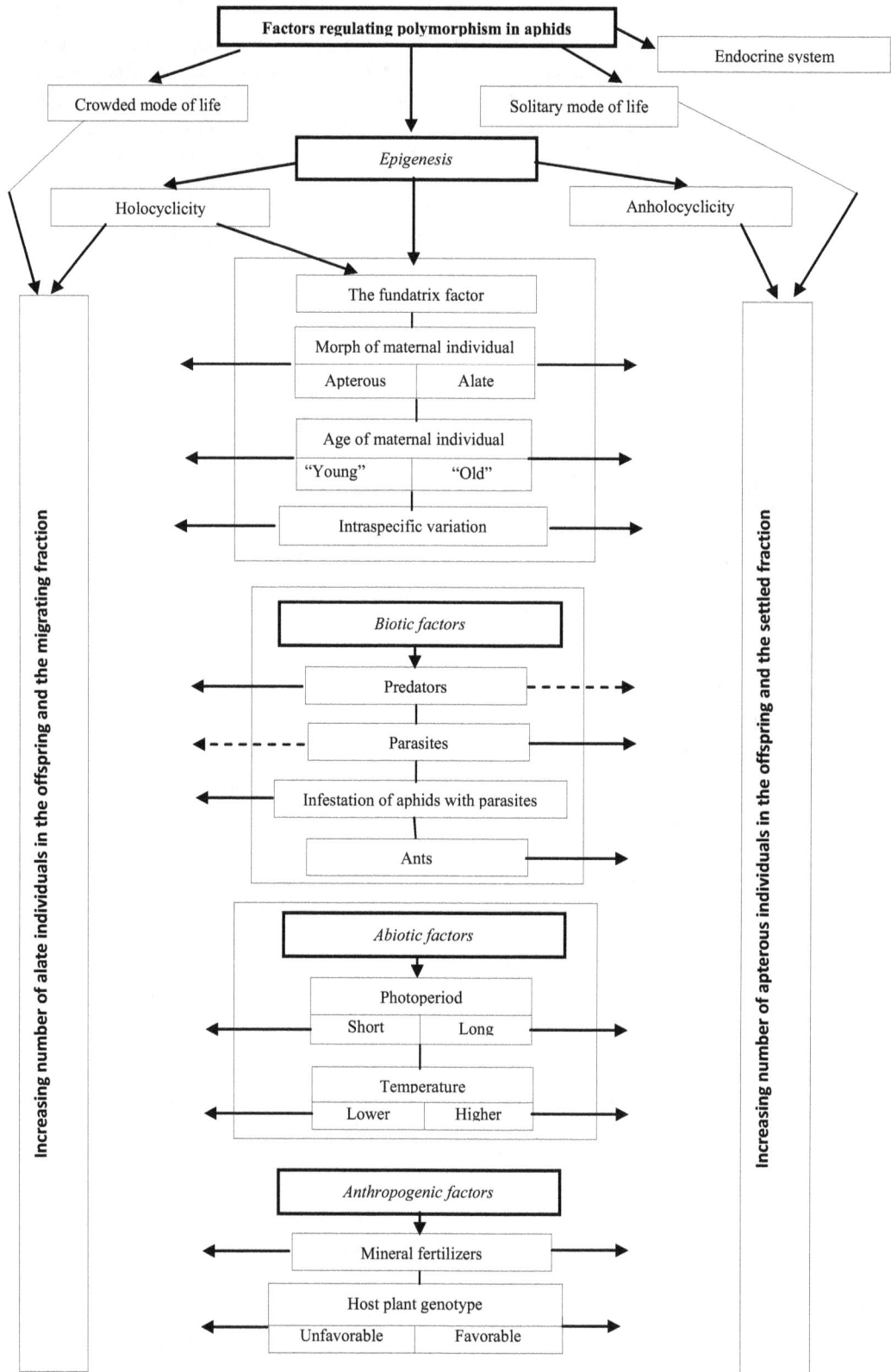

Figure 1. Factors regulating polymorphism in aphids

Note: ——▶ less probable action of the factor

5.4 Specific Traits of the Aphid Flight and Transmission of Viral Diseases

In order to monitor the dispersal and distribution of aphids over cultivars, one has to know not only the factors determining wing induction but also the specific traits of the aphid flight. The behavior of aphids in flight is primarily determined by biological motivations that underlie the behavioral responses of phytophages (Vilkova, 1979).

Aphids have two types of flight: the migration (long-distance) flight typical of spring emigrants, autumn remigrants, and some alate individuals in the summer population, and the trivial (short-distance) flight typical of alate viviparae. The migration flight is obligatory and takes place even when the host plant is favorable, while the trivial flight largely depends on external factors. The specificity of the two types of flight was covered in detail in the reviews (Müller, 1962; Vereschagina & Vereschagin, 2013, etc.). The composition of emigrants and alate viviparae by their tendency to long- or short-distance flights or settled existence may vary within the clone depending on the host plant (Vereschagina & Vereschagin, 2013). The flight motivation is related to the reproductive states of alate aphids which vary even among individuals in conspecific clones. Aphids having large gonads show a tendency to trivial flights, whereas those with small gonads but considerable fat reserves usually fly to long distances (Walters & Dixon, 1983). Some alate individuals may be sedentary while some apterous ones may be migratory. The apterous adults and larvae migrate to short distances. According to our data, the migrating fraction of the summer generations of *Rh. padi* includes 60% or more of alate morphs, 19% of apterous adults, and up to 5% of larvae. Nymphs rarely participate in migration. In the beet root aphid *Pemphigus fuscicornis* and some other species the function of dispersal is performed by the I instar larvae referred to as "crawlers." In *A. fuscicornis* such larvae have a low sailing capacity and can be carried by wind to small distances only; they disperse actively via soil fissures and infest the roots of beet and some wild representatives of *Chenopodiaceae* (Gorbatyuk, 1969).

The migration flight is obligatory and takes place even when the host plant is favorable. Aphids in general are poor fliers but they can remain in the air for up to 14 h and can be carried to hundreds of kilometers by wind as part of the anemoplankton. The flight velocity of aphids varies from 0.8 to 3.3 km/h depending on the wind. Aphids need light to fly and take off at an illumination of 1000 lx. Bright sunlight (10700 lx) decreases the flight activity of aphids but the illumination thresholds are species-specific. The most active flight is usually observed in the morning and 1–2 h before sunset but some species also fly in the night. The mean lower temperature threshold of the aphid flight is $+ 13$–$15°C$, and the upper one, $+ 31°C$ (Robert, 1987). The flight peaks of summer viviparae depend on the air temperature and the species' abundance during the most favorable phases of the host plants. For example, the bird cherry-oat and corn aphids have two flight peaks during the season: on corn and sorghum the first peak is observed near the end of stage VIII of the plant organogenesis, whereas the second peak starts with stage X and lasts until stage XII. Some aphids have a unimodal flight (Guidelines, 1983).

The dynamics of aphid flight is monitored by the network of suction traps established in Europe, which helps to protect plants from damage. Each trap provides a representative sample of aphids within the radius of 20–50 km (Dixon, 1987). The modern global climatic changes affect the phenology of aphids and correlation of their development with that of their host plants. The unequal changes in the rates of development of plants and aphids increase the probability of dissociation of their relations and changes in their dynamics and dispersal (Harrington et al., 1999). As shown by the cited authors, the dispersal of migrants allows them to "scan" large territories. The flight is followed by the search for host plants which is accomplished, similar to other insects, by distant, contact, and gustatory orientation (Vilkova, 1979). Distant orientation of emigrants and viviparae is limited by their weak sensitivity to low concentrations of volatile compounds (Anderson & Bromley, 1987; Webster, 2012), therefore during the host search aphids often land on unfavorable plants, mostly on wild herbaceous ones. Although aphids leave such plants quickly, after a few shallow punctures (contact and gustatory orientation), noncirculative phytopathogenic viruses get attached to the tips of their stylets and can be transferred by the insect onto other plants. Circulative viruses can be transmitted only several days after feeding on the infested plant, because they have to multiply first and reach the insect's digestive system (Zykin, 1970; Keldysh & Pomazkov, 2003).

The intensity of aphid flight is particularly significant for transmission of viral diseases; for example, the potato virus Y can be transmitted by many aphid species which do not colonize the potato plants themselves, such as *Anoecia corni, Aphis rumicis, Phorodon humuli, Rh. padi*, and others (Heimbach et al., 1998). The barley yellow dwarf virus is known to be widely distributed on representatives of the family *Poaceae*. It is therefore not surprising that the virus itself and its vectors, aphids of the genus *Rhopalosiphum*, have a common center of origin located in North America (Halbert & Voegtlin, 1998). This historically established trophic association allows these aphids to transmit the virus to cultured grasses during their long- and short-distance migrations.

According to our observations, of special interest are the aphid species transmitting the non-persistent plum pox virus distributed in Moldova, where plum trees occupy the second largest fraction of orchard territories, yielding only to apple trees. This virus is also present in Russia: in Krasnodar and Stavropol Territories, Rostov, Volgograd, Voronezh, and other provinces (Kuleshova & Rynza, 2010). The virus vectors developing on plum trees in Moldova are *Hyalopterus pruni* subsp. *pruni*, *Brachycaudus cardui*, *B. helichrysi*, and *Myzus persicae*. In Russia, this list is supplemented with *Aphis craccivora*, *M. varians*, and *Ph. humili* (Kuleshova & Rynza, 2010).

In spring, after the imaginal molt on plum trees, the aphids lose the virus and migrate onto herbaceous plants; however, in autumn the plum trees get infected by the gynoparae and males remigrating from one tree to another (Verderevskaja et al., 1983). Transmission of the virus from infected plants to healthy ones can be explained by the fact that the gynoparae and males, like the corresponding morphs of *Rh. padi*, still consume some sap from the primary host plants (Walters et al., 1984). So far, the only known ways to decrease infestation in the plum pox virus foci are the application of aphicides after harvesting in order to kill the alate remigrants, and also removal of weeds on which the vector aphid species may develop.

6. Can Aphids be Beneficial?

However, most species of aphids do not form outbreaks and do not transmit viral diseases of plants, therefore they do not act as agricultural pests at low abundance (Miles, 1999). The other side of the biology of aphids is their positive role in the functioning of biogeocenoses. Many forest aphids are consumed by the same entomophages which destroy aphid pests. The aphid honeydew serves as food for ants and some nectar-feeding insects, including the adults of parasitic and predatory entomophages, and also bees and other pollinators. In forests, aphids act as a factor stabilizing the abundance of insects. The sugars of the honeydew get into the soil, increasing the abundance of nitrogen-fixing bacteria and improving the soil composition (Vereschagin et al., 1985).

7. Conclusion

The relation between polymorphism in aphids and their damage is characterized. The key factors controlling polymorphism in aphids are classified for the first time and a diagram showing their action is proposed.

The damage of aphids is shown to be related to the ability of apterous viviparae to rapidly increase their abundance. Due to high fecundity and development rate, feeding on phloem sap in the zones of high content of assimilates, and the action of aggregation attractant, apterous viviparae can form large aggregations. On the one hand, group action of aphids on the host plant optimizes their feeding and provides the means of contact communication, which are particularly important in case of danger. On the other hand, crowded aphids reduce the amount of nutrients in the plant, contaminate it with excrements, and cause various deformations and metabolic disturbances which may eventually kill the plant.

The damage of aphids is also determined by their wide ecological plasticity and the formation and dispersal of aggressive intrapopulation forms. Besides, it is related to the ability of most alate morphs: emigrants, alate viviparae, gynoparae, and males, to transmit viral diseases of plants and to disperse onto new territories, spreading the harmful intrapopulation forms.

The alate morphs appear at certain stages of the life cycle in accordance with endogenous rhythms. Wing induction in summer aphid populations is primarily affected by the colony density and deterioration of the trophic conditions, and also by the presence of entomophages, parasites, and pathogens.

Thus, the importance of alate morph formation as a factor of aphid damage has been demonstrated; this factor should be taken into account during development of new strategies of plant protection, including selection of resistant cultivars. The positive role of aphids in ecosystems is one of the aspects of the biocenotic approach to plant protection.

References

An, C., Fei, X., Chen, W., & Zhao, Z. (2012). The integrative effects of population density, photoperiod, temperature and host plant on the induction of alate aphids in *Schizaphis graminum*. *Archives of insect biochemistry and physiology, 79*(4-5), 198-206. http://dx.doi.org/10.1002/arch.21005

Anderson, M., & Bromley, A. K. (1987). Sensory system. In A. K. Minks, & P. Harrewijn (Eds.), *Aphids: Their Biology, Natural Enemies and Control* (pp. 153-162). Amsterdam: Elsevier.

Ankersmit, G. W., & Dijkman, H. (1983). Alatae production in the cereal aphid *Sitobion avenae*. *Netherlands Journal of Plant Pathology, 89*, 105-112. http://dx.doi.org/10.1007/BF01976349

Auclair, J. L. (1963). Aphid feeding and nutrition. *Annual Review of Entomology, 8*, 439-490. http://dx.doi.org/10.1146/annurev.en.08.010163.002255

Ban, L., Ahmed, E., Ninkovic, V., Delp, G., & Glinwood, R. (2008). Infection with an insect virus affects olfactory behavior and interactions with host plant and natural enemies in an aphid. *Entomologia experimentalis et applicata, 127*(2), 108-117. http://dx.doi.org/10.1111/eea.2008.127.issue-2

Blackman, R. L. (1981). Aphid genetics and host plant resistance. International organization for Biological Control of Noxious Animals and Weeds. *West Palearctic Regional Section Bulletin, IV*, 13-19.

Bonnemaison, L. (1967). L'effect de group chez les aphids. *Colloques internationaux du centre National de la Recherche Scientifique, 173*, 213-236.

Bonner, A. B., & Ford, J. B. (1972). Some effects of crowding on the biology of *Megoura viciae*. *Annals of Applied Biology, 71*(2), 91-98. http://dx.doi.org/10.1111/j.1744-7348.1972.tb02943.x

Brown, P. A., & Blackman, R. L. (1988). Karyotype variation in the corn leaf aphid Rhopalosiphum maidis species complex (Hemiptera: Aphididae) in relation to host plant and morphology. *Bulletin of Entomological Research, 78*, 351-363. http://dx.doi.org/10.1017/S0007485300013110

Campbell, B. C., & Dreyer, D. L. (1985). Host-plant resistance of sorghum: differential hydrolysis of sorghum pectic substances by polysaccharases of greenbug biotipes (Schizaphis graminum: Homoptera, Aphididae). *Archives of Insect Biochemistry and Physiology, 2*, 213-215.

Chakrabarti, S. (2001). Host plant association and utilization of gall forming aphids (Homoptera: Aphididae) in western and northwest Himalayas. In J. C. Simon, C. A. Dedryver, C. Rispe, & M. Hulle (Eds). *Aphids in a New Millennium: Proceedings VI International Symposium on Aphids* (pp. 125-130). Rennes 3-7 Sept., France, 2001. Paris: INRA.

Delisle, J., Clauter, C., & McNeil, J. N. (1983). Precocene II-induced alate production in isolated and crowded alate and apterous virginoparae of the aphid *Macrosiphum euphorbiae*. *Journal of Insect Physiology, 29*, 477-484. http://dx.doi.org/10.1016/0022-1910(83)90078-1

Dixon, A. F. G. (1977). Aphid ecology: life cycles, polymorphism and population dynamics. *Annual Review of Ecology and Systematics, 8*, 329-353. http://dx.doi.org/10.1146/annurev.es.08.110177.001553

Dixon, A. F. G. (1985). Structure of Aphid Populations. *Annual Review of Entomology, 30*, 155-174. http://dx.doi.org/10.1146/annurev.en.30.010185.001103

Dixon, A. F. G. (1987). Cereal Aphids as an Applied Problem. *Agricultural Zoology Reviews, 2*(November), 1-57.

Dudnik, G. F. (1985). Population dynamics and damage of the grain aphid in Vinnitsa Province. *Biological and chemical plant protection from pests, diseases and weeds in the Ukrainian SSR. Kiev*, 150-156.

Dyakonov, K. P. (2000). Insects as a factor of dispersal of phytopathogenic viruses in the Russian Far East. A.I. Kurentsov Memorial Lectures. *Vladivostok: Biology and Soil Institute, XI*, 15-26.

Ehrhardt, P. (1968). Einfluss von Ernährungsfaktoren auf die Entwicklung von Säfte saugenden Insekten unter besonderer Berücksiehtigung von Symbionten. *Zeitschrift für Parasitenkunde, 31*(1), 38-36. http://dx.doi.org/10.1007/BF00716427

El Naghy, M. A., & Shaw, M. (1966). Correlation between Resistance to Stem Rust and the Concentration of a Glucoside in Wheat. *Nature, 210*(5034), 417-418. http://dx.doi.org/10.1038/210417a0

Eleftherianos, I., Foster, S., Goodson, S., Williamson, M., & Denholm, I. (2004). Toxicological and molecular characterization of pyrethroid knockdown resistance (kdr) in the peach-potato aphid Myzus persicae (Sulzer). J.-Ch. Simon, Ch.-A Dedryver, C. Rispe, M. Hulle (Eds). *Aphids in a New Millennium: Proceedings VI International Symposium on Aphids* (pp. 213-218). Rennes 3-7 Sept., France, 2001. Paris: INRA.

Evert, R. F., Eschrich, W., Meddler, J. T., & Alfieri, F. J. (1968). Observations on penetration of linden branches by stylets of the aphid Longistigma caryae. *American Journal of Botany, 55*(7), 860-874. http://dx.doi.org/10.2307/2440974

Figueroa, C. C., Simon, J. C., Dedryver, C. A., & Niemeyer, H. M. (2004). Genetic variability of the recently introduced Sitobion avenae in Chile. In J. C. Simon, C. A. Dedryver, C. Rispe, & M. Hulle (Eds), *Aphids in a New Millennium: Proceedings VI International Symposium on Aphids* (pp. 219-225). Rennes 3-7 Sept., France, 2001. Paris: INRA.

Forrest, J. M. S. (1974). The effect of crowding on morph determination of the aphid Dysaphis devecta. *Journal of Entomology, A48*(2), 171-175.

Garratt, M. P. D., Wright, D. J., & Leather, S. R. (2010). The effects of organic and conventional fertilizers on cereal aphids and their natural enemies. *Agricultural and Forest Entomology, 12*(3), http://dx.doi.org/307-318.10.1111/j.1461-9563.2010.00480.x

Gildow, F. E. (1983). Influence of barley yellow dwarf virus-infected oats and barley on morphology of aphid vectors. *Phytopathology, 73*(73), 1196-1199. http://dx.doi.org/10.1094/Phyto-73-1196.

Gliwood, R. T., & Petterson, J. (2000). Host choice and host leaving in *Rhopalosiphum padi (Hemiptera: Aphididae)* emigrants and repellency of aphid colonies on the winter host. *Bulletin of Entomological Research, 90*, 57-61. http://dx.doi.org/10.1017/S0007485300000717

Gorbatyuk, N. M. (1969). *The beet root aphid (Pemphigus fuscicornis Koch) in Moldavia. Candidate's dissertation in biology.* Kishinev: Institute for Zoology. 20 pp.

Goszczynski, W., & Cichocka, E. (1998). Effects of aphids on their host plants. J. M. N. Nafria, A.F.G. Dixon (Eds), *Aphids in Natural and Managed Ecosystems: Proceedings V International Symposium on Aphids* (pp. 197-203). 15-19 Sept., Leon, 1997. Leon: Universidad de Leon.

Gutierres, A. P., Summers, C. G., & Baumgaertner, J. (1980). The phenology and distribution of aphids in California on alfalfa as modified by ladybird beetle predation (*Coleoptera: Coccinellidae*). *Canadian Entomologist, 112*, 489-495. http://dx.doi.org/10.4039/Ent112489-5

Halbert, S. E., & Voegtlin, D. J. (1998). Evidence for the North American origin of Rhopalosiphum and Barley Yellow Dwarf Virus. In J. M. N. Nafria, & A.F.G. Dixon (Eds), *Aphids in Natural and Managed Ecosystems: Proceedings V International Symposium on Aphids* (pp. 351-356). 15-19 Sept., Leon, 1997. Leon: Universidad de Leon.

Hardie, J. (1981). Juvenile hormone and photoperiodically controlled polymorphism in *Aphis fabae*: postnatal effects on presumptive gynoparae. *Journal of Insect Physiology, 27*(5), 347-355. http://dx.doi.org/10.1016/0022-1910(81)90059-7

Harrewijn, P. (1978). The role of Plant Substances in Polymorphism of the Aphid Myzus persicae. *Entomologia experimentalis et applicata, 24*(3), 198-214.

Harrington, R., Woiwod, I., & Sparks I. (1999). Climate change and trophic interactions. *Tree, 14*(4), 146-150. http://dx.doi.org/10.1016/S0169-5347(99)01604-3.

Hatano, E., Baverstock, J., Kunert, G., Pell, J., K., & Weisser, W. W. (2012). Entomopathogenic fungi stimulate transgenerational wing induction in pea aphids, Acyrthosiphon pisum (Hemiptera: Aphididae). *Ecological Entomology, 37*, 75-82. http://dx.doi.org/10.1111/j.1365-2311.2011.01336.x

Heimbach, U., Thieme, T., Weidemann, H. L., & Thieme, R. (1998). Transmission of potato virus Y by aphid species which do not colonize potatoes. In J. M. N. Nafria, & A. F. G. Dixon (Eds), *Aphids in Natural and Managed Ecosystems: Proceedings V International Symposium on Aphids* (pp. 555-559). 15-19 Sept., Leon, 1997. Leon: Universidad de Leon.

Keldysh, M. A., Pomazkov, Y. I. (2003). *Viruses, viroids and mycoplasmata of plants: A brief study manual.* Moscow: Peoples' Friendship University of Russia. 157 pp.

Kidd, N. A. C. (1976). Aggregation in the lime aphid (*Eucallipterous tiliae* L.) 2. Social aggregation. *Oecologia, 25*(2), 175-185. http://dx.doi.org/10.1007/BF00368852

Kuleshova, Yu. G., & Rynza, E. T. (2010). The plum pox virus in the Russian Federation. *Zashchita i karantin rastenii.* Retrieved October 10, 2010, from http://rostov.vniikr.ru/articles/4231/

Kunert, G., Otto, S., Röse, Ursula S. R., Gershenzon, J., Weisser, & Wolfgang W. (2005). Alarm pheromone mediates production of winged dispersial morphs in aphids. *Ecology letters, 8*, 596-603. http://dx.doi.org/10.1111/j.1461-0248.2005.00754.x

Leather, S. R., & Dixon, A. F. G. (1981). Growth, survival and reproduction of the bird-cherry oat aphid, Rhopalosiphum padi, on its primary host. *Annals of Applied Biology, 99*, 115-118. http://dx.doi.org/10.1111/j.1744-7348.1981.tb05136.x

Lees, A. D. (1966). The control of polymorphism in Aphids. Advances *in* Insect *Physiology, 3*, 207-277. http://dx.doi.org/10.1016/S0065-2806(08)60188-5

Lees, A. D. (1967). The production of the apterous and alate forms in the aphid Megoura viciae Buckton, with special reference to the role of crowding. *Journal of Insect Physiology, 13*, 289-318. http://dx.doi.org/10.1016/0022-1910(67)90155-2

Llewellin, K. S., Loxdale, H. D., Harrington, R., Brookes, C. P., Clark, S. J., & Sunnucks, P. (2003). Migration and genetic structure of the grain aphid *(Sitobion avenae)* in Britain related climate and clonal fluctuation as revealed using microsatellites. *Molecular Ecology, 12*, 21-34. http://dx.doi.org/10.1046/j.136 5-294X.2003.01703.x

Loxdale, H. D., Hardie, J., Halbert, S., Foottit, R., & Kidd, N. A. C. (1993). The relative importance of short- and long-range movement of flying aphids. *Biological Reviews, 68*(2), 291-311. http://dx.doi.org/10.1111/j.1469-185X.1993.tb00998.x

Lukasik, I., Leszczynski, B., & Dixon, A.F.G. (2004). Changes in bird cherry-oat aphid metabolism while occurring on primary host. In J.-Ch. Simon, Ch.-A Dedryver, C. Rispe, M. Hulle (Eds.), *Aphids in a New Millennium: Proceedings VI International Symposium on Aphids* (pp. 463-469). Rennes 3-7 Sept., France, 2001. Paris: INRA.

Medvedev, S. S. (2004). *Plant physiology.* Saint-Petersburg State Univ. 335 pp.

Michaud, J. P., Jyoti, J. L., & Qureshi, J. A. (2006). Positive correlation of fitness with group size in two biotypes of Russian Wheat Aphid (Homoptera: Aphididae). *Journal of Economic Entomology, 99*(4), 1214-1224. http://dx.doi.org/10.1603/0022-0493-99.4.1214

Miles, P. W. (1972). The saliva of Hemiptera. *Advances of Insect Physiology, 9*, 183-255. http://dx.doi.org/10.1016/s0065-2806(08)60277-5

Miles, P. W. (1998). Aphid salivary functions: the physiology of deception. In J. M. N. Nafria, A. F. G. Dixon (Eds.), *Aphids in Natural and Managed Ecosystems: Proceedings V International Symposium on Aphids* (pp. 255-263). 15-19 Sept., Leon, 1997. Leon: Universidad de Leon.

Miles, P.W. (1999). Aphid saliva. *Biological Reviews, 74*(1), 41-85. http://dx.doi.org/10.1017/S00063231980 05271

Mittler, T. E., & Kleinjan, J. E. (1970). Effect of artificial diet composition on wing-production by the aphid *Myzus persicae. Journal of Insect Physiology, 16*(5), 833-850.

Müller, H. J. (1962). Moderne Vorstellungen über Biologie und Ökologie des Blattlausflüges und seine Bedeutung für die Virusausbreitung. *Zeitschrift für Pflanzenkrankheiten und Pflanzenschutz, 69*(7), 385-392. http://dx.doi.org/10.1016/0022-1910(70)90217-9

Novgorodova, T. A. (2005). Specific traits of mutualistic relations with aphids in two ant species of the genus *Lasius (Formicinae). Uspekhi sovremennoi biologii, 125*(2), 199-205.

Peng, Z., & Miles, P. W. (1988). Studies on the salivary physiology of plant bugs: Function of the catechol oxidase of the rose aphid. *Journal of Insect Physiology, 11*, 1027-1033. http://dx.doi.org/10.1016/0022-1910(88) 90202-8

Popova, A. A. (1967). *Adaptations of aphids to feeding on their host plants.* Leningrad: Nauka. 292 pp.

Powel G., & Hardie J. (2001). The chemical ecology of aphid host alternation: how do return migrants find the primary host plant? *Applied Entomology and Zoology, 36*, 259-267. http://dx.doi.org/10.1303/aez.2001.259

Robert, Y. (1987). Aphids and their environment. Dispersion and migration. In A. K. Minks, P. Harrewijn (Eds.), *Aphids: Their Biology, Natural Enemies and Control* (pp. 299-313). Amsterdam: Elsevier, A.

Ryabov, E. V., Keane , G., Naish, N., Evered, C., & Winstanley, D. (2009). Densovirus induces winged morphs in asexual clones of the rosy apple aphid, *Dysaphis plantaginea. Proceedings of the National Academy of Sciences of the United States of America, 106*, 8465-8470. http://dx.doi.org/10.1073/pnas.0901389106

Sandström, J. (1995). Temporal changes in host adaptation in the pea aphid, Acyrthosiphon pisum. In J. Sandström (Ed.), *Host adaptation in the pea aphid. Temporal changes and nutrition.* Dissertation. Department of Entomology Swedish University of Agricultural Sciences: Uppsala, IV, 1-17.

Shaposhnikov, G. C. (1981). *Populations and species in aphids and need for a Universal species concept.* Special Publication Research Branch Agriculture. Canada. P. 61.

Shaposhnikov, G. C. (1986). New species of the genus Dysaphis Börrn. (Homoptera, Aphidinea) and peculiarities of the taxonomic work on aphids. *Entomologicheskoe obozrenie, 36*(1), 535-550.

Sloggett, J. J., & Weisser, W. W. (2004). A general mechanism for predator and parasitoid - induced dispersal in the pea aphid Acyrthosiphon pisum. In J.-Ch. Simon, Ch.-A Dedryver, C. Rispe, & M. Hulle (Eds). *Aphids in a New Millennium. Proceedings VI International Symposium on Aphids* (pp. 79-85). Rennes 3-7 Sept., France, 2001. Paris: INRA.

Sobetsky, L. A. (1967). *Some specific traits of trophic physiology of aphids. Candidate's dissertation in biology.* Kishinev: Institute for Zoology. 20 pp.

Stadler, B., Dixon, & A. F. G. (2005). Ecology and Evolution of Aphid-Ant Interactions. *Annual Review of Ecology, Evolution and Systematics, 36*, 345-372. http://dx.doi.org/10.1146/annurev.ecolsys.36.091704.175531

Stafford, D. B., Tariq, M., Wright, D. J., Rossiter, J. T., Kazana, E., Leather, S. R., Ali, M., & Staley J. T. (2012) Opposing effects of organic and conventional fertilizers on the performance of a generalist and a specialist aphid species. *Agricultural and Forest Entomology, 14*(3), 270-275. http://dx.doi.org/10.1111/j.1461-9563.2011.00565.x

Sudd, J. H. (1987). Ant aphid mutualism. A. K. Minks, & P. Harrewijn (EDds.), *Aphids: Their Biology, Natural Enemies and Control* (Vol. A, pp. 355-365). Amsterdam: Elsevier.

Sutherland, O. R. W. (1969). The role of crowding in the production on wing forms by two strains of the pea aphid *Acyrthosiphon pisum. Journal of Insect Physiology, 15*, 1385-1410. http://dx.doi.org/10.1016/0022-1910(69)90199-1

Tsitsipis, J. A., & Mittler, T. E. (1977). Influence of daylength on the production of parthenogenetic and sexual females of *Aphis fabae* at 17.5°. *Entomologia experimentalis et applicata, 21*(2), 163-173. http://dx.doi.org/10.1007/BF00287091

Urbanska, A., Leszcsynski, B, Matok, H., & Dixon, A. F. G. (2004). Hydrolysis of plant glycosides by cereal aphids. In J.-Ch. Simon, Ch.-A Dedryver, C. Rispe, & M. Hulle (Eds.), *Aphids in a New Millennium. Proceedings VI International Symposium on Aphids* (pp. 521-527). Rennes 3-7 Sept., France, 2001. Paris: INRA.

Urbanska, A., Leszczynski, B., Laskowska, I., & Matok, H. (1998). Enzymatic defense of grain aphid against plant phenolics. In J. M. N. Nafria, & A. F. G. Dixon (Eds.), *Aphids in Natural and Managed Ecosystems: Proceedings V International Symposium on Aphids* (pp. 119-124). 15-19 Sept., Leon, 1997. Leon: Universidad de Leon.

Verderevskaja, T. D, Vereschjagin, B. V., Andreev, A. V., & Polynkovsky, A. I. (1983). The epidemiology of plum pox virus in Moldavia. *Acta Horticulturae (Canada), 130*, 317.

Vereschagin, B. V., Andreev, A. V., & Vereshchagina, A. B. (1985). *The Aphids of Moldavia.* Kishinev: Shtiintsa.

Vereschagina, A. B. (1982). *The physiological peculiarities of trophic specialization of aphids as related to plant resistance. Candidate's dissertation in biology.* Leningrad: All Union Institute for Plant Protection.

Vereschagina, A. B. (2001). The effect of the host plant genotype and crowding on development and mortality rates of cereal aphids *(Homoptera: Aphididae). Trudy Rossiiskogo entomologicheskogo obshchestva, 72*, 83-88.

Vereschagina, A. B. (2007). Adaptive variation of Aphids as the Basis for Criteria of Phenetic Monitoring of Their Populations. In V. A. Pavljushin (Ed.), *International Organization for Biological Control of Noxious Animals and Plants* (pp. 56-60). East Palearctic Regions Section. Newsletter. Saint-Petersburg: All Russian Institute for Plant Protection.

Vereschagina, A. B. (2008). Ecological mechanisms of phenotypic and genotypic variation of the aphid population structure. *Modern problems of plant immunity to harmful organisms: Collected works of the 2nd All-Russian Conference* (pp. 197-200). *September 29 - October 2, 2008.* Saint-Petersburg.

Vereschagina, A. B. (2012). Development of clones of the bird cherry-oat aphid Rhopalosiphum padi (L.) (Homoptera: Aphididae) as related to the resistance mechanisms of soft spring wheat of Delfi 400 cultivar. *Proc. of XIV Congress of Russian Entomological Society*, August 27 - September 1, 2012. Saint-Petersburg.

Vereschagina, A. B., & Shaposhnikov, G. C. (1998). Influence of crowding and host-plant on development of winged and apterous aphids. In J. M. N. Nafria, & A. F. G. Dixon (Eds.), *Aphids in Natural and Managed Ecosystems: Proceedings V International Symposium on Aphids* (pp. 642-645). 15-19 Sept., Leon, 1997. Leon: Universidad de Leon.

Vereschagina, A. B., & Vereschagin, B. V. (2013). Classification of host plants of aphids (Homoptera, Aphidoidea) as related to their selection and use by aphids under the recent conditions of biogeocenosis transformation. *Entomologicheskoe obozrenie, XLII*(2), 265-281.

Vereshchagina, A. B., & Gandrabur, S. I. (1988). Behavioral responses of cereal aphids feeding on corn plants with different degrees of resistance. *Ekologiya, 2*, 35-39.

Vilkova, N. A. (1983). *Guidelines for estimation of pest resistance of cereal cultivars* (1983). N.A. Vilkova (Ed.), Leningrad: All-Union Institute for Plant Protection. 46 pp.

Vilkova, N. A. (1979). Plant immunity to pests and its relation to the trophic specialization of phytophagous insects. *N.A. Kholodkovsky Memorial Lectures, 11 April 1978*. Saint-Petersburg: *Zoological Institute, 31*, 68-103.

Walters, K. F. A., & Dixon, A. F. G. (1983) Migratory urge and reproductive investment in aphids: Variation within clones. *Oecologia, 58*, 70-75. http://dx.doi.org/10.1007/BF00384544

Walters, K. F. A., Dixon, A. F. G., & Eagles, G. (1984). Non-feeding by adult gynoparae of *Rhopalosiphum padi* and its bearing on the limiting resource in the production of sexual females on host-alternating aphids. *Entomologia Experimentalis et Applicata, 36*, 9-12. http://dx.doi.org/10.1111/j.1570-7458.1984.tb03398.x

Way, M. J., & Banks, C. J. (1967). Intraspecific mechanisms in relation to the natural regulation of numbers of *Aphis fabae Scop. Annals of Applied Biology, 59*(2), 189-205. http://dx.doi.org/10.1111/j.1744-73 48.1967.tb04428.x

Way, M. J., & Cammell, M. (1970). Aggregation behavior in relation to food utilization by aphids. In A. Watson (Ed.), *Animal Populations in Relation to their Food resources* (pp. 229-246). Oxford-Edinburgh.

Webster, B. (2012). The role of olfaction in aphid host location. *Physiological Entomology, 37*, 10-18. http://dx.doi.org/10.1111/j.1365-3032.2011.00791.x

Zykin, A. G. (1970). *Aphids as vectors of potato viruses*. Leningrad: Kolos.

14

Occurrence of *Trichograma* Parasitoids in Eggs of Soybean Lepidopteran Pests in Mato Grosso, Brazil

Angélica Massaroli[1], Alessandra Regina Butnariu[1] & Augusta Karkow Doetzer[1]

[1] Department of Biological Sciences, University of Mato Grosso State, Brazil

Correspondence: Augusta Karkow Doetzer, Rodovia MT 358, Km 07, Jardim Aeroporto, Tangará da Serra, MT, CEP: 78300-000, Brazil. E-mail: alebut@unemat.com; gutaunemat@gmail.com

Abstract

Occurrence of parasitoids in eggs of soybean lepidopteran pests, such as velvetbean caterpillar, soybean loopers and others the genus *Spodoptera* and *Helicoverpa* in Mato Grosso, Brazil was studied. Due to the scarcity of information on the natural occurrence of parasitoids in lepidopteran eggs in soybean fields in Mato Grosso surveys were conducted in Tangará da Serra to verify the occurrence these parasitoids. Surveys were conducted in two areas in Tangará da Serra, with a total of 591 eggs, of which 138 and 77 eggs were collected in area I and 274 and 102 eggs in area II, during the 2008/2009 and 2009/2010 soybean seasons, respectively. From the eggs collected in area I, 26.6% were parasitized in the first year and 30.0% in the second season. In area II, the parasitism rates were 40.2% and 27.5% for the first and second season respectively. The species of parasitoid collected was *Trichogramma pretiosum* Riley, 1879 (Hymenoptera: Trichogrammatidae) totaling 354 parasitoids. We conclude that there is a natural occurrence of *T. pretiosum*, this being the species of egg parasitoid found, and that parasitism rates were higher when compared with other studies under natural conditions, especially considering the large number of chemical products used in the fields. The identification of the eggs parasitoids in Mato Grosso State is important to development of biological control programs for lepidopteran pests.

Keywords: insecta, natural enemies, Noctuidae, biocontrol, pesticides, Brazil

1. Introdution

Brazil is a major producer of soybeans (*Glycine max* L.), and the state of Mato Grosso is the major producer of this grain, with a planted area of 6.16 million hectares, accounting for 27.5% of the national production (CONAB, 2010). However, the large size of the area occupied by soybeans favors the incidence of insect pests in the crop (Corrêa-Ferreira & Panizzi, 1980). Among the pests that attack soybeans defoliating caterpillars, such as velvetbean caterpillar, soybean loopers and others the genus *Spodoptera* and *Helicoverpa* stand out for decreasing leaf area, and consequently lowering the photosynthetic rate which decreases productivity, while leaving the plant susceptible to disease and sometimes causing its death. In most cases these pests are controlled with wide spectrum chemicals that reduce the action of natural enemies. However, the pests eventually become resistant, while the chemicals pose health hazards and environmental contamination.

For these reasons, states such as Rio Grande do Sul (Costa & Link, 1974), Paraná (Corrêa-Ferreira et al., 1998; Foerster & Avanci, 1999), São Paulo (Campos et al., 1997), Mato Grosso do Sul (Godoy et al., 2005), Minas Gerais (Venzon et al., 2000), Distrito Federal (Medeiros et al., 1997) and Acre (Thomazini, 2001) have developed research on species of insect pests and natural enemies, aiming at the implementation of biological control programs.

Among the natural enemies that act in reducing pest populations, egg parasitoids are very effective, since they prevent the development of the pest while still in the egg stage and thus avoid damage to the crop. Egg parasitoids of the genus *Trichogramma* (Hymenoptera: Trichogrammatidae) are important because they parasitize eggs of Lepidoptera, Hemiptera and Coleoptera that attack various crops (Parra & Zucchi, 2004) and, for this reason, are used in biological control programs worldwide (Beserra & Parra, 2004). Currently, there are 190 known species of *Trichogramma*, and 38 of these occur in South America, 28 of which have been recorded in Brazil (Moreira et al., 2009). Despite the importance of soybean production in the state of Mato Grosso, no research has been carried out in the state regarding the occurrence of egg parasitoids, although their effectiveness has been proven by previous research (Foerster & Avanci, 1999; Foerster & Butnariu, 2004).

Due to the scarcity of information on the natural occurrence of parasitoids in lepidopteran eggs in soybean fields in Mato Grosso, surveys were conducted in two areas in Tangará da Serra with the aim to verify the occurrence of the *Trichogramma* parasitoids and assess the rates of parasitism. The occurrence of the phytophagous Lepidoptera was also assessed, as well as the pesticides used in the study areas.

2. Materials and Methods

2.1 Study Areas

The study was conducted in two areas located in the municipality of Tangará da Serra, Mato Grosso, Brazil, cultivated with soybean. The first area (area I) is cultivated with conventional soybean and a dense submontane forest reserve (14°39′53″S and 57°24′30″W). The second area (area II), was planted with organic soybean until the previous year and then conventionally during the sampling season. This area has Cerrado reserves (14°18′44″S and 57°45′18″W) and is located 90 km away from area I. Temperature and relative humidity (RH) of the areas were obtained using a digital thermohygrometer on the sampling days. The average temperature and RH were 28.6 °C and 59.8% in area I and 26.8 °C and 67.5% in area II.

2.2 Sampling of Lepidopteran Eggs

Eggs of Lepidoptera were collected during the crop seasons of 2008/2009 and 2009/2010 in the months of November, December and January (vegetative period and early reproductive period). Two samplings were performed per month for 1.5 hours each, except for the second crop in the area II, where monthly samples were collected.

The sampling methodology used was described by Foerster and Avanci (1999): soybean leaves containing eggs were collected and taken to the laboratory where the eggs were removed from the leaves with the help of a fine brush under a stereomicroscope and individualized in 1.5 ml microtubes which were kept in conditions controlled of BOD at 25 °C and 70% relative humidity.

The occurrence of parasitism was checked daily and the number of emerged parasitoids per host egg and sex of the individuals determined by the morphological characteristics of the antennae (Querino & Zucchi, 2003). When parasitoids emerged, females were placed in microtubes containing 70% ethanol and males were used to identify the species.

2.3 Identification of Parasitoids

Specimens were identified according to the morphological characters of the male genitalia (Nagaraja & Nagarkatti, 1973). Parasitoids were cleared in acetic acid for 24 hours, and mounted individually on glass slides in a micro-droplet of Hoyer's medium. Specimens were mounted with their ventral side up and legs pushed to the side, leaving the genitals free for observation. Another micro-droplet was placed on the parasitoid followed by the coverslip. After confirming the sex, the slides were sent for identification to Dr. Ranyse B. Querino of EMBRAPA Meio-Norte.

Identified specimens were deposited in the entomological collection of the laboratory of Zoology at the Centro de Pesquisas, Estudos e Desenvolvimento Agro-ambientais (CPEDA) at the Universidade do Estado de Mato Grosso campus de Tangará da Serra (UNEMAT/CUTS).

2.4 Sampling of Lepidopteran Larvae

All larvae found at the sampling points were collected manually in order to record the lepidopteran species found in the area. These were taken to the laboratory for identification using books and field guides. The identification of adults by an expert was not possible, since chemical products that were applied in the areas prevented the collected larvae from reaching the adult stage.

2.5 Pesticides Used

Using a predefined form, an interview was conducted with the persons responsible for applying the pesticides in the fields to obtain the names of the products used.

In both areas studied chemical management is the only method of pest control used. The main insecticides used were Lambda-cyhalothrin (Karate Zeon®), Thiodicarb (Larvin®), Thiomethoxam (Engeo Pleno®), Metal phosphides (Fosfito®), Organophosphates (Metamidofós®). The herbicides used were Chlorimuron (Classic®), Fomesafen (Flex®), Fluazifop (Fusilade®), Glyphosate (Roundup®), and the fungicides were Azoxystrobin and Cyproconazole (Priori Xtra®) and Triactol (Alto 100®). All pesticides used were recommended by agronomists.

2.6 Data Analysis

We calculated the mean percentage of parasitism in the study areas and analyzed the data using a one way analysis of variance (p ≤ 0.05) (ANOVA). The sex ratio was calculated for the parasitoids collected in the 2009/2010 season, using the formula: rs = №♀ / (№♀ + №♂). The average number of parasitoids emerged per egg was also calculated.

3. Results and Discussion

A total of 591 eggs were collected in the two areas, of which 138 and 77 eggs were collected in area I and 274 and 102 eggs area II, during the 2008/2009 and 2009/2010 soybean seasons, respectively. From the eggs collected in area I, 26.6% were parasitized in the first year and 30.0% in the second season. In area II, the parasitism rates were 40.2% and 27.5% for the first and second season, respectively (Figure 1). The average parasitism in two areas was 35.7%, without significant difference between the two areas (F = 0.49 and P = 0.68 (p ≤ 0.05).

Despite the fact that the parasitoids were influenced by the host and the chemical management employed in the areas (Zago et al., 2010), percentages of parasitism in Mato Grosso were higher than those in Paraná, where Avanci et al. (2005) found that parasitism in eggs of *Anticarsia gemmatalis* Hübner, 1818 (Lepidoptera: Noctuidae) varied from 3.7% to 11.5% during four seasons.

The species of parasitoid collected was *Trichogramma pretiosum* Riley, 1879 (Hymenoptera: Trichogrammatidae) totaling 354 parasitoids. According to Gonçalves et al. (2003), this species is the most widely distributed in Brazil and, according to Zucchi and Monteiro (1997), it is the most generalist and is associated with 26 host species (Potrich et al., 2009).

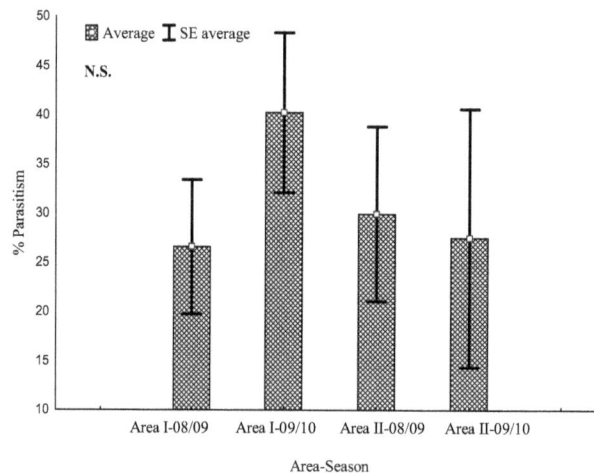

Figure 1. Percentage of eggs of noctuid pests on soybean parasitized by *Trichograma pretiosum*. The two areas were sampled in Mato Grosso, Brazil during the 2008/2009 and 2009/2010 growing seasons. The rates of parasitism were not significantly different according to ANOVA (F = 0.49; P = 0.68; p ≤ 0.05)

Avanci et al. (2005) also recorded a predominance of *T. pretiosum*, which accounted for 80% of the parasitized eggs of *A. gemmatalis*. In another study in Paraná, during five consecutive soybean seasons Foerster and Avanci, (1999) recorded 90% of the eggs of *A. gemmatalis* parasitized by *T. pretiosum*. In maize (*Zea mays* L.) crops in Minas Gerais, Tironi and Ciociola (1994) recorded more than 90% of the eggs of *Helicoverpa zea* (Boddie, 1850) (Lepidoptera: Noctuidae) parasitized by *T. pretiosum*.

According to Moreira et al. (2009), in Brazil, *T. pretiosum* is the most frequently occuring parasitoid and is associated with various hosts in different cultures, such as *Tuta absoluta* (Meyrick, 1917) in tomato (Pratissoli et al., 2005), *Spodoptera frugiperda* (Smith, 1797) (Bessera & Parra, 2003) and *H. zea* (Pratissoli & Oliveira, 1999) in maize, *Plutella xylostella* (Linnaeus, 1758) in cabbage (Pereira et al., 2004; Zago et al., 2010), *Heliothis virescens* (Fabricius, 1781) in cotton (Zucchi et al., 1989), *A. gemmatalis* and *Chrysodeixis* (= *Pseudoplusia*) *includens* (Walker, 1858) in soybean (Bueno et al., 2009).

Three species of phytophagous Lepidoptera were found in area I, during two seasons, of which 74.9% was *C.*

includens, 20.5% *A. gemmatalis* and 4.6% *Spodoptera* spp. In area II there was an inversion of these data, being 69.4% *Spodoptera* spp., 19.2% *A. gemmatalis* and 11.4% *C. includens*.

This inversion can be related to the type of management used as in area I, where regular insecticide applications were made, since the area has always been planted with conventional soybean. This may have favored the emergence of resistant strains of *C. includens*, justifying the large number of this species. The high number of *Spodoptera* spp. Can be related to the conventional maize has been cultived after last organic soybean cultivation. Maize cultivastion was not appropriately carried out, leaving residues on the crop.

Considering the presence of larvae, we concluded that the parasitized eggs belong to these species since the identification of eggs is difficult due to the darkening of the egg as the parasitoid develops inside the eggs. Thus, all eggs were included in the count. However, the method of oviposition found in the areas (individual eggs on leaves) which is characteristic of *A. gemmatalis* and *C. includens*, and leads us to infer that these species were predominant when compared to eggs of *Spodoptera* spp., which are laid in masses of more than 100 eggs.

The sex ratio is an important feature of biological control programs, since the greater the production of females, the greater the number of parasitized eggs and parasitoids in the next generation. In this study, the sex ratio of the parasitoids was 0.5 ± 0.10 for eggs collected in area I and 0.6 ± 0.07 for eggs collected in area II during the 2009/2010 season. Avanci et al. (2005) obtained similar results in eggs of *A. gemmatalis* under natural conditions as the ones recorded in the area II (0.65 ± 0.02) for *T. pretiosum*. Bueno et al. (2009) observed, under laboratory conditions, similar results in eggs of *C. includens* (0.70 ± 0.02). Thus we can infer that environmental differences do not affect the sex ratio of the parasitoids.

Of the eggs collected, the average number of adults emerged per egg was (average $\pm$ SE) 1.4 ± 0.46 and 1.6 ± 0.36 in I area (area with predominance of larvae of *C. includens*) and 1.6 ± 0, 11 and 1.3 ± 0.06 in II area (where occurrence of *A. gemmatalis* was higher than that of *C. includens*) in the 2008/2009 and 2009/2010 seasons, respectively. Bueno et al. (2009) obtained lower emergence rates for *C. includens* (1.0 ± 0.0) whereas Avanci et al. (2005) recorded higher emergence of parasitoids in eggs of *A. gemmatalis* (2.2 ± 0.06).

Differences in the number of adults emerged per egg can be associated with the host species. Nutritional characteristics, egg size, and adaptations of the parasitoid are very important in this process (Bueno et al., 2009).

As explained above, pesticides were utilized during sampling periods, and the use of chemicals also influenced the viability of parasitoids. Bueno et al. (2008) showed a 50% reduction in the viability of eggs parasitized by *T. pretiosum* in laboratory with the use of products such as Lambda-cyhalothrin and Thiodicarb. Moreover, other products such as glyphosate and chlorimuron reduced by 100% the viability of parasitized eggs in laboratory tests, and all these products were used in the areas sampled in our study.

Despite the use of these products in the study areas, parasitoids were present throughout the period the collections were made. Figures 2A and 2B show the percentage of parasitism for each sampling.

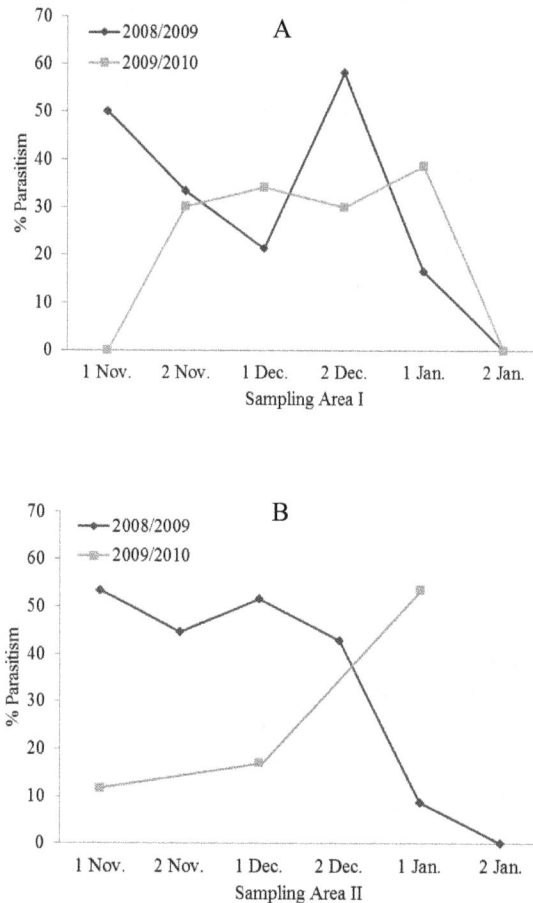

Figure 2. Population fluctuation of *Trichogramma pretiosum* egg parasitoids during the 2008/2009 and 2009/2010 soybean seasons in the two areas sampled in Mato Grosso, Brazil

During the first year, in area I, the parasitoids were present from the first sampling (Figure 2A), and the level of parasitism was higher when compared to the second season. Declines in the percentage of parasitism, as seen in the samplings made in November (2 Nov.) and December (1 Dec.), may be related to the use of pesticides, increasing mortality of caterpillars and unavailability of eggs.

In area II (Figure 2B), during the first season, since the beginning of tsampling, parasitism remained similar and, in January, when soybeans entered the stage of pre-harvest, there was a decrease in the percentage of parasitism. In the second soybean season, due to difficulties in accessing the area, there were only three samples taken., They showed inverse relation to the first year, which may also be related to the use of chemical products as this was the second year of chemical use in the area.

4. Conclusions

Concluded that *Trichogramma pretiosum* was the only species parasitizing lepidopteran eggs on soybeans in our sampling areas, and that the parasitism rate was higher than recorded in other Brazilian regions. The lepidopteran pests the most abundant lepidopteran species in the areas investigated are *Chrysodeixis includes*, *Anticarsia gemmatalis* and *Spodoptera* sp.

Acknowledgments

We thank to Dr. Ranyse B. Querino of EMBRAPA Meio-Norte (Teresina-PI), for the identification of parasitoids, to Dr. Luis Amilton Foerster and the entire staff of the Laboratorio de Controle Integrado de Insetos (UFPR) for the suggestions on collection and management of parasitoids. To the producers who allowed us to collect in their areas and to FAPEMAT for the Scientific Initiation scholarship in 2008-2009.

References

Avanci, M. R. F., Foerster, L. A., & Cañete, C. L. (2005). Natural parasitism in eggs of *Anticarsia gemmatalis* Hübner (Lepidoptera, Noctuidae) by *Trichogramma* spp. (Hymenoptera, Trichogrammatidae) in Brazil. *Revista Brasileira de Entomologia, 49*, 148-151. http://dx.doi.org/10.1590/S0085-56262005000100017

Beserra, E. B., & Parra, J. R. P. (2003). Comportamento de parasitismo de *Trichogramma atopovirilia* Oatman & Platner e *Trichogramma pretiosum* Riley (Hymenoptera, Trichogrammatidae) em posturas de *Spodoptera frugiperda* (J. E. Smith) (Lepidoptera, Noctuidae). *Revista Brasileira de Entomologia, 47*, 205-209. http://dx.doi.org/10.1590/S0085-56262003000200009

Beserra, E. B., & Parra, J. R. P. (2004). Biologia e parasitismo de *Trichogramma atopovirilia* Oatman & Platner e *Trichogramma pretiosum* Riley (Hymenoptera, Trichogrammatidae) em ovos de *Spodoptera frugiperda* (J.E. Smith) (Lepidoptera, Noctuidae). *Revista Brasileira de Entomologia, 48*, 119-126. http://dx.doi.org/10.1590/S0085-56262004000100020

Bueno, A. F., Bueno, R. C. O. F., Parra, J. R., & Vieira, S. S. (2008). Efeitos dos agroquímicos utilizados na cultura da soja ao parasitoide de ovos *Trichogramma pretiosum*. *Ciência Rural, 38*, 1495-1503. http://dx.doi.org/10.1590/S0103-84782008000600001

Bueno, R. C. O. F., Parra, J. R. P., Bueno, A. F., & Haddad, M. L. (2009). Desempenho de tricogramatídeos como potenciais agentes de controle de *Pseudoplusia includens* Walker (Lepidoptera: Noctuidae). *Neotropical Entomology, 38*, 389-394. http://dx.doi.org/10.1590/S1519-566X2009000300015

Campos, O. R., Campos, A. R., & Lara, F. M. (1997). Ocorrência sazonal de insetos pragas e predadores entomófagos em duas variedades de soja. *Cultura Agronômica, 6*, 1-11.

Companhia Nacional de Abastecimento (CONAB). (2010). *Acompanhamento de safra brasileira: grãos, quarto levantamento, janeiro 2010 – Brasília, DF.* Retrieved from http://www.conab.gov.br/conabweb/download/safra/3graos_09.12.pdf

Costa, E. C., & Link, D. (1974). Incidência de percevejos em soja. *Revista Centro de Ciências Rurais, 4*, 397-400. Retrieved from http://cascavel.ufsm.br/revista_new/ojs/index.php/RCCCR/article/viewFile/226/226

Corrêa-Ferreira, B. S., Nunes, M. C., & Uguccioni, L. D. (1998). Ocorrência do parasitoide *Hexacladia smithii* Ashmead em Adultos de *Euschistus heros* (F.) no Brasil. *Anais da Sociedade Entomológica do Brasil, 27*, 495-498.

Corrêa-Ferreira, B. S., & Panizzi, A. R. (1980). Geometrídeos em soja: Flutuação estacional e ressurgência após o uso de inseticidas. *Pesquisa Agropecuária Brasileira, 15*, 159-161.

Foerster, L. A., & Avanci, M. R. F. (1999). Egg parasitoids of *Anticarsia gemmatalis* Hübner (Lepidoptera: Noctuidae) in soybeans. *Anais da Sociedade Entomológica do Brasil, 28*, 545-548. http://dx.doi.org/10.1590/S0301-80591999000300025

Foerster, L. A., & Butnariu, A. R. (2004). Development, reproduction, and longevity of *Telenomus cyamophylax*, egg parasitoid of the velvetbean caterpillar *Anticarsia gemmatalis*, in relation to temperature. *Biological Control, 29*, 1-4. http://dx.doi.org/10.1016/S1049-9644(03)00133-6

Godoy, K. B., Galli, J. C., & Ávila, C. J. (2005). Parasitismo em ovos de percevejos da soja *Euschistus heros* (Fabricius) e *Piezodorus guildinii* (Westwood) (Hemíptera: Pentatomidae) em São Gabriel do Oeste, MS. *Ciência Rural, 35*, 455-458. Retrieved from http://www.redalyc.org/pdf/331/33135234.pdf

Gonçalves, J, R., Holtz, A. M., Pratissoli, D., & Guedes, R. N. C. (2003). Avaliação da qualidade de *Trichogramma pretiosum* (Hymenoptera:Trichogrammatidae) em ovos de *Sitotroga cerealella* (Lepidoptera: Gelechiidae). *Acta Scientiarum Agronomy, 25*, 485-489. http://dx.doi.org/10.4025/actasciagron.v25i2.2328

Medeiros, M. A., Schimidt, F. V. G., Loiácono, M. S., Carvalho, V. F., & Borges, M. (1997). Parasitismo e predação em ovos de *Euschistus heros* (Fab.) (Heteroptera: Pentatomidae) no Distrito Federal, Brasil. *Anais da Sociedade Entomológica do Brasil, 26*, 397-401. http://dx.doi.org/10.1590/S0301-80591997000200026

Moreira, M. D., Santos, M. C. F., Beserra, E. B., Torres, J. B., & Almeida, R. P. (2009). Parasitismo e Superparasitismo de *Trichogramma pretiosum* Riley (Hymenoptera: Trichogrammatidae) em Ovos de *Sitotroga cerealella* (Oliver) (Lepidoptera: Gelechiidae). *Neotropical Entomology, 38*, 237-242. http://dx.doi.org/10.1590/S1519-566X2009000200011

Nagaraja, H., & Nagarkatti, S. (1973). A Key to some New World species of *Trichogramma* (Hymenoptera: Trichogrammatidae), with descriptions of four new species. *Proceedings of the Entomological Society of Washington, 85,* 710-713.

Parra, J. R. P., & Zucchi, R. A. (2004). *Trichogramma* in Brazil: Feasibility of use after twenty years of research. *Neotropical Entomology, 33,* 271-281.

Pereira, F. F., Barros, R., Pratissoli, D., & Parra, J. R. P. (2004). Biologia e exigências térmicas de *Trichogramma pretiosum* Riley e *T. exiguum* Pinto & Platner (Hymenoptera: Trichogrammatidae) criados em ovos de *Plutella xylostella* (Lepidoptera: Plutelidae). *Neotropical Entomology, 33,* 231-236. http://dx.doi.org/10.1590/S1519-566X2004000200014

Potrich, M., Alves, L. F. A., Haas, J., Silva, E. R. L., Daros, A., Pietrowski, V., & Neves, P. M. O. J. (2009). Seletividade de *Beauveria bassiana* e *Metarhizium anisopliae* a *Trichogramma pretiosum* Riley (Hymenoptera: Trichogrammatidae). *Neotropical Entomology, 38,* 822-826. http://dx.doi.org/10.1590/S1519-566X2009000600016

Pratissoli, D., Thuler, R. T., Andrade, G. S., Zanotti, L. C. M., & Silva, A. F. (2005). Estimativa de *Trichogramma pretiosum* para o controle de *Tuta absoluta* em tomateiro estaqueado. *Pesquisa Agropecuária Brasileira, 40,* 715-718. http://dx.doi.org/10.1590/S0100-204X2005000700013

Pratissoli, D., & Oliveira, H. N. (1999). Influência da idade dos ovos de *Helicoverpa zea* (Boddie) no parasitismo de *Trichogramma pretiosum* Riley. *Pesquisa Agropecuária Brasileira, 34,* 231-236. http://dx.doi.org/10.1590/S0100-204X1999000500023

Querino, R, B., & Zucchi, R. A. (2003). Caracterização morfológica de dez espécies de *Trichogramma* (Hymenoptera: Trichogrammatidae) registradas na América do Sul. *Neotropical Entomology, 32,* 597-613. http://dx.doi.org/10.1590/S1519-566X2003000400010

Tironi, P., & Ciociola, A. I. (1994). Parasitismo natural de ovos de *Helicoverpa zea* (Boddie, 1850) (Lepidoptera: Noctuidae) por duas espécies de *Trichogramma* (Hymenoptera: Trichogrammatidae) em culturas de milho em Lavras, MG. *Ciência e Prática, 18,* 61-67.

Thomazini, M. J. (2001). Insetos associados à cultura da soja no estado do Acre, Brasil. *Acta Amazônica, 31,* 673-681.

Venzon, M., Ferreira, J. A. M., & Ripposati, J. G. (2000). Parasitism of stink bug eggs (Hemíptera: Pentatomidae) of soybean fields in the Triangulo Mineiro. *Revista de Biologia tropical, 47,* 1007-1013.

Zago, H. B., Barros, R., Torres, J. B., & Pratissoli, D. (2010). Distribuição de ovos de *Plutella xylostella* (L.) (Lepidoptera: Plutellidae) e o parasitismo por *Trichogramma pretiosum* Riley (Hymenoptera: Trichogrammatidae). *Neotropical Entomology, 39,* 241-247. http://dx.doi.org/10.1590/S1519-566X2010000200015

Zucchi, O. L. A. D., Parra, J. R. P., Silveira Neto, S., & Zucchi, R. A. (1989). Desenvolvimento de um modelo determinístico comportamental para simular o controle de *Heliothis virescens* (Fabr., 1781) através de *Trichogramma* spp. *Anais da Sociedade Entomológica do Brasil, 2,* 357-365.

Zucchi, R. A., & Monteiro, R. C. (1997). O gênero *Trichogramma* na América do Sul. In J. R. P. Parra, R. A. Zucchi (Eds.). *Trichogramma e o controle biológico aplicado* (Cap. 3, pp. 41-66). Piracicaba: FEALQ.

15

In Vitro Culture of Fibroblast-Like Cells From Sheep Ear Skin Stored at 25-26°C for 10 Days After Animal Death

Mahipal Singh[1] & Xiaoling Ma[1]

[1] Animal Biotechnology Laboratory, Agricultural Research Station, Fort Valley State University, Fort Valley, GA, USA

Correspondence: Mahipal Singh, Animal Biotechnology Laboratory, Agricultural Research Station, Fort Valley State University, Fort Valley, GA 31030, USA. E-mail: singhm@fvsu.edu

Abstract

Successful somatic cell nuclear transfer *aka* cloning requires good quality undamaged nuclear DNA from desired cell types. *In vitro* culture of cells is one way of ensuring nuclear integrity. Cellular contents including nucleus gradually decompose postmortem, if not preserved within a reasonable time, leading to cell and ultimately nuclear DNA damage. The goal of this study was to determine time limits within which live and culturable cells can be obtained, after death of an animal, using sheep as a model. How long the somatic cells are alive and have potential to replicate after the animal death is not precisely known. Here we show, for the first time, that the sheep ear skin stored at 25-26°C after animal death can be cultured up to 10 days postmortem. The culture confluence is inversely correlated with increasing postmortem time interval. The cultured fibroblast-like cells have 95±5.2 % post cryopreservation cell-viability; have normal karyotype, and a comparable growth profile to that of fresh tissue derived cells. This study shows that sheep skin has potential for *in vitro* culture of its cells up to 10 days postmortem. Cultured cells can be successfully used for preservation of biodiversity for possible future cloning of animals.

Keywords: fibroblast, postmortem recovery, skin stem cells, cryopreservation, skin banking, animal cloning

1. Introduction

Cryopreservation of tissues from various breeds/species and/or animals with superior traits, both domesticated and wild relatives, has been suggested to conserve genetic diversity, before it is too late. These preserved genetic resources could be utilized globally, to meet the climatic and/or other challenges in future, to meet the global demand for food, especially protein source. However, obtaining viable and culturable tissues/cells from live animals for cryopreservation presents ethical challenges. Recent studies show effective preservation of postmortem tissues and producing live animals from these tissues by somatic cell nuclear transfer *aka* cloning. However, for success of a cloning experiment, nuclear integrity of donor cells is a key requirement (Hoshino et al., 2009). *In vitro* culture of cells ensures nuclear integrity and enhances the success rate of animal cloning (Mastromonaco et al., 2006). *In vitro* culture of cells from live or dead animal tissues, preserved at sub-zero temperatures, has been discussed in several studies (Palmer et al., 2001; Viel et al., 2001; Wakayama et al., 2008; Hoshino et al., 2009; Erker et al., 2010). The nucleus from these cultured cells from preserved tissues has been used to clone the animals even after many years of their death (Hoshino et al., 2009; Wakayama et al., 2008). However, in all these cases the tissues used to obtain viable cells as a source of nucleus were preserved within few hours of animal death. Delays in preservation of tissues within a reasonable time may compromise *in vitro* culture potential, and thus the cloning efficiency, due to cellular death. How long live and culturable cells can be recovered from postmortem tissues in mammalian species is not precisely known. Hasegawa and his colleagues have shown that 2% neurosphere-forming activity remained in a postmortem rat brain even after 25 h of storage at 25°C (Hasegawa et al., 2009). Neural stem and progenitor cells were recovered from postmortem rat brains which were stored at 4°C for a week (Xu et al., 2003). Cells in postmortem tissues stored at room temperature remain culturable up to 4 days after animal death in goat and sheep (Silvestre et al., 2004) and up to 6 days in goat (Singh et al., 2012) and rabbits and pigs (Silvestre et al., 2003). Do cells remain culturable beyond 6 days after animal death in carcasses lying at room temperature (25-26°C) is not known? Here we show, *in vitro* culture of cells after

10 days of animal death, in sheep skin tissues stored at 25-26°C, and demonstrate that these cell cultures are genetically stable and have similar growth profile as that of fresh tissue derived cells.

2. Material and Methods

2.1 Tissue Sample Collection:

The tissue samples were procured from the University slaughter house. Ears of individual healthy Katahdin sheep (1.0-1.5 years of age) were excised from the animal head and brought to the laboratory within an hour. The ears were cleaned with 70% alcohol swabs, wrapped in clean sterile paper towels, and stored in plastic bags in the laboratory at room temperature (25-26°C).

2.2 Explant Preparation, Primary Cell Cultures, and Data Collection:

The explants were prepared and cultured as described earlier (Singh et al., 2011). In brief, inner side of the ear was cleaned with 70% ethanol swabs, and the skin samples excised aseptically after 0, 2, 4, 6, 8, and 10 days of postmortem storage. Excised samples were chopped into 2-3 mm^2 size pieces (explants) and adhered onto 60 mm diameter dishes (Falcon, B. D. Biosciences, Oxnard, CA, USA). Each dish contained 5 explants placed at a uniform distance. Adhered explants were cultured in P116, a fibroblast culture media (Cell Applications Inc., San Diego, CA) at 37°C, 5% CO_2 in a humidified environment. The media in dishes was changed on Mondays, Wednesdays and Fridays. At the same time the dishes were also observed for any microbial or fungal contamination, explant dislodging, and for outgrowth of cells around explants, under an inverted microscope. In our experience, we never observed outgrowth around explants that dislodged within first 3-4 days of culture and, therefore, such explants were removed from the dish as soon as observed. Explants dislodged after day 4 of culture were observed for any outgrowth at their foot-prints and the results included in final analysis. Presence or absence of the outgrowth of cells around explants (clusters of > 50 cells) was recorded and the level of confluence was compared on day 10-12 of culture for different postmortem time intervals.

2.3 Establishing Secondary Cultures and Cell Storage:

Primary outgrowing cells (80-90% confluence) around the explants were trypsinized and secondary cultures established as described (Singh et al., 2011). Briefly, the cells in dishes were washed twice with 2.0 mL of the balanced salt solution without calcium and magnesium (Gibco@ Life Technologies, Grand Island, NY, USA), and incubated with 2.0 mL of 0.125% trypsin for 5-10 min at 37°C. The trypsinized cells were neutralized with 5 volumes of complete growth media, counted to assess cell-viability using Trypan Blue Dye Exclusion Method (Strober, 2001), and pelleted at 200 X g for 7 min. The pellets were suspended in Synth-a-Freeze® (Life Technologies Corp., Carlsbad, CA) media, aliquoted into cryogenic storage vials (1.0 X 10^5 cells / vial) and frozen at -80°C deep freezer o/n using Nalgene™ Cryo 1°C Freezing Container (Nalgene, Rochester, NY). Next day the vials were transferred to liquid nitrogen tank and stored till used for further experiments. To establish secondary cultures, and to expand cell numbers, the frozen vials were quickly thawed at 37°C, mixed slowly with 9.0 mL of the media, pelleted at 200 X g for 7 min, dissolved in complete growth media, and cultured in appropriate (T25 or T75) culture flasks. To determine post cryopreservation cell-viability, representative vials in triplicate were thawed after 10 months of cryopreservation and viable cells were counted manually using a hemocytometer. Cell-viability was expressed as % of live cells from total cell count.

2.4 Generating Growth Curves:

Growth curves were generated from passage 4 cultures using a 24-well micro titer plate format as described earlier (Singh et al., 2012). Briefly, 20, 000 cells were inoculated in each well in 0.5 mL of growth media to initiate the culture. The wells were trypsinized (in triplicate) after day 1, 3, 5, 7, 9 and 11 of culture. The viable cells were counted in each well using a hemocytometer. Mean and standard deviation of the cells / mL in 3 wells were plotted against time using Excel program to generate growth curves.

2.5 Cytogenetic Analysis:

Sheep cells cultured from 10 days postmortem tissues stored at 25-26°C were processed for cytogenetic analysis, at passage 2 level, using previously established methodologies (Meisner & Johnson, 2008) by Cell Line Genetics (Madison, WI; www.clgenetics.com). The chromosome assignments were made according to the Atlas of Mammalian Cytogenetics (O'Brien, Menninger, & Nash, 2006).

3. Results and Discussion

3.1 In Vitro Culture of Cells After 10 Days of Animal Death:

We studied healthy sheep of Katahdin breed for postmortem survival of their skin cells, using an in vitro culture approach. In vitro cell culture ensures nuclear integrity, a key requirement of successful animal cloning. Sheep ears were chosen as a source of skin sample because they are easily accessible. Ear skin stored in the laboratory at 25-26°C postmortem was sliced into small explants and cultured in complete growth media. Forty explants were cultured in 8 dishes of 60 mm diameter (5 explants / dish) after 0, 2, 4, 6, 8, and 10 days of animal death. Outgrowth of fibroblast-like cells around the explants was scored after 10-12 days of culture for each time point. Our results show outgrowth of fibroblast-like cells around all explants (except one) that adhered to dish surface, even from 10 days postmortem tissues (Table 1). To our knowledge this is the first report of *in vitro* culture of postmortem tissues after 10 days of animal death in mammals from their carcass exposed to 25-26°C. Earlier studies have shown outgrowth of fibroblast-like cells up to 4 days of animal death in goats and sheep (Silvestre et al., 2004), six days in goats (Singh et al., 2012), and up to 6 days in rabbits and pigs (Silvestre et al., 2003) when the skin tissue was exposed to around 25°C. It should be noted that in all these earlier studies, outgrowth was observed in all the time points studied, including the last time point, although the number of explants that responded varied. In fact, that inspired us to explore, if the postmortem tissues exposed to room temperature beyond six days, still have live cells that can be cultured *in vitro*. It is presumed that the individual cells in skin tissues die gradually over time. It is perhaps correlated with gradual reduction in live adult stem cells in skin tissues with increasing postmortem time interval. What made these skin cells to survive for such a long time of 10 days postmortem is not clear. It is possible that diffusion of environmental oxygen into skin tissues kept them alive for such a longer period at room temperature (Fife et al., 2009). The time taken to have complete death of all cells in a tissue may depend upon environmental temperature, humidity, age, and health of an animal. Although for normal practical purposes, it is unlikely that the animal will remain in field without notice for such a long time after its death, this study shows that the live cells can be recovered after several days postmortem, and effectively cryopreserved for future use.

Table 1. Outgrowth of fibroblast-like cells around skin explants after different days postmortem

Sheep	Breed / age / weight	Number of outgrowth positive / total explants adhered to dish surface					
		0-dpm	2-dpm	4-dpm	6-dpm	8-dpm	10-dpm
S1	Katahdin / 1.5 year / 120 lb	25/25	20/20	16/16	15/15	17/17	10/10
S2	Katahdin / 1.0 year / 100 lb	15/15	10/10	15/15	10/10	15/15	9/10
Total		40/40	30/30	31/31	25/25	32/32	19/20
% outgrowth		100	100	100	100	100	95

Note: Forty explants were used for each time point. The explants floated within first 3 days of culture (non-adhered explants) and the contaminated dishes during experimentation were removed from the study.

3.2 Comparison of Confluence of Outgrowth Around Tissues Cultured After Different Postmortem Time Intervals:

Confluence in cell biology is defined as the number of cells per unit area in a culture dish or flask. We show in Figure 1 that the confluence level of outgrowing cells is decreasing with increasing postmortem time interval on a given day. For example, the number of cells around skin explants of 0-dpm (first panel in Figure 1) is more than 2-dpm, which intern has more cells than 4-dpm. It is known that oxygen deprivation, due to cessation of blood circulation, progressively leads to decomposition of tissues (Zdravkovic, 2010). This happens due to action of lysosomal enzymes present in the cells which ultimately results in cellular death. It is conceivable that longer the interval from animal death to cell culture, lesser the number of live skin adult stem cells available to produce outgrowth. This could explain the observed reduction in outgrowth with increasing postmortem time interval in this study. Our results are in agreement with similar reduction in cells with extended postmortem time interval in neurosphere cell cultures in adult human retina (Carter et al., 2007), and fibroblast-like cell cultures in goat ear skin (Singh et al., 2012).

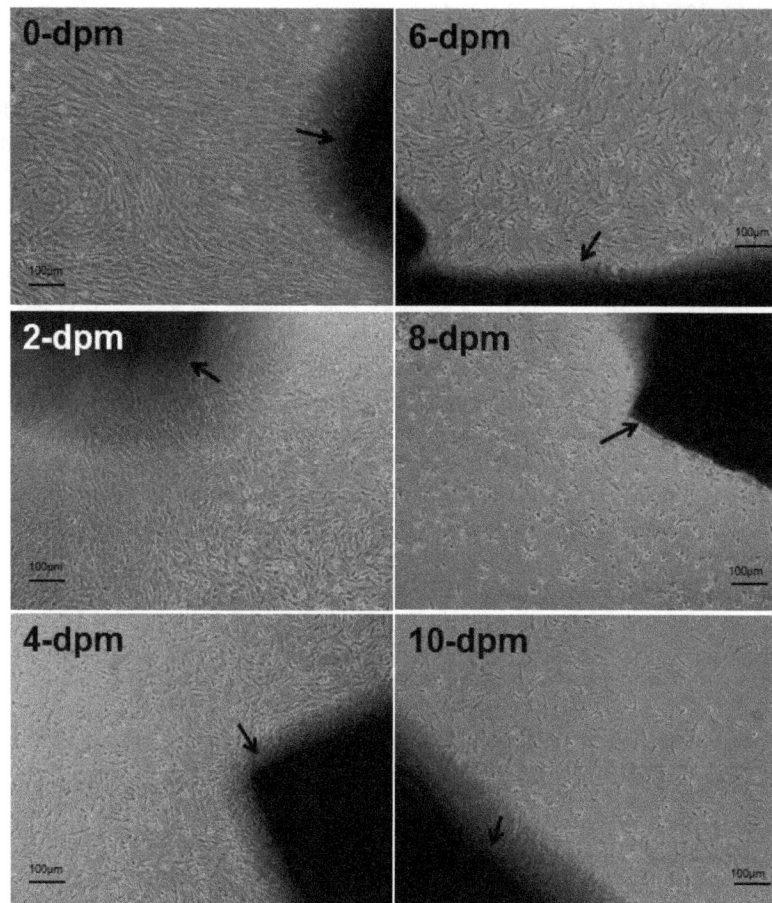

Figure 1. A representation of comparative cell confluence of different days postmortem tissues after a fixed interval of 10-12 days of *in vitro* culture (note that except 6-dpm, rest images here were after 10 days of culture). Arrow marked dark areas are skin explants adhered on dish surface. Nikon Eclipse TS100 inverted microscope was used to capture the images. Dpm = days postmortem

3.3 Establishment of Secondary Cultures, Their Cryopreservation, and Growth Profiles:

Although, the confluence level of primary outgrowth reduces with increasing postmortem time interval, how the cell populations in subsequent passages differ, with respect to their growth profile, is not known. To find out the differences between these cell populations, we established secondary cultures from the primary outgrowth in 0-dpm (control) and 10-dpm tissues, by serial passaging. It was observed that secondary cultures of these cell populations grow much faster as compared to the primary outgrowth. These cultures became 50-60% confluent in 3-4 days (Figure 2; upper panels) as opposed to the primary outgrowth (Figure 1), which took 2-5 times longer in reaching same level of confluence. Faster growth of secondary cultures may be related to their adaptation to the culture environment.

The cells harvested from passage 2 and passage 3 cultures of 0-dpm and 10-dpm tissues were preserved in liquid nitrogen in freezing media. Representative Cryovials, in triplicate, were tested for their post-freezing cell-viability and *in vitro* culture ability after 10 months of cryopreservation. Results show 96.67±1.53 % and 95.00±5.20 % post-freezing cell-viability for 0-dpm and 10-dpm cell cultures, respectively (Figure 2; lower right panel). All tested cryovials showed normal growth in *in vitro* cultures, which was indistinguishable for 0-dpm and 10-dpm cell populations.

To determine the similarities / differences between two cell populations (*i.e.* 0-dpm and 10-dpm), we performed sequential passaging of cells so as to obtain pure fibroblast-like cells (note that some epithelial cell colonies also appear in primary outgrowth which are eliminated by partial trypsinization during sequential passages). Subsequently, these cell populations were compared at passage 4 level for their, a) growth curves, and b) growth morphology. Growth curves were generated using a 24-well micro titer plate format. The results show typical mammalian growth curves, which are similar but not identical, for both 0-dpm and 10-dpm cell populations

(Figure 2; lower left panel). Comparative images of growing cell cultures on day 4 were captured to study their morphology. As shown in Figure 2 (upper panels), images display typical elongated, bipolar fibroblast-like morphology which is indistinguishable for both cell populations. It suggests their similarity, despite of the time lapse between animal death and the cell culture. It would be interesting to see reprogramming of their nuclear DNA by enucleated oocyte cytoplasm, embryo development, and ultimately cloning of sheep that was dead for 10 days.

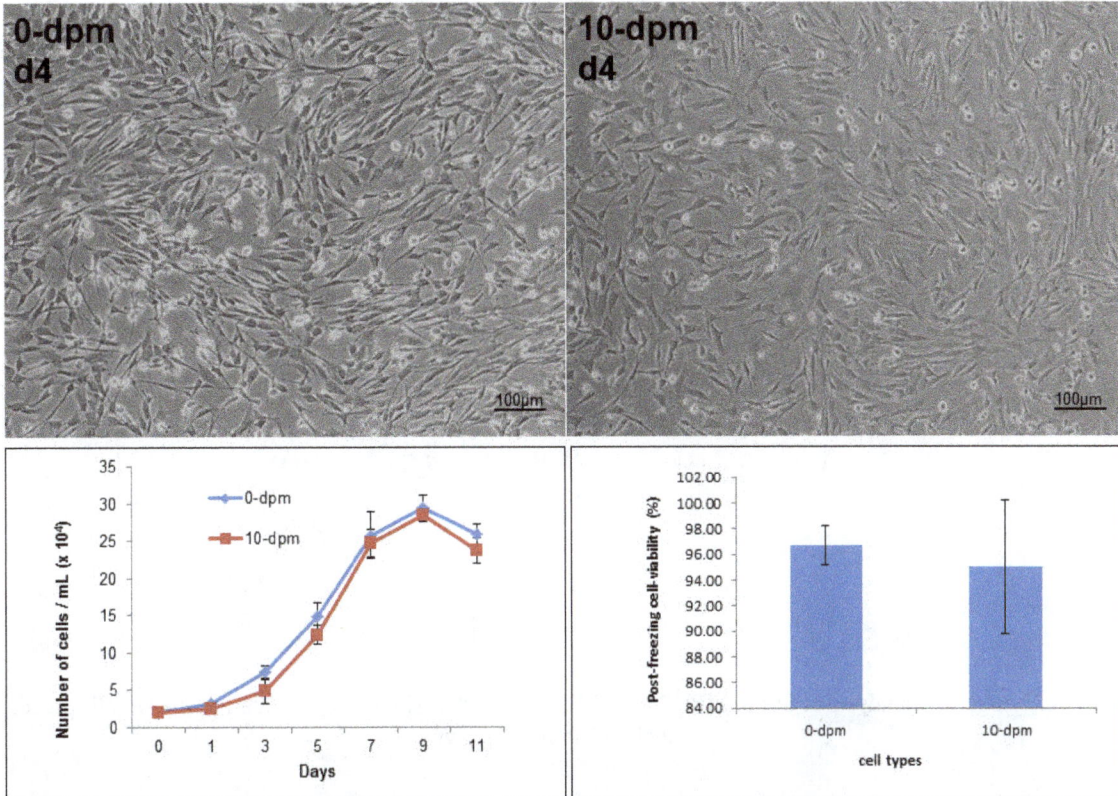

Figure 2. Figure shows comparative cell morphology and confluence of 0-dpm (control) and 10-dpm cell-lines at passage 4 level on day 4 of culture (upper panels); growth curves show mean ± SD of cells in triplicate wells (lower left panel); and post-freezing cell-viability show mean ± SD of 3 experiments (lower right panel)

3.4 Cytogenetic Stability of 10 Days Postmortem Tissue Derived Cell-Line:

To determine chromosomal stability of the cells cultured from 10 days postmortem tissues, a cytogenetic analysis was performed. The diploid (2x) number of the chromosomes in the cell-line was determined to be 54, XY chromosomes. As can be seen in Figure 3 (panel B), it consisted of 52 autosomes and two (X and Y) sex chromosomes, which is consistent with earlier studies on sheep cytogenetics (Ansari et al., 1996; Iannuzzi et al., 2009). Cytogenetic analysis was performed on twenty G-banded metaphase cells of 10-dpm (SSFRT-10DPM, p2) cell-line. Nineteen cells exhibited an apparently normal male karyotype, while one cell showed a non-clonal chromosomal aberration which is most likely an artifact of culture. Thus these results are apparently consistent with a normal male sheep karyotype.

In conclusion, we have shown for the first time that sheep ear skin stored at 25-26°C has *in vitro* culture potential up to 10 days of animal death. The postmortem tissue-derived cells are effectively cryopreserved with > 95% of post-freezing cell-viability, are indistinguishable from normal control cells, with respect to their growth profiles, and are cytogenetically stable. Future studies will determine their potential for reprogramming and animal cloning.

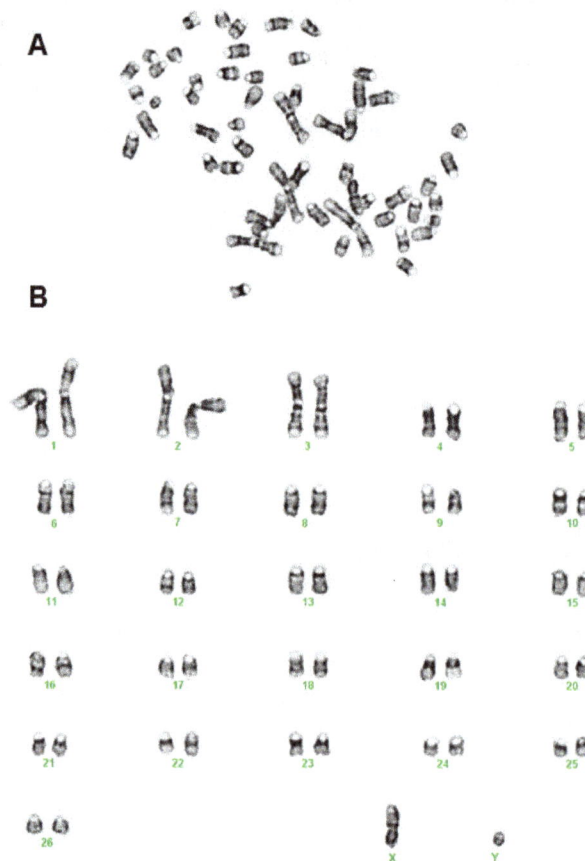

Figure 3. Cytogenetic analysis of SSFRT-10DPM cell-line: Panel "A" show a representative single cell spread, and panel "B" show karyotype

Acknowledgements

Part of this work was presented in the *Society for In Vitro Biology* meeting at Providence, RI, as an interactive poster in June 2013. This work was partly supported by a USDA-NIFA Capacity Building Grant No: 2011-38821-30910 to MS. We would like to thank Drs. Bharat Singh and Seyedmehdi Mobini for the critical comments on the manuscript.

References

Ansari, H. A., Maher, D. W., Pearce, P. D., & Broad, T. E. (1996). Resolving ambiguities in the karyotype of domestic sheep (ovis aries). II. G-, Q-, and R-banded idiograms, and chromosome-specific molecular markers. *Chromosoma, 105*, 62-67. http://dx.doi.org/10.1007/bf02510040

Carter, D. A., Mayer, E. J., & Dick, A. D. (2007). The effect of postmortem time, donor age and sex on the generation of neurospheres from adult human retina. *Br J Ophthalmol, 91*, 1216-1218. http://dx.doi.org/10.1136/bjo.2007.118141

Erker, L., Azuma, H., Lee, A. Y., Guo, C., Orloff, S., Eaton, L., … Grompe, M. (2010). Therapeutic liver reconstitution with murine cells isolated long after death. *Gastroenterology, 139*, 1019-1029. http://dx.doi.org/10.1053/j.gastro.2010.05.082

Fife, C. E., Smart, D. R., Sheffield, P. J., Hopf, H. W., Hawkins, G., & Clarke, D. (2009). Transcutaneous oximetry in clinical practice: Consensus statements from an expert panel based on evidence. *Undersea Hyperb Med, 36*, 43-53.

Hasegawa, A., Yamada, C., Tani, M., Hirano, S., Tokumoto, Y., & Miyake, J. (2009). Caspase inhibitors increase the rate of recovery of neural stem/progenitor cells from post-mortem rat brains stored at room temperature. *J Biosci Bioeng, 107*, 652-657. http://dx.doi.org/10.1016/j.jbiosc.2009.01.021

Hoshino, Y., Hayashi, N., Taniguchi, S., Kobayashi, N., Sakai, K., Otani, T., Iritani, A., & Saeki, K. (2009). Resurrection of a bull by cloning from organs frozen without cryoprotectant in a -80 degrees C freezer for a decade. *PLoS One, 4*, e4142. http://dx.doi.org/10.1371/journal.pone.0004142

Iannuzzi, L., King, W. A., & Di Berardino, D. (2009). Chromosome evolution in domestic bovids as revealed by chromosome banding and FISH-mapping techniques. *Cytogenet Genome Res, 126*, 49-62. http://dx.doi.org/10.1159/000245906

Mastromonaco, G. F., Perrault, S. D., Betts, D. H., & King, W. A. (2006). Role of chromosome stability and telomere length in the production of viable cell lines for somatic cell nuclear transfer. *BMC Dev Biol, 6*, 41. http://dx.doi.org/10.1186/1471-213x-6-41

Meisner, L. F. & Johnson, J. A. (2008). Protocols for cytogenetic studies of human embryonic stem cells. *Methods, 45*, 133-141. http://dx.doi.org/10.1016/j.ymeth.2008.03.005

O'Brien, S. J., Menninger, J. C., & Nash, W. G. (2006). *Atlas of mammalian chromosomes* (pp. 653-655). http://dx.doi.org/10.1002/0471779059

Palmer, T. D., Schwartz, P. H., Taupin, P., Kaspar, B., Stein, S. A., & Gage, F. H. (2001). Cell culture. progenitor cells from human brain after death. *Nature, 411*, 42-43. http://dx.doi.org/10.1038/35075141

Silvestre, M. A., Saeed, A. M., Cervera, R. P., Escriba, M. J., & Garcia-Ximenez, F. (2003). Rabbit and pig ear skin sample cryobanking: Effects of storage time and temperature of the whole ear extirpated immediately after death. *Theriogenology, 59*, 1469-1477. http://dx.doi.org/10.1016/s0093-691x(02)01185-8

Silvestre, M. A., Sanchez, J. P., & Gomez, E. A. (2004). Vitrification of goat, sheep, and cattle skin samples from whole ear extirpated after death and maintained at different storage times and temperatures. *Cryobiology, 49*, 221-229. http://dx.doi.org/10.1016/j.cryobiol.2004.08.001

Singh, M., Ma, X., & Sharma, A. (2012). Effect of postmortem time interval on in vitro culture potential of goat skin tissues stored at room temperature. *In Vitro Cell Dev Biol Anim, 48*, 478-482. http://dx.doi.org/10.1007/s11626-012-9539-3

Singh, M., Ma, X., Amoah, E., & Kannan, G. (2011). In vitro culture of fibroblast-like cells from postmortem skin of katahdin sheep stored at 4 degrees C for different time intervals. *In Vitro Cell Dev Biol Anim, 47*, 290-293. http://dx.doi.org/10.1007/s11626-011-9395-6

Strober, W. (2001). Trypan blue exclusion test of cell viability. *Curr Protoc Immunol* Appendix 3: Appendix 3B. http://dx.doi.org/10.1002/0471142735.ima03bs21

Viel, J. J., McManus, D. Q., Cady, C., Evans, M. S., & Brewer, G. J. (2001). Temperature and time interval for culture of postmortem neurons from adult rat cortex. *J Neurosci Res, 64*, 311-321. http://dx.doi.org/10.1002/jnr.1081

Wakayama, S., Ohta, H., Hikichi, T., Mizutani, E., Iwaki, T., Kanagawa, O., & Wakayama, T. (2008). Production of healthy cloned mice from bodies frozen at -20 degrees C for 16 years. *Proc Natl Acad Sci (U S A), 105*, 17318-17322. http://dx.doi.org/10.1073/pnas.0806166105

Xu, Y., Kimura, K., Matsumoto, N., & Ide, C. (2003). Isolation of neural stem cells from the forebrain of deceased early postnatal and adult rats with protracted post-mortem intervals. *J Neurosci Res, 74*, 533-540. http://dx.doi.org/10.1002/jnr.10769 Actions

Zdravkovic, M. (2010). Ultrastructural changes of renal epithelial cells during postmortem autolysis--experimental work. *Med Pregl, 63*, 15-20. http://dx.doi.org/10.2298/mpns1002015z

16

Studying the Effect of the *Ziziphora tenuior* L. Plant on Some Biochemical Factors of Serum in Rats

Hamed Soleyman Dehkordi[1,3], Hamid Iranpour Mobarakeh[1], Mohsen Jafarian Dehkordi[2] & Faham Khamesipour[1,3]

[1] Under Graduated Student of Veterinary Medicine, College of Veterinary Medicine, Islamic Azad University, Shahrekord Branch, Shahrekord, Iran

[2] Department of Pathology, Faculty of Veterinary Medicine, Islamic Azad University, Shahrekord Branch, Shahrekord, Iran

[3] Young Researchers and Elite Club, Islamic Azad University, Shahrekord Branch, Shahrekord, Iran

Correspondence: Hamed Soleyman Dehkordi, Faculty of Veterinary Medicine, Islamic Azad University, Shahrekord Branch, Shahrekord, PO box 166,Iran.E-mail: hamed.soleyman.dvm@gmail.com

Abstract

The *Ziziphora tenuior* L. is an herbaceous, plant rich in essential oil. *Ziziphora tenuior* L. has antibacterial and disinfection effect and its vapour is used to treat respiratory tract problems and removing symptoms of the common cold. This plant is carminative, promotes expelling of phlegm and sputum and is a tonic for the stomach. To perform this study some chemical factors (triglyceride, cholesterol, HDL, LDL, ALT, AST and total protein) of the rat's serum were measured in the clinical laboratory. In this study, number of 24 white male wistar rats with the weight range of 215 ± 15 g and at 10 weeks of age were prepared from the Laboratory Animals Breeding Center of the University. Rats were divided into three classes of 8 and 200 mg/kg and 400 mg/kg of the *Ziziphora tenuior* L. plant essence was used in groups 1 and 2 respectively and no plant compound was used in the third group as the control group. The plant was prescribed for 21 days in groups. Amount of the cholesterol, ALT and triglyceride factors showed the significant difference compared to the control group.

Keywords: *Ziziphora tenuior* L., biochemical factors, serum, rats

1. Introduction

The *Ziziphora tenuior* L. (named as Kakoti in Persian) is a herbaceous, seasonal, annual, flocculent, slender and erect plant with a height of 5-15 cm (In vitro propagation of the medicinal plant *Ziziphora tenuior* L. and evaluation of its antioxidant activity). This plant has limited growth and is not able to resist dry condition of the summer. The flowering season of this plant is April-Jun (Verdian-Rivi, 2008; Al-Rawashdeh, 2011; Ghassemi et al., 2013).

There is abundant oil essence in the *Ziziphora tenuior* L. Researchers have concluded in studying compounds of the *Ziziphora tenuior* L. plant essence that three compounds of Pulegone (82.6%), Limonene (6.8%) and Cineol (1.9%) have had the highest percent and totally have constituted 91.36% of the essence (Al-Rawashdeh, 2011; Ghassemi et al., 2013; Karimi et al., 2013). Leaves and blooming branches of the plant (fresh or dried) are consumed as infused or sodden with water or tea (Amanlou et al., 2005).

Ziziphora tenuior L. is known as a medicinal herb, because its extract shows antifungal and antibacterial effects. Even a lot research studies have been performed on *Ziziphora tenuiro*, the research assay described in this manuscript is quite novel. Measuring the concentrations of chemical factors in serum is easy and relatively cheap. It might be helpful to find new usage of *Ziziphora tenuior*. *Ziziphora tenuior* L. has antibacterial and disinfection effect and its vapour is used to treat respiratory tract problems and removing symptoms of the common cold. Also, it is used as antispasmoic in gastrointestinal disorders. This plant is carminative, promotes expelling of phlegm and sputum and is a tonic for the stomach. Mixture of its used and powder is used in some areas in remedy of diarrhea. Its seed in used to treat fever in India. It is also useful to increase the sexual force (Naeini et al., 2005; Talebi et al., 2012; Ghassemi et al., 2013).

Infusion of the *Ziziphora tenuior* L. is useful in treatment of the rheumatoid arthritis and *rickets*. Also it relieves the chronic tonsillitis. This plant is a common antiepileptic agent and increases secretion of bile therefore, it is useful for liver. Infusing some grams of this plant in the hot water is useful to treat Stomachache, Bronchitis, Pertussis, bloodshed, etc. (Sharafzadeh & Alizadeh, 2012; Talebi et al., 2012; Ghassemi et al., 2013; Karimi et al., 2013; Tabatabaei et al., 2013). Habitats of this species have high potential to apiculture the due to the suitable nectar (Al-Rawashdeh, 2011; Darbandi et al., 2013; Ghassemi et al., 2013). To perform this study some chemical factors of the rat's serum were measured in the clinical laboratory.

2. Material and Method

This study is experimental and is conducted in 2014 in the Islamic Azad University of Shahrekord Branch. To perform this study some chemical factors (triglyceride, cholesterol, HDL, LDL, ALT, AST and total protein) of the rat's serum were measured in the Al-Mahdi clinical laboratory of Shahrekord.

Samples were collected from leaves of the *Ziziphora tenuior* L. plant and essence making was performed in the Islamic Azad University of Shahrekord Branch laboratory and they were accurately identified and confirmed by means of adapting the herbarium specimens. Then they were dried at 25-35 °C temperature for 3 hours. After drying they were crushed and extracted using BP (British Pharmacopoeia, 1988). It should be noted that 1 cc essence were obtained per 100 gr plant (British Pharmacopoeia, 1988).

In this study, number of 24 white male wistar rats with the weight range of 215 ± 15 gr and at 10 weeks of age were prepared from the Laboratory Animals Breeding Center of the Islamic Azad University of Shahrekord Branch and then were maintained in standard cages and ready access to water and food was provided for them. Rats were divided into three classes of 8 and 200 mg/kg and 400 mg/kg of the *Ziziphora tenuior* L. plant essence was used in groups 1 and 2 respectively and no plant compound was used in the third group as the control group.

The plant was prescribed for 21 days in groups. After 21 days animals were anesthetized using the chloroform in special containers provided for this purpose and blood sample was collected using the cardiac puncture technique (Barreto et al., 2008) and its serum was isolated using centrifuge apparatus at speed of 2000 cycle per minute and then AST, ALT, HDL, total protein, cholesterol and triglyceride factors were measured. Data were statistically analyzed using the SPSS software and significant levels ($p < 0.05$).

3. Results

Results were compared with each other using the Dunnet test. Amount of serum triglyceride was 45.3 ± 20.9 mg/dl in the group 1 that did not show significant difference compared to the control group (111.7 ± 21 mg/dl). Also amount of triglyceride in the group 2 (41.2 ± 18.2 mg/dl) was that did not show significant difference compared to the group control (111.7 ± 21 mg/dl) (Table 1).

Amount of total protein in the group 1 was (7.6-0.40 mg/dl) that has not significant difference compared to the control group (7.5 ± 1.7 mg/dl). Amount of total protein in the group 2 was (6.7 ± 1.3 mg/dl) that has not significant difference compared to the control group (7.5 ± 1.7 mg/dl) (Table 1).

Amount of the cholesterol in the group 1 was (65.4 ± 9.5 mg/dl) that showed the significant difference compared to the control group (142.2 ± 24.13 mg/dl). Also amount of the cholesterol in the group 2 was (60.6 ± 8.5 mg/dl) that showed the significant difference compared to the control group (142.2 ± 24.13 mg/dl) (Table 1).

Amount of the AST in the group 1 was (166.2 ± 74.6 mg/dl) that did not show significant difference compared to the group control (173.7 ± 55.8 mg/dl). Amount of the AST in the group 2 was (124.2 ± 15 mg/dl) that did not show significant difference compared to the group control (173.7 ± 55.8 mg/dl) but in two groups show a relative decrease compared to the control group (Table 2).

Amount of the LDL in the group 1 was (5.6 ± 3.4 mg/dl) that did not show significant difference compared to the group control (4.3 ± 1 mg/dl). Amount of the LDL in the group 2 was (3.3 ± 0.17 mg/dl) that did not show significant difference compared to the group control (4.3 ± 1 mg/dl) (Table 1).

Amount of the HDL in the group 1 was (9.7 ± 2.01 mg/dl) that did not show significant difference compared to the group control (12 ± 1.4 mg/dl). Amount of the HDL in the group 2 was (10.2 ± 1.7 mg/dl) that did not show significant difference compared to the group control (12 ± 1.4 mg/dl) (Table 1).

Amount of the ALT in the group 1 was (36.3 ± 17.1 mg/dl) that showed the significant difference compared to the control group (80 ± 10.1 mg/dl). Also amount of the ALT in the group 2 was (33.6 ± 18.3 mg/dl) that showed the significant difference compared to the control group (80 ± 10.1 mg/dl) (Table 2).

Amount of the AST in the group 1 was (166.3 ± 74.6 mg/dl) that did not show significant difference compared to the group control (173.7 ± 55.8 mg/dl). Amount of the AST in the group 2 was (124.2 ± 15 mg/dl) that did not show significant difference compared to the group control (173.7 ± 55.8 mg/dl) (Table 2).

Table 1. Studying effect of the *Ziziphora tenuior* L. plant on biochemical factors of the serum (LDL, HDL, total protein, cholesterol, total protein and triglyceride) in rat for 21 days

Groups	*Ziziphora tenuior* L. plant essence was used (SD±Mean)	Total Protein (SD±Mean)	Cholesterol (SD±Mean)	Triglyceride (SD±Mean)	LDL (SD±Mean)	HDL (SD±Mean)
Control	-	7.5±1.7[a]	142±24.13[a]	111.7±21[a]	4.3±1[a]	12±1.4[a]
Group 1	200 mg/kg	7.6±0.4[a]	65.4±9.5[b]	45.3±20.9[b]	5.6±3.4[a]	2.7±2.01[a]
Group 2	400 mg/kg	6.7±1.3[a]	60.6±8.5[b]	41.2±18.2[b]	3.2±0.17[a]	10.2±1.7[a]

There is no significant difference in numbers with similar letters (p < 0.05).

Table 2. Studying effect of the *Ziziphora tenuior* L. plant on liver enzymes (ALT, AST) in rat for 21 days

Groups	*Ziziphora tenuior* L. plant essence was used (SD±Mean)	ALT (SD±Mean)	AST (SD±Mean)
Control	-	80±10.1[a]	173.7±55.8[a]
Group 1	200 mg/kg	36.3±17.1[b]	166.3±74.6[a]
Group 2	400 mg/kg	33.6±18.3[b]	124.2±15[a]

There is no significant difference in numbers with similar letters (p < 0.05).

4. Discussion

With regard to obtained results, the triglyceride and cholesterol levels in treatment groups showed significant decrease than the control group (p < 0.05) that this decrease was more significant in the group that received more amount of the essence. Therefore it can be concluded that the effect of the *Ziziphora tenuior* L. essence on level of the triglyceride and cholesterol factors depends on the dosage. The *Ziziphora tenuior* L. plant contains the essential oil that its ingredients are thymol and Carvacrol. Carvacrol decreases the plasma triglyceride concentration (Lee et al., 2003). Case et al. (1995) showed that Thymol and Carvacrol at doses of 0.15 (V/V) decreased the serum cholesterol in Leghorn chickens (Case et al., 1995). Elson and Qureshi (1995) showed that hypocholesterolemic effects of the Thymol and Carvacrol is related to the inhibition of the enzyme HMG_COA *reductase* (Elson & Qureshi, 1955). Tsherich (2000) reported that using the Carvacrol stimulates the lactobacillus reproduction and growth. Lactobacillus has the important role in improvement of blood factors and decreasing serum lipids (Tschirch, 2000). Also perkival reported that lactobacilli can metabolize and absorb the cholesterol in the Small intestine and decrease its absorption through the blood (Percival, 2001). Cerig (1999) also reported role of medicinal plants and effective oils in decreasing cholesterol and protecting against the cancer (Craig, 1999). Also the serum cholesterol concentration was shown by adding the effective oils in the livery of broilers (Case et al., 1995) that obtained results in conducted research are consistent with the present study. Youshika et al. (2000) also reported that plant essential oils can decrease the abdominal fat deposition. This decrease can be caused by reduction of serum lipids (Yoshioka et al., 2000). Phenolic compounds available in medicinal plants control activity of the (HMG_COA) enzyme and in result the cholesterol synthesis is controlled. This act increases LDL receptors in surface of liver cells and consequently the LDL catabolism is accelerated. Inhibitors of HMG_COA *reductase* decrease the LDL and the plasma triglyceride concentration to a lower concentration (Barreto et al., 2008). In the present study a relative decrease was observed in the LDL amount in the group receiving the 400 mg/kg doses of the *Ziziphora tenuior* L. essence. In this study enzymes serum ALT and AST were used to investigate the liver performance biochemistry. These enzymes naturally are existed in liver cells and these enzymes are released into the blood due to disorder in the plasma membrane or cells deposition when liver cells are damaged and so serum levels of these enzymes are increased. Therefore increased concentration of these two enzymes is considered as a suitable criterion for evaluating the level of the damage to liver cells (Chidambarama & Carani, 2010). In the conducted study the ALT level in both treatment groups showed the significant decrease than the control group that this decrease was more in the second group that had received 400 mg/kg of the plant essence. So it can be said that the plant essence depends on the dose in causing the decrease in ALT and has no toxic effects on liver cells in doses of 200 and 400 mg/kg but improve the liver activity. Also the AST level in both treatment groups showed the relative decrease than the control group.

References

Al-Rawashdeh, I. M. (2011). Molecular Taxonomy Among Mentha spicata, Mentha longifolia and Ziziphora tenuior Populations using the RAPD Technique. *Jordan Journal of Biological Sciences, 4*(2), 63-70.

Amanlou, M. D., Dadkhah, F., Salehniak, A., & Farsam. H. (2005). An antinocieptive effect of hydro alcoholic extract of satureje khuzestanica jamzad extract. *The Journal of Pharmaceutical Sciences, 8*(1), 102-106.

Barreto, M. S. R., Menten, J. F. M., Racanicci, A. M. C., Pereira, P. W. Z., & Rizzo, P. V. (2008). Plant extracts used as Growth promoters in Broilers . *Brazilian Journal of Poultry Science, 10*(2), 109-115.

British Pharmacopoeia. (1988). *British pharmacopoeia* (Vol. 2, pp. 137-138). London: HMSO.

Case, G. L., He, L., Mo, H., & Elson, C. E. (1995). Induction of geranyl pyrophosphate pyrophsphatase activity by cholesterol suppressive iso pre noids. *Lipids, 30*, 357-359. http://dx.doi.org/10.1007/BF02536045

Chidambarama, J., & Carani, V. A. (2010). Cissus quadrangularis stem alleviates insulin resistance, oxidative injury and fatty liver disease in rats fed high fat plus fructose diet. *Food and Chemical Toxicology, 48*(8-9), 2021-2029. http://dx.doi.org/10.1016/j.fct.2010.04.044

Craig, W. G. (1999). Health promoting properties of common herbs. *The American Journal of Clinical Nutrition, 70*, 491-499.

Darbandia, T., Honarvarb, B., Sinaei, N. M., & Rezaei, A. (2013). Extraction of Ziziphora tenuior essential oil using supercritical CO_2. *European Journal of Experimental Biology, 3*(3), 687-695.

Elson, C. E., & Qureshi, A. A. (1955). Coupling the cholesterol and tumor suppressive actions of palm oil to the impact of this minor Constituents on 3-hydroxy-3-methyl glutaryl coenzymeA reductase activity. *prostaglandins Leukorrienes and Essential Fatty Acids, 852*, 205-280.

Ghassemi, N., Ghanadian, M., Ghaemmaghami, L., & Kiani, H. (2013). Development of a Validated HPLC/Photodiode Array Method for the Determination of Isomenthone in the Aerial Parts of Ziziphora tenuior L. *Jundishapur J Nat Pharm Prod, 8*(4), 180-186.

Karimi, I., Hayatgheybi, H., Motamedi, S., Naseri, D., Shamspur, T., Afzali, D., & Hassanpour, A. A. (2013). Chemical Composition and Hypolipidemic Effects of an Aromatic Water of *Ziziphora tenuior* L. in Cholesterol-fed Rabbits. *Journal of Applied Biological Sciences, 7*(3), 61-67.

Lee, K. W., Evert, H., Kappert, H. G., Ferehner, M., Losa, R., & Beynen, A. C. (2003). Effect of dietary essential oils on growth performance, digestive enzymens and lipid metabolism in femaile broiler chiken. *British Poultry Science, 44*, 450-457. http://dx.doi.org/10.1080/0007166031000085508

Naeini, A., Khosravi, A. R., Chitsaz, M., Shokri, H., & Kamlnejad, M. (2009). Anti- Candida albicans activity of some Iranian plants used in traditional medicine. *Journal of Medical Mycology, 19*(3), 168-72. http://dx.doi.org/10.1016/j.mycmed.2009.04.004

Percival, M. (2001). Choosing a probiotic supplement. Clinical nutrition Insights. *Advances in Nutrition, 6*, 1-9.

Sharafzadeh, S., & Alizadeh, O. (2012). Some Medicinal Plants Cultivated in Iran. *Journal of Applied Pharmaceutical Science, 2*(1), 134-137.

Tabatabaei, Y. F., Mortazavi, A., Koocheki, A., Afsharian, S. H., & Alizadeh, B. B. (2013). Antimicrobial properties of plant extracts of *Thymus vulgaris* L., *Ziziphora tenuior* L. and *Mentha Spicata* L., against important foodborne pathogens in vitro. *Scientific Journal of Microbiology, 2*(2), 23-30.

Talebi, S. M., Rezakhanlou, A., & Salahi, I. G. (2012). Trichomes Plasticity in *Ziziphora tenuior* L. (Labiatae) in Iran: An ecological review. *Annals of Biological Research, 3*(1), 668-672.

Tschirch, H. (2000). The use of natural plants extracts as production inhansers in modern animal rearing practiees. Zeszyty Naukowe Akademii Rolniczej Wroclaw, *Zootechnika, XXV376*, 25-39.

Verdian-Rivi, M. (2008). Effect of the Essential Oil Composition and biological activity of Ziziphora clinopodiodes Lam. on the against Anopheles Stephensi and Culex pipiens Parva from Iran. *Saudi Journal of Biological Sciences, 15*(1), 185-188.

Yoshioka, M., Matsuo, T., Link, T. A., & Suzuki, M. (2000). Effect of capsaicin on abdominal fat and serum free fatty acids in exercise_trained rats. *Nutrition Research, 20,* 1041-1045. http://dx.doi.org/10.1016/S0271-5317(00)00180-9

Green Roof Performance Towards Good Habitat for Butterflies in the Compact City

Lee-Hsueh Lee[1] & Jun-Cheng Lin[1]

[1] Department of Landscape Architecture, Chung Hua University, Hsinchu, Taiwan

Correspondence: Lee-Hsueh Lee, Department of Landscape Architecture, Chung Hua University, 707, Sec.2, WuFu Rd., Hsinchu, Taiwan, R.O.C. E-mail: lslee@chu.edu.tw

Abstract

Urban ecology is threatened by habitat loss and fragmentation due to increasing urbanization. Green roofs may act as habitats to compensate for loss of green space at the ground level. Here, we assessed greening variables of 11 green roofs for butterflies in Taipei City. Butterfly number, species, and richness on green roofs were lower than parks, but some less common species were observed on green roofs. The nectar plant area, number of nectar plant species and age of green roof were the main positive effectors of butterfly number. The height above ground of green roof had not impact on butterfly survival; supposed high rise buildings are spreading out over city that would be a potential good habitat for butterflies on skyscape. However, since the scale of green roofs was small in Taipei, we adopted a habitat suitability index (HSI) method to determine optimal value of selected greening variables to attract more butterfly number on green roofs. HSI curves' findings suggested that achieving a nectar plant area of more than 25 m^2 and not less than 10 nectar plant species would greatly benefit butterfly number; meanwhile, the age of green roof was higher than 38 months, butterfly number is expected to increase rapidly. We confirmed that carefully design green roof could play a good habitat for butterflies in Taipei city.

Keywords: habitat suitability index curve, green roofs, nectar plants, urbanization, greening approach

1. Introduction

The increase of human populations in city centers and urban sprawl has eliminated or severely restricted green space and threatens habitats. Since large amounts of land area are covered by various kinds of buildings makes the urban green space area at ground level is limited, many countries have begun to consider the green roof as an important habitat for wildlife in the city (Oberndorfer et al., 2007). Green roofs in urban centers provide other benefits such as mitigation of urban heat islands, reduced energy use, reduced air pollution, enhanced storm water management, and the creation of natural habitats for animals and plants (Carter & Fowler, 2008; Clark, Adriaens, & Talbot, 2008; USGSA, 2013).

Green roofs in large cities have high potential as habitat for species negatively impacted by land-use changes (Brenneisen, 2006). Although green roofs support a lower abundance of invertebrate than nearby urban green space (Brenneisen, 2006; Gedge & Kadas, 2005; Kadas, 2006; Snep, WallisDeVries, & Opdam, 2011), green roofs can be designed to mitigate habitat loss for insects and wildlife in urban areas (Colla, Willis,, & Packer, 2009; Ksiazek, Fant, & Skogen, 2012; MacIvor & Lundholm 2011; Tonietto, Fant, Ascher, Ellis, & Larkin, 2011).

Butterflies are highly appreciated by the public because of their beautiful appearance and daytime activities, which can be easily observed. The survival of butterflies is affected by various factors of urbanization (Dennis & Hardy 2001; Wood & Pullin, 2002). Urban development, habitat loss, and fragmentation have a negative impact on butterfly species distributions, species richness, and Shannon diversity (Blair & Launer, 1997; Di Mauro, Dietz, & Rockwood, 2007). Blair & Launer (1997) asserted that some butterfly species can survive in urban areas that have suitable habitats. And butterfly diversity in urban areas can be preserved by constructing suitable habitats in these areas (Chikamatsu, Natuhara, Mizutani, & Nakamura,2002; Collier, Mackay, Benkendorff, Austin, & Carthew, 2006; Smallidge & Leopold, 1997; Wood & Pullin, 2002).

We was sensible to deliberately determined six greening variables had effect on butterflies survival on green roof, which include total greening area, nectar plant area, number of nectar plant species, age of green roof, height above ground and surrounding green space ratio. Insects may be more attracted to suitable green roof vegetation due to the provision of appropriate food sources or pollen resources (Coffman & Waite, 2010; Everaars, Strohbach, Gruber, & Dormann, 2011; Gedge &d Kadas, 2005; Lundholm 2006), meanwhile nectar plants are the major plants that attract butterflies in metropolitan areas (Sharp, Parks, & Ehrlich, 1974). Insect survival on green roof was influenced by vegetal structures (Coffman & Waite, 2010; Madre, Vergnes, Machon, & Clergeau, 2013), which can offer suitable habitats for the foraging or nesting of a variety of insects (Colla et al., 2009; Everaars et al., 2011; Ksiazek et al., 2012).

The number of arthropod species was positively correlated with living vegetation cover; but was not significantly correlated with the distance of green roof from the ground (Schindler, Griffith, & Jones, 2011), that may higher sun exposure between buildings provide a good environment for insects (Everaars et al., 2011). Thus, carefully designed green roofs have the potential to be used to support more species in urban area (Bates, Sadler, & Mackay, 2013; Molles, 2005; Smallidge & Leopold 1997).

The presence of wild species was influenced by geographic location (Coffman & Waite, 2011). Kadlec, Benes, Jarosik, and Konvicka (2008) asserted that fragmentary habitats could support small butterfly survival. Butterflies move from patch to patch, their dispersal range is typically no more than 2 km (Maes & Bonte, 2006), and 1 km has been described as an appropriate "stepping stone" distance between butterfly habitats, green roofs can act as stepping stones and refuges for butterflies and insects in urban (Dramstad, Olson, & Forman, 1996; Snep et al., 2011).

Taipei city is the capital of Taiwan, has a population over three million people within a total area of 272 km^2, with population density nearly10, 000 people/km^2 and is the seventh most populous city in the world. The population density is even more than 27, 000 people/km^2 in city center. Parks and green space areas are less than 5% of total area (Taipei City Government, 2014); it would be difficult to increase green space on the ground level. In Taipei City center area, there were more than 110 types of butterflies identified in 1936, 69 from 1960 to 1970, and 50 in 2008 (Zhou, 2009); however, many butterflies have disappeared due to habitat loss from urbanization (Hiura, 1973; Jiang, 2012; Zhou, 2009). Therefore, green roofs may mitigate habitat loss in this urban area (Colla et al, 2007), green roofs represent bioengineering (Matteson & Langellotto, 2011) and carefully design green roof can offer suitable habitat for butterflies.

In our study, we used an interdisciplinary approach that combined several methods for answer the following questions: (1) we would compare green roofs with parks to verify that green roof is a potential habitat for butterflies in Taipei city, (2) tested for the critical greening variables of green roof affect butterfly survival, (3) determine the optimal value of selected greening variables could attracted more butterfly number on green roof, (4) made the suggestion of construct sufficient habitats on green roof and planning green roof the location for butterflies in compact city.

2. Materials and Methods

2.1 Study Sites and Greening Variables Survey

We selected 11 public buildings were located in Taipei city, Taiwan (Figure 1), and carry out the greening variables survey from August 2011 to May 2012; which were set up extensive green roofs that were under maintenance, with planting mediums at depths below 20 cm.

For each green roof, we measured 7 greening variables, i.e. site area of roof, greening area, nectar plant area, number of nectar plant species, height above ground, age of green roof, and surrounding green space ratio.

Greening area was figured on total greening area of roof. Nectar plant was defined the plant attract butterflies; height above ground was estimated the height from ground level to green roof location; age of green roof was calculated the period from green roof was established to investigation timing. Finally, we calculated the surrounding green space ratio that was the proportion of all area of parks and green space to all land area within a 1000-m radius of green roof, the park and green space was according to land-use regulations of Taipei urban planning, and verified by 2011 aerial photograph.

We surveyed these greening variables on each site twice per month from August 2011 to May 2012, total 20 times. All 11 buildings were constructed more than 20 years ago; roof area was from 142 to 610 m^2, the entire green roof area ranged from 95 to 590 m^2. The height above ground of green roof was form 7 to 34 m. The ages of the green roofs ranged from 13 to 46 months old from green roof was established to August 2011. The surrounding green space ratio was from 3.84% to 50.68%.

The plants and nectar plants were identified by the Catalogue of Life in Taiwan (TaiBNET, 2011), the Taiwan Biodiversity Information Facility (TAIBIF, 2011), the Taiwan Flowers and Plants Illustrated Handbook (Xue, 2003) and Taiwan Butterfly Host and Nectar Plant Illustrated Handbook (Lin, 2008). During the study period, we recorded 34 butterfly nectar plants species belonging to 16 families. Statistics for all nectar plants from 11 green roofs were calculated; these green roofs contained primarily Asteraceae, Brassicaceae, Rubiaceae and Verbenaceae family plants; the most five common nectar plant species were *Belamcanda chinensis* (*Belamcanda* family), *Begonia semperflorens Link. & Otto* (Begoniaceae family), *Youngia japonica (L.) DC.* (Youngia family), *Oxalis corymbosa DC.* (Oxalis family), and *Bidens pilosa L. var. radiata* (Bidens family).

ST1	ChengDe Junior High School 25°01'32.68"N 121°34'7.41"E	ST2	Xinyi District office 25°01'59.71"N 121°34'1.03"E
ST3	Xinyi Junior High School 25°01'43.56"N 121°34'3.95"E	ST4	Department of Civil Servant Development Building 25°0'11.89"N 121°33'39.52" E
ST5	Wenshan School of Special Education 24°59'29.53"N 121°34'4.12"E	ST6	North Site ChengGhi University 24°59'11.28"N 121°34'24.92"E
ST7	South Site, ChengGhi University 24°59'11.28"N 121°34'24.92"E	ST8	Tsai Lecture Hall, Taiwan University 25°01'14.85"N 121°32'37.12"E
ST9	LinSen community centre 25°02'54.52"N 121°31'29.52"E	ST10	BaiLing Elementary School 25°05'7.38"N 121°31'13.09"E
ST11	ChungShih community centre 25°05'05.45"N 121°32'16.95"E		

Daan Park 25°1'46"N 121°32'6"E.   Biha Park 25°4'56"N 121°35'0"E

Figure 1. The map shows the research sites location, included 11 green roofs and two parks in Taipei city

Table 1. Description of 11 green roofs and parks

Green roof	roof area[a] (m^2)	greening area[b] (m^2)	height above ground[c] (m)	nectar plant area[d] (m^2)	number of nectar plant[e] species	age of green roof[f] (months)	surrounding green space ratio[g] (%)
ST1	180	124	3	15.20	4	21	18.30
ST2	600	410	11	29.73	16	46	8.92
ST3	637	590	3	13.18	3	41	20.33
ST4	234	234	2	5.20	8	29	50.68
ST5	188	93	6	1.90	1	17	30.19
S56	160	105	6	16.56	6	18	40.53
ST7	152	127	7	10.70	8	18	40.53
ST8	610	330	3	8.30	4	13	7.05
ST9	142	120	4	7.75	5	18	3.84
ST10	214	135	3	8.90	6	30	5.67
ST11	250	192	3	41.10	13	45	45.80

Park	area (ha)	greening area[b] (m^2)	height above ground[c] (m)	nectar plant area[d] (m^2)	number of nectar plant[e] species	age of park (to 2011) (year)	surrounding green space ratio[g] (%)
Daan Park	26	-	-	-	20	18	-
Bihu Park	13	-	-	-	45	25	-

(a) the whole roof area.

(b) total greening area on roof.

(c) estimated the height from ground level to green roof location.

(d) total area of butterfly-attractive plants.

(e) total butterfly-attractive plant species.

(f) total months form green roof was establish to August, 2011.

(g) the proportion of all area of parks and green space to all land area within a 1000-m radius of green roof.

2.2 Butterfly Sampling

Due to the area of 11 green roofs were small, butterfly sampling was performed by the main method of observation, and the net-capture method was subordinate. Each site was surveyed from 9:00 to 11:00 and 14:00 to 16:00 at each survey time. Survey period was from August 2011 to May 2012, and twice per month, total 20 times.

We recorded 514 butterflies belonging to 5 families (12 species) during the investigation period on 11 green roofs. The most frequent species were *Zizeeria maha okinawana* (Zizeeria family; Matsumura, 1929; 69.8% at 11 sites), followed by *Pieris canidia* (Pieris family; Linnaeus, 1768 ; 21.8% at 10 sites). The remaining 10 species, observed at lower frequencies, were *Abraximorpha davidii ermasis* (Abraximorpha family; ermasis Fruhstorfer, 1914), *Catochrysops panormus exiguu*s (Catochrysops family;Distant, 1886), *Danaus genutia* (Danaus family; Cramer, 1779), *Eurema blanda arsakia* (Eurema family; Fruhstorfer, 1910), *Graphium sarpedon connecten* (Graphium family; Fruhstorfer, 1906), *Hebomoia glaucippe formosana* (Hebomoia family; formosana Fruhstorfer, 1908), *Jamides bochus formosanus* (Jamides family; formosana Fruhstorfer, 1908), *Tirumala limniace* (Tirumala family; Fruhstorfer, 1910), *Vanessa indica* (Vanessa family; Herbst, 1794), and *Ypthima multistriata* (Ypthima family; Butler, 1883).

2.3 Data analysis methods

We used correlation analysis to test the greening variables (i.e., site area of roof, greening area, nectar plant area, number of nectar plant species, height above ground, age of green roof, and surrounding green space ratio) with the distribution of butterfly number, and the results would used to determine the significant impact of the greening variables on butterfly number.

HSI is based on hypothesized species-habitat relationships rather than statements of proven cause and effect relationships, which was employed to develop optimal habitat environment level of a certain habitat factor for that given species (Fish and wildlife Service, 1981; Schamberger & O'neil, 1986). HSI curve was plotted by the x-axis is a habitat variable and the y-axis is a suitability score with 1 being optimum habitat and 0 being least suitable habitat. Below that hypothetical plots for given species where the curves describe suitability of habitat variables for that species existence. We utilized the normalized butterfly number as vertical axis and the individual selected greening variable as the horizontal axis, and then drew the envelope, that was the HSI for butterfly and greening variable. Then we could use the HSI curve to determine the optimal level of greening variables for butterflies on green roof.

3. Results

3.1 Compared of Butterflies of Green Roofs and Park

To show the green roof was a potential habitat for butterflies, we processed to compare butterflies species, number, species richness and Shannon-Wiener diversity index of 11 green roofs and Zhou (2009) and Jiang (2012) research finding.

Zhou (2009) research finding of the change of butterfly fauna at Daan Park, which location in Taipei city center, with area is 26 ha. 50 species belong to 7 families total 3141 butterflies were recorded from July 2008 to June 2009. The most frequent species was *Pieris spp.* (41.1%; Pieridae family) and *Zizeeria maha okinawaana* (23.5%; Zizeeria family; Matsumura, 1929).

Jiang (2012) research finding of butterfly fauna at Bihu Park, which located in the end of WuZhi mountain chain in Taipei city suburban; and with an 8 ha lake, total area is 13 ha. 109 species belong to 5 families total 8,882 butterflies were recorded from March 2011 to February 2012. The most frequent species was *Pieris spp.* (20.5%; Pieridae family), *Jamides spp.*(9.9%; Lycaenidae family), and *Zizeeria maha okinawaana* (8.5%; Zizeeria family; Matsumura, 1929).

Both Zizeeria maha okinawana and *Pieris canidia* were dominant species at green roofs and parks. We also observed these butterflies species were nectar-based on green roofs, that was similar to Zhou's findings. There were fewer butterflies visiting green roofs than park (p=0.000). Butterflies species richness was significant different between green roof and park (p=0.019); but there was not a significant different between Shannon-Wiener diversity index at green roof and park.

3.2 Dominant Influence of Greening Variables on Butterflies

We performed correlation analysis using the 7 greening variables versus butterfly number to assess the relationships between greening variables and butterfly. The butterfly number was significantly correlated with the nectar plant area (r=0.825, p=0.003), number of nectar plant species (r=0.804,p=0.002), and age of green roof (r=0.895,p=0.000); all of these were positive correlations. No significant correlation was observed between butterfly number and site area of roof, greening area, height above ground, and surrounding green space ratio. Meanwhile, due to the butterflies species was very low, which was negligible in the step of correlation analysis.

3.3 Habitat Suitability Index Curve of Greening Variables for Butterflies

Our research main objectives was to determine the optimal value of greening variables, which was estimated by HSI method, and the results could be used to created a suitable habitat attract more butterflies.

When establishing HSI curve, the properties of greening variables should be determined, first. The correlation analysis results showed that nectar plant area, number of nectar plant species, and age of green roof were significant correlation with butterfly number, which were further described using the HSI method. Based on the HSI concept, we utilized the value of normalized butterfly number on the vertical axis and the values of individual greening variable on the horizontal axis; we then established an envelope based on the scatter plots of normalized butterfly number and of a particular greening variable, i.e., HSI curve for butterfly number and nectar plant area, number of nectar plant species and age of green roof, as shown in Figures 2–4. As shown in Figure 2, the nectar plant area rose with increasing butterfly number. When the nectar plant area was about 25 m^2, there were many butterflies; after the increased trend become slowly. Therefore, we determined the optimal value of nectar plant area for butterflies on green roofs was not less than 25 m². When the number of nectar plant species was about 10, there were many butterflies, the optimal value of number of nectar plant species for butterflies on green roofs was not less than 10 (Figure 3). Butterfly number continue to show an upward tendency from the green roof was established to nearly 38 months, after the trend become stability; and at this time, there were many butterflies. We believed that attracted more butterflies on green roof, which optimal value of age of green roof was not less than 38 months (Figure 4).

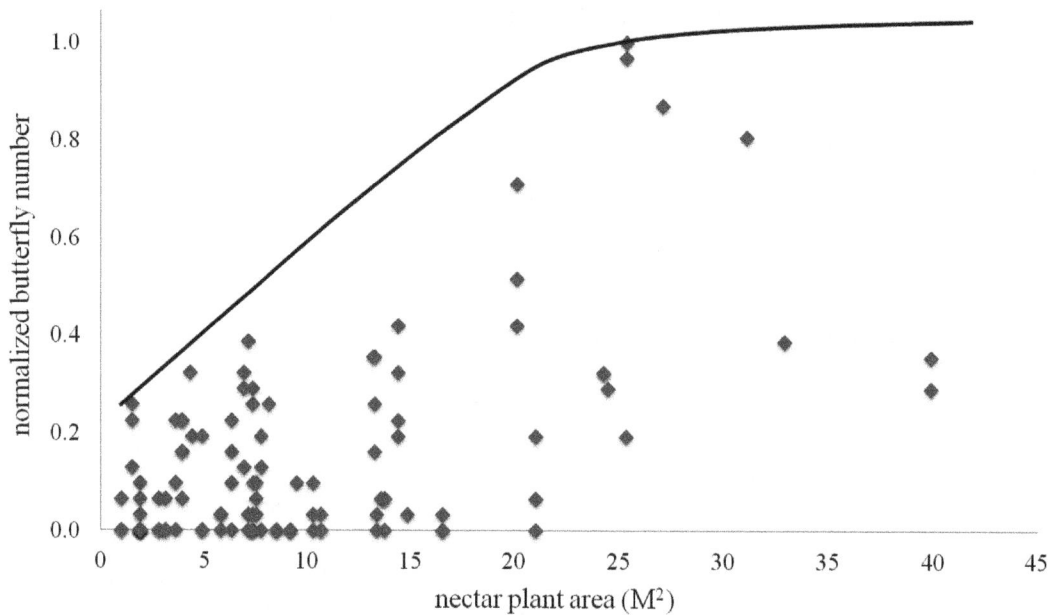

Figure 2. HSI curve for normalized butterfly number versus the nectar plant area. The curve used to determine the effect of nectar plant area on butterfly number. Nectar plant area was below 25m², butterfly number has a fast growth trend; once nectar plant area was over 25m², and the growth trend of butterfly number became stability

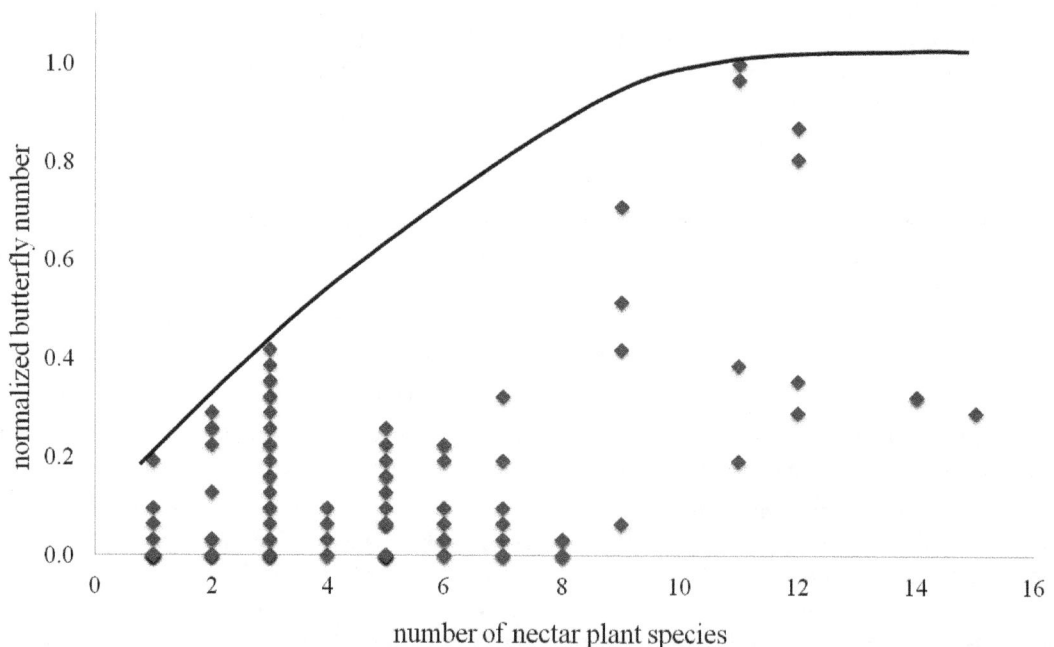

Figure 3. HSI curve for normalized butterfly number versus number of nectar plant species. Effect of the number of nectar plant species on butterfly number was determined by the curve. When the number of nectar plant species was less than 10, growth in butterfly number was increasing; after the growth trend of butterfly number was slower

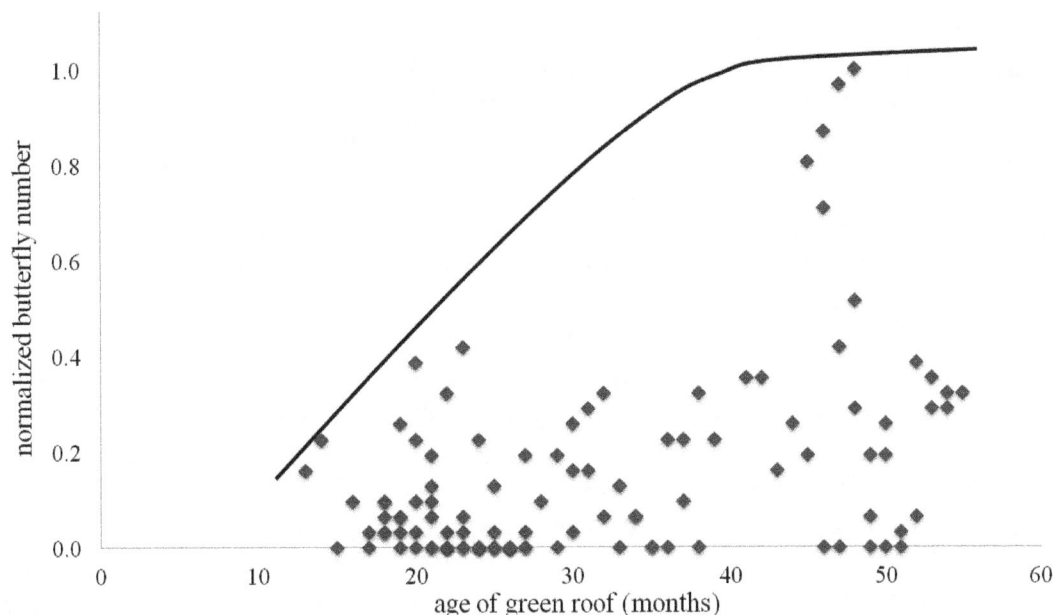

Figure 4. HSI curve for normalized butterfly number versus age of green roof. That used to determine the effect of green roof age on butterfly number. If green roof was established more than 38 months, butterfly number continued to show an upward tendency, after the growth trend became stability

4. Discussion and Conclusions

Butterflies were present on green roofs in Taipei city but were represented by fewer species and individuals than in parks. Few less common species were observed on green roof, although which support a lower abundance of bees than nearby urban park (Brenneisen, 2006; Gedge & Kadas, 2005; Kadas, 2006; Snep et al., 2011).

In our research findings demonstrated that butterfly number were strongly influenced by nectar plant area and number of nectar plant species, which suggested that carefully design of green roofs would attract more butterflies (Bates et al., 2013; Coffman & Waite ,2010; Collinge, Prudic, & Oliver, 2003; Everaars et al., 2011; Feltwell, 1993; Lundholm, 2006; Molles, 2005; Saarinen, Valtonen, Jantunen, &Saarnio, 2005; Vickery, 1995; Wood & Samways, 1991). Meanwhile, butterfly number increased with the established time of green roof went on. Schrader & Böning (2006) asserted that older green roofs had more stable environments for insects.

There are no significant correlations between butterfly number and height above ground of green roof (Snep et al., 2011). Sunlight is one of the major factors limiting butterfly diversity on the rooftops of heavily urbanized areas (Matteson & Langellotto, 2010). That implied high-rise buildings are prevalent in large cities and can become good habitats for butterflies. Butterfly number and surrounding green space ratio of green roof was also no significant correlations. Madre et al. (2013) found the surrounding environment exhibited minor influences on arthropods survival. Moreover, the green space was very limited on ground level in Taipei city and the surrounding green space ratio of selected green roofs in our research was less than the research of Snep et al. (2011); that may be possible to cause no significant correlations between butterfly number and surrounding green space ratio.

In Taipei City, there are over 230,000 buildings, the average building area is less than 800 m^2 (Taipei City Government, 2013), that could offer greening area was small. Green roofs are thought to function as novel habitat islands in urban settings, providing resources for butterflies (Fernandez-Canero & Gonzalez-Redondo, 2010; Tonietto et al., 2011). We are sure the carefully designed green roofs had the better support potential for insect survival (Bates et al., 2013; Coffman & Waite,2010; Davies, Simcock, & Toft, 2010; Everaars et al., 2011; Gedge & Kadas, 2005; Lundholm, 2006; Tonietto et al., 2011); the establishment of green roofs is critical to reducing habitat loss. Due to the habitat suitability index (HSI) method was being used to determine the greening variables appropriate condition for butterfly survival; we adopted HSI method to determine the optimal value of selected greening variables and confirm that with a small scale green roof, one can create a good habitat for butterflies in urban area. Our suggestions, for attract more butterflies, we must plant nectar plant area more than

$25m^2$ and at least 11 species of nectar plant species on green roof; which greening approach was suitable for the compact city of Taipei.

The dominant species appeared by an overwhelming majority on green roofs, but a few of less common butterfly species were observed. Suitable planting of green roofs may support more species (Bates et al., 2013; Everaars et al., 2011), and the diversity of blooming plants could increase insects diversity (Tonietto et al., 2011). Due to few butterfly species were observed, our research findings were useful to attract more butterfly number on green roofs. Clearly there is some extra restriction of our research. Such is the influence of some variables that it can promote number of butterfly species on green roof that must be explored. In the face of climate change impacts on urban areas, our research approach and methodology could easily be applied and modified to test the ecological benefit of green roof.

References

Bates, A. J., Sadler, J. P., & Mackay, R. (2013). Vegetation development over four years on two green roofs in the UK. *Urban Forestry & Urban Greening, 12*(1), 98-108. http://dx.doi.org/10.1016/j.ufug.2012.12.003

Blair, R. B., & Launer, A. E. (1997). Butterfly diversity and human land use: Species assemblages along an urban grandient. *Biological Conservation, 80*(1), 113-125. http://dx.doi.org/10.1016/S0006-3207(96)00056-0

Brenneisen, S. (2006). Space for urban wildlife: designing green roofs as habitats in Switzerland. *Urban Habitats, 4*(1), 27-36.

Carter, T., & Fowler, L. (2008). Establishing green roof infrastructure through environmental policy instruments. *Environmental Management, 42*(1), 151-164. http://dx.doi.org/10.1007/s00267-008-9095-5

Chikamatsu, M., Natuhara, Y., Mizutani, Y., & Nakamura, A. (2002). Effect of artificial gaps on the butterfly assemblage in urban woods. *Journal of the Japanese Society of Revegetation Technology, 28*(1), 97-102. http://dx.doi.org/10.7211/jjsrt.28.97

Clark, C., Adriaens, P., & Talbot, F. B. (2008). Green roof valuation: a probabilistic economic analysis of environmental benefits. *Environ. Sci. Tech., 42*(6), 2155-2161. http://dx.doi.org/10.1021/es0706652

Coffman, R. R., & Waite, T. (2010). Vegetated roofs as reconciled habitats: Rapid assays beyond mere species counts. *Urban Habitats 6*(1). http://urbanhabitats.org/v06n01/vegetatedroofs_full.html

Colla, S. R., Willis, E., & Packer, L. (2009). Can green roofs provide habitat for urban bees (Hymenoptera: Apidae)? *Cities and the Environment, 2*(1), 4. http://digitalcommons.lmu.edu/cgi/viewcontent.cgi?article= 1017&context=cate

Collier, N., Mackay, D. A., Benkendorff, K., Austin, A. D., & Carthew, S. M. (2006). Butterfly communities in South Australian urban reserves: estimating abundance and diversity using the Pollard walk. *Austral Ecology, 31*(2), 282-290. http://dx.doi.org/10.1111/j.1442-9993.2006.01577.x

Collinge, S. K., Prudic, K. L., & Oliver, J. C. (2003). Effects of local habitat characteristics and landscape context on grassland butterfly diversity. *Conservation Biology, 17*(1), 178-187. http://dx.doi.org/10.1046/j. 1523-1739.2003.01315.x

Davies, R., Simcock, R., & Toft, R., (2010). Islands in the sky, urban biodiversity enhancement in NZ on indigenous living roof landscapes. *44th Annual Conference of the Architectural Science Association*, ANZAScA 2010, Unitec Institute of Technology.

Dennis, R. L., & Hardy, P. B. (2001). Loss rates of butterfly species with urban development. A test of atlas data and sampling artefacts at a fine scale. *Biodiversity & Conservation, 10*(11), 1831-1837. http://dx.doi.org/10. 1023/A:1013161522916

Di Mauro, D., Dietz, T., & Rockwood, L. (2007). Determining the effect of urbanization on generalist butterfly species diversity in butterfly gardens. *Urban ecosystems, 10*(4), 427-439. http://dx.doi.org/10.1007/s1125 2-007-0039-2

Dramstad, W. E., Olson, J. D., & Forman, R. T. T. (1996). *Landscape Ecology Principles in Landscape Architecture and Land-use Planning*. Harvard University Graduate School of Design.

Everaars, J., Strohbach, M. W., Gruber, B., & Dormann, C. F. (2011). Microsite conditions dominate habitat selection of the red mason bee (*Osmia bicornis*, Hymenoptera: Megachilidae) in an urban environment: A case study from Leipzig, Germany. *Landscape and Urban Planning, 103*(1), 15-23. http://dx.doi.org/10. 1016/j.landurbplan.2011.05.008

Feltwell, J. (1993). *The illustrated encyclopedia of butterflies*. Blandford a Cassell imprin, London.

Fernandez-Canero, R., & Gonzalez-Redondo, P. (2010). Green roofs as a habitat for birds: a review. *Journal of Animal and Veterinary Advances, 9*(15), 2041-2052. http://dx.doi.org/10.3923/javaa.2010.2041.2052

Fish and Wildlife Service. (1981). *Standards for the development of habitat suitability index models for use in the habitat evaluation procedures*. U.S.D.I. Fish and Wildlife Service, Division of Ecological Services. ESM 103.

Gedge, D., & Kadas, G. (2005). Green roofs and biodiversity. *Biologist, 52*(3), 161-169.

Hardy, P. B., & Dennis, R. L. (1999). The impact of urban development on butterflies within a city region. *Biodiversity & Conservation, 8*(9), 1261-1279. http://dx.doi.org/10.1023/A:1008984905413

Hiura, I. (1973). *Butterfly across the sea*. Aoki Shobo, Tokyo.

Jiang, S. L. (2012). *Studies on the butterfly fauna and ecology of Biha Park in Taipei City*. Master's thesis, University of Taipei.

Kadas, G. (2006). Rare invertebrates colonizing green roofs in London. *Urban Habitats, 4*(1), 66-86

Kadlec, T., Benes, J., Jarosik, V., & Konvicka, M. (2008). Revisiting urban refuges: changes of butterfly and burnet fauna in Prague reserves over three decades. *Landscape and Urban Planning, 85*(1), 1-11. http://dx.doi.org/10.1016/j.landurbplan.2007.07.007

Ksiazek, K., Fant, J., & Skogen, K. (2012). An assessment of pollen limitation on Chicago green roofs. *Landscape and Urban Planning, 107*(4), 401-408. http://dx.doi.org/10.1016/j.landurbplan.2012.07.008

Lin, C. J. (2008). *Taiwan Butterfly host and nectar plant illustrated handbook*. Commonwealth Publishing Co., Ltd, Taipei City, Taiwan, ROC.

Lundholm, J. T. (2006). Green roofs and facades: a habitat template approach. *Urban habitats, 4*(1). http://www.urbanhabitats.org/v04n01/habitat_full.html

MacIvor, J. S., & Lundholm, J. (2011). Insect species composition and diversity on intensive green roofs and adjacent level-ground habitats. *Urban Ecosystems, 14*(2), 225-241. http://dx.doi.org/10.1007/s11252-010-0149-0

Madre, F., Vergnes, A., Machon, N., & Clergeau, P. A. (2013). Comparison of 3 types of green roof as habitats for arthropods. *Ecological Engineering, 57,* 109-117. http://dx.doi.org/10.1016/j.ecoleng.2013.04.029

Maes, D., & Bonte, D. (2006). Using distribution patterns of five threatened invertebrates in a highly fragmented dune landscape to develop a multispecies conservation approach. *Biological Conservation, 133*(4), 490-499. http://dx.doi.org/10.1016/j.biocon.2006.08.001

Matteson, K. C., & Langellotto, G. A. (2010). Determinates of inner city butterfly and bee species richness. *Urban Ecosystems, 13*(3), 333-347. http://dx.doi.org/10.1007/s11252-010-0122-y

Molles, M. C. (2005). *Ecology: concepts and applications*. McGraw-Hill, New York

Oberndorfer, E., Lundholm, J., Bass, B., Coffman, R. R., Doshi, H., Dunnett, N., ... Rowe, B. (2007). Green roofs as urban ecosystems: ecological structures, functions, and services. *BioScience, 57*(10), 823-833. http://dx.doi.org/10.1641/B571005

Saarinen, K., Valtonen, A., Jantunen, J., & Saarnio, S. (2005). Butterflies and diurnal moths along road verges: does road type affect diversity and abundance? *Biological Conservation, 123*(3), 403-412. http://dx.doi.org/10.1016/j.biocon.2004.12.012

Schamberger, M. L., & O'Neil, L. J. (1986). Concepts and constraints of habitat-model testing. In: Vemer J, Momson ML, Ralph CJ (eds) *Wildlife 2000*. The University of Wisconsin Press, Madison, Wisconsin, pp 177-182.

Schindler, B. Y., Griffith, A. B., & Jones, K. N. (2011). Factors influencing arthropod diversity on green roofs. *Cities and the Environment, 4*(1). Retrieved from http://digitalcommons.lmu.edu/cgi/viewcontent.cgi?article=1085&context=cate.

Schrader, S., & Böning, M. (2006). Soil formation on green roofs and its contribution to urban biodiversity with emphasis on Collembolans. *Pedobiologia, 50*(4), 347-356. http://dx.doi.org/10.1016/j.pedobi.2006.06.003

Sharp, M. A., Parks, D. R., & Ehrlich, P. R. (1974). Plant resources and butterfly habitat selection. *Ecology, 55,* 870-875. http://dx.doi.org/10.2307/1934423

Smallidge, P. J., & Leopold, D. J. (1997) .Vegetation management for the maintenance and conservation of butterfly habitats in temperate human-dominated landscapes. *Landscape and Urban Planning, 38*(3), 259-280. http://dx.doi.org/10.1016/S0169-2046(97)00038-8

Snep, R. P., WallisDeVries, M. F., & Opdam, P. (2011). Conservation where people work: A role for business districts and industrial areas in enhancing endangered butterfly populations? *Landscape and Urban Planning, 103*(1), 94-101. http://dx.doi.org/10.1016/j.landurbplan.2011.07.002

TaiBNET. (2011). *Catalogue of life in Taiwan.* Biodiversity Research Center, Academia Sinica, Taiwan. Retrieved from http://taibnet.sinica.edu.tw/home.php

Taiwan Biodiversity Information Facility (TAIBIF). (2014). *The Catalogue of Life of Taiwan Biodiversity.* Academia Sinica, Taiwan, ROC. Retrieved from http://www.taibif.tw/en/catalogue_of_life/browse.

Taipei City Government. (2013, 2014). *Taipei city statistical yearbook 2012, 2013.* Taipei City Government, ROC.

Tonietto, R., Fant, J., Ascher, J., Ellis, K., & Larkin, D. (2011). A comparison of bee communities of Chicago green roofs, parks and prairies. *Landscape and Urban Planning, 103*(1), 102-108. http://dx.doi.org/10.1016/j.landurbplan.2011.07.004

US General Services Administration (USGSA). (2014). *The benefits and challenges of green roofs on public and commercial buildings: A report of the United States General Services Administration.* ARUP Study. Washington: General Services Administration; 2013. Retrieved from http://www.gsa.gov/portal/getMedia Data?mediaId=158783

Vickery, M. L. (1995). Gardens: the neglected habitat. In: Pullin A (ed) *Ecology and Conservation of Butterflies.* Springer Netherlands, pp 123-134. http://dx.doi.org/10.1007/978-94-011-1282-6_9

Wood, B. C., & Pullin, A. S. (2002). Persistence of species in a fragmented urban landscape: the importance of dispersal ability and habitat availability for grassland butterflies. *Biodiversity & Conservation, 11*(8), 1451-1468. http://dx.doi.org/10.1023/A:1016223907962

Wood, P. A., & Samways, M. J. (1991). Landscape element pattern and continuity of butterfly flight paths in an ecologically landscaped botanic garden, Natal, South Africa. *Biological Conservation, 58*(2), 149-166. http://dx.doi.org/10.1016/0006-3207(91)90117-R

Xue, C. X. (2003). *Taiwan flowers and plants illustrated handbook.* Taiwan: Pu-Lu Publishing Co., Ltd.

Zhou, M. Y. (2009). *Studies on the chane of butterfly fauna at Daan Park in Taipei City.* Ms thesis, University of Taipei. Master's thesis, University of Taipei.

18

Microbiological Assay of Folic Acid Content in Some Selected Bangladeshi Food Stuffs

Tanjina Rahman[1], Mohammed Mehadi Hassan Chowdhury[2], Md. Tazul Islam[1] & M. Akhtaruzzaman[3]

[1] Department of Food Technology and Nutrition Science, Noakhali Science and Technology University, Noakhali 3814, Bangladesh

[2] Department of Microbiology, Noakhali Science and Technology University, Noakhali 3814, Bangladesh

[3] Institute of Nutrition and Food Science, University of Dhaka, Dhaka-1000, Bangladesh

Correspondence: Mohammed Mehadi Hassan Chowdhury, Department of Microbiology, Noakhali Science and Technology University, Noakhali 3814, Bangladesh. E-mail: md.mehadihassanchy@yahoo.com

Abstract

Folic acid concentration in six common food samples of Bangladesh including lentil (*Lens culinaris Medik*) and bengal gram(*Cicer arietinum*), representatives of legumes; spinach (*Spinacia oleracea)* and basil (*Ocimum basilicum*), representatives of green leafy vegetables; milk, representative of a common animal source food and topa boro rice (*Oryza sativa*), a representative of cereal-grains were measured in the present study by microbiological assay using a new trienzyme folic acid extraction method. In this experiment, we estimated that, lentil, bengal gram, spinach, basil, milk and topa boro rice contained about 63 µg, 48 µg, 195 µg, 131 µg,10 µg and 42 µg of folic acid per 100 g of food sample respectively. For the assay of this water-soluble vitamin, *Lactobacillus casei* (ATCC 7469) was employed as test organism. Trienzyme treatment was performed to release bound folic acid using protease, and α-amylase with chicken pancreases as the conjugase. The highest folic acid content was recorded for spinach *(Spinacia oleracea)*, the well-known richest source of folic acid, followed by basil (*Ocimum basilicum*). The lowest value was recorded for milk. Topa boro rice, a kind of parboiled rice was also found to be a fair source of folic acid. For these six types of Bangladeshi foods studied, the content of folic acid ranged from 10 to 195 µg/ 100 g, indicating a very wide range of folic acid content in foods.

Keywords: folic acid, *lactobacillus casei*, trienzyme, chicken pancrease

1. Introduction

Folic acid is an important member of the water-soluble B-group vitamins. Folic acid (PGA) is an orange-yellow crystal or powder with a molecular weight of 441.4; Hydrate, 477.4. At pH 7.0, absorption spectrum, λ_{max} is 282-350 nm and at pH 13.0, λ_{max} is 256-365 nm (Rex et al., 1986). It has limited solubility in water, but is soluble in acidic and alkaline solutions and insoluble in organic solvents (Dick et al., 1948). Folic acid is more stable in alkaline than in acidic conditions, thus standards for folic acid derivatives are prepared in basic solution (Keagy, 1986). It is involved in the formation of new cells, the metabolism of ribonucleic acids (RNA) and deoxyribonucleic acids (DNA), essential for protein synthesis, formation of blood and transmission of genetic code. It is essential during pregnancy to reduce the risk of neural tube defects (birth defects affecting the brain and/or spinal cord) essential for the normal growth and development of the fetus. Folates, as various forms of tetrahydrofolate (THF), are substrates and coenzymes in the acquisition, transport, and enzymatic processing of one-carbon units for amino acid and nucleic acid metabolism and metabolic regulation (Cook et al., 2001). There is strong evidence for a relationship among inadequate folic acid status, elevated homocysteine concentration, and risk of coronary heart disease, venous thrombosis, carotid artery stenosis, and other forms of vascular disease (Rimm et al., 1998; Robinson et al., 1998). Mothers with inadequate folic acid status are at increased risk for having children with neural tube defects or other forms of birth defects (Botto et al., 1999). The risk of certain forms of cancer (e.g., colon, cervical, and breast) also increases when folic acid intake is inadequate (Kim, 1999; Choi & Manson, 2000). Folic acid deficiency can also contribute to depression (Fava et al., 1997; Green & Miller, 1997), impaired immune response, and neural and neurological damage (Houseton et al., 1988; Snowdon et al., 2000). The method of food folic acid measurement has improved over the years. As food, folic acid tables are notoriously unreliable for estimating accurate dietary intake, to improve the method of food folic acid analysis, the tri-enzyme extraction method was developed about 15 years ago by researchers in Georgia, United States. For decades, the traditional food folic acid extraction method involved two steps including heat treatment, to release folic acid from its binding proteins, and folate conjugase treatment, to hydrolyze polyglutamyl folate to monoglutamyl folate. However, a trienzyme extraction method of food folic acid was developed in the mid 1990s. This method involves the use of α-amylase, protease and folic acid conjugase and allows for a more extraction of folic acid trapped in carbohydrate or protein matrices in food than the traditional method. In the last several years, this extraction method became widely used. In Bangladesh, it is the first attempt to determine food folic acid content by microbiological assay with the help of tri-enzume extraction method. The method is based on the observation that certain organisms require specific vitamins for growth, using the basal medium containing all nutrients except that to be assayed. Growth responses of the organism are then compared quantitatively with standard of known concentration.

The microbiological assay of folic acid started with the findings of Stokstad (Stokstad, 1943) that growth of lactic acid bacteria such as *Lactobacillus rhamnosus* and *Streptococcus lactis* is influenced by liver and yeast extracts which are rich in folic acid (Bird et al., 1945). The most common detection method is a microbiological assay relying on the turbidimetric bacterial growth of *Lactobacillus casei* (ATCC 7469) (Hawkes et al., 1989). If anyone of the required growth factors is completely left out of the medium and then an unknown amount added, the final level of cell growth is directly related to the amount of the growth factor present in the medium. Bird et al reported that among the different methods, microbiological assay is the most reliable and accurate even at the fractions levels (Bird et al., 1969). Today the microbiological method is the only food folic acid method given Official Status by AOAC (AOAC, 2006) and AACC (AACC, 2000).

The biologically active compounds derived from folic acid and present in biological materials are folic acid, formyl-FAH$_4$(tetra-hydrofolic acid), 5-formyl-FAH$_4$, 5,10-methenyl-FAH$_4$, 5,10-methylene-FAH$_4$, 5-formyl-FAH$_4$. All these compounds promote the growth of *Lactobacillus casei* but not *Streptococcus fecalis*. For this reason, *Lactobacillus casei* is the most suitable organism for determining the free folate activity of biological materials (Swaminathan, 1985) because it responds to the widest variety of folate derivatives (Krumdeick et al., 1983).

In the 1980s, a number of researchers reported that treatment with folic acid conjugase alone is usually not effective to liberate food-bound folate. The use of additional enzymes, proteolytic or amylolytic, was shown to liberate folic acid from the foods, thereby maximizing the folic acid activity in certain foods (Cerna & Kas, 1983). Later DE.Souza and colleagues reported a method of folic acid extraction where, in addition to the traditional treatment with folic acid conjugase, protease and α-amylase were also used (De Souza & Eitenmiller, 1990). The extraction method was named as the 'tri-enzyme treatment'. Although the order of enzyme addition was found to differ with investigators, the more common order appeared to be protease, α-amylase and finally conjugase (Tamura et al., 1997). It is now realized that conditions of the enzyme treatment might be different for each type of food and therefore, the identification of the optimum pH and a suitable incubation time for each food must be done prior to folic acid analysis (Aiso &Tamura, 1998). It is our hope that, the wide use of an appropriate procedure of the tri-enzyme extraction method, in combination with a reasonable detection method, help in establishing accurate and reliable food- folic acid tables, preparing a complete Bangladeshi Food Composition Table and this, in turn, makes it possible to accurately assess folic acid intake in the general population of Bangladesh. Then adequate folic acid nutritional status will be ensured among the mass people through consumption of folic acid -rich food items that will contribute for maintaining good health of people, especially for the adolescent girls and the pregnant and lactating women.

To determine folic acid content of some selected Bangladeshi Food stuffs employing the organism *Lactobacillus casei subsp. rhamnosus* (ATCC no. 7469) is used to determine the amount of folic acid present in these foods.

2. Materials and Methods

2.1 Collection of Sample

The present investigation was conducted on six types of food samples for the determination of folic acid content together with nutritive value. Here we used the following samples including lentil and bengal gram, representatives of legumes; spinach and basil, representatives of green leafy vegetables; topa boro rice, representatives of cereal grains and milk, a representative of animal source food from six different parts of Bangladesh. These six foods were mixed into three pairs of food. Each pair was mixed completely to make a composite mixture. Then folic acid concentration of the samples were estimated from each composite mixture using the method of Tamura et al. (1997).

2.2 Sample Preparation

2.2.1 Sample Homogenization

The collected samples were thoroughly washed for 3-4 times with distilled water to remove excess dirts, dust particles, unnecessary microbes and other extraneous materials (except milk). Then the samples, namely spinach and basil cut into small pieces and mixed homogenizely. For the samples, namely lentils, bengal gram, boro rice and milk, the above steps were not needed. Next a certain portion of each sample was put in mortar-pastel and was ground vigorously to make fine paste. The procedures of food folic acid assay generally begin with homogenization of a single food or food mixtures in a buffer with an appropriate pH, containing ascorbic acid or 2-mercaptoethanol. Here we used the working buffer with a pH of 6.1 containing ascorbic acid. About 100 ml of this buffer was added with each 1-2 g of sample and the solution was shaken properly in an earlenmeyer flask to make a homogenized sample solution.

2.2.2 Sample Extraction

2.2.2.1 Preparation of Stock Buffer:

For the convenient of this assay, we formerly prepared two types of stock buffer and stored them in a refrigerator for later use. Buffer A was prepared by dissolving 31.199 g of NaH_2PO_4. $2H_2O$ in 1000ml of de-ionized water and Buffer B was prepared by dissolving 71.59 g of Na_2HPO_4. $12H_2O$ in another 1000ml of de-ionized water. These buffers were then used to prepare different types of working buffers having different pH.

2.2.2.2 Preparation of Working Buffer:

It was usually prepared freshly before use by diluting 212.5 ml of Buffer A and 35.5 ml of Buffer B. About 5 g of ascorbic acid was added with it. Then the pH was adjusted up to 6.1 by using Buffer B. Finally the volume was made up to 1000 ml.

2.2.2.3 Application of Working Buffer in Sample Extraction:

About 1 to 2 grams of test sample was mixed with 100 ml of working buffer, pH 6.1. Then the solution was autoclaved at 120°C and 15lbs pressure for 10 min. Then the extracted sample was cooled.

2.2.2.4 Enzymes-Sources of Enzymes:

Protease: It was prepared from the strains of *Streptomyces griseus* and sold from Sigma Chemical Co., USA.

α-Amylase: it was prepared from *Aspergillus orygae* and also sold from Sigma Chemical co., USA.

Folate conjugase: It was prepared in laboratory by using the pancreas of the freshly slaughtered chicken.

2.2.2.5 Preparation of Enzyme Folate Conjugase

2.2.2.5.1 Preparation of Extraction Buffer, pH 7.8: 1.42 g of sodium phosphate dibasic and 1.0 g of ascorbic acid were dissolved in de-ionized water and diluted up to 100 ml. Then its pH was adjusted to 7.8 with 4N NaOH

2.2.2.5.2 Chicken Pancreases Solution (5 mg/ml): 0.5g of fresh chicken pancreases was weighed and washed with acetone. It was then chopped into small pieces using a sharp knife and was ground in a mortar-pastel. 100ml of pH 7.8 Phosphate buffer was added with it. Then the solution was stirred vigorously for 10 minutes at 3000 rpm capacity centrifuging machine and squeezed through glass wool.

2.3 Enzyme Treatment

After cooling, 1 ml of protease (2 mg/ml) was added per gram of sample and incubated at 37°C, for 3 hours. Then the enzyme was inactivated in water bath at 100°C for 5 min and cooled. Secondly, 20 mg of amylase powder per gram of sample was added and incubated at 37°C for 2 hours. Then the pH of the sample was adjusted up to 7.2. Finally, the enzyme folate conjugase was added in the sample along with 0.5 ml toluene and incubated at 37°C for 16 hours. The activity of this enzyme was then stopped by autoclaving at 120°C for 5 min (or boiling in a water bath at 100°C for 15 min) (Ashok et al., 2000)

2.4 Sample Dilution and Adjustment of pH

After cooling, the solution was diluted up to 200 ml and filtered. The pH of a portion of the clear filtrate was adjusted up to pH 6.2 and diluted to concentration of about 0.25-0.3 ng folic acid/ ml.

2.5 Preparation of Inoculums

2.5.1 Microorganism: Here we used *Lactobacillus casei* ATCC 7469 for the assay method.

2.5.2 Stock Medium: Bacto-Micro Assay Culture Agar (MACA). Micro Vitamin Test Media were recommended for cultivation and maintenance of stock cultures of Lactobacilli used in microbiological assays of vitamins.

2.5.3 Culture Medium: Bacto-Micro Assay Culture Agar (MACA) for stock culture, Micro-Inoculums Broth Media for the preparation of inoculums and Folic acid Casei Media for Folic acid assay. All the three media were collected from Hi-Media Laboratories Limited, Mumbai, India.

2.5.4 Stock Culture: Stock cultures were prepared by stab inoculation in triplicates. One was used for the preparation of stock cultures while others were used for inoculums preparation for assays, followed by incubation at 35-37°C for 24-48 hours. Transfer of cultures should be made at weekly or bi-weekly intervals, culture (3 tubes) in MACA. The tubes were stored in refrigerator.

2.5.5 Preparation of Micro Vitamin Test Culture Agar for Stock culture: 52.1 gram of medium was suspended in 1000 ml of distilled water. 15 g of agar was added with it. Then it was heated if necessary to dissolve the medium completely. pH 6.7±0.2 was adjusted. Then it was dispensed and sterilized by autoclaving at 15 lbs pressure (121°C) for 15 minutes.

2.5.6 Preparation of Micro Vitamin Test Inoculums Broth for Inocula: 37.1 gram of media was suspended in 1000 ml of distilled water. Then it was heated if necessary to dissolve the medium completely. pH 6.7±0.2 was adjusted. Then it was dispensed and sterilized by autoclaving at 15 lbs pressure (121°C) for 15 minutes.

2.5.7 Inoculums: Subculture of *L casei* was used to form a stock culture for micro-inoculum broth. After 18-24 hours, the stock culture was incubated at 35-37°C, under aseptic condition. Then the cells were washed with 3×10 ml (5 ml) portions sterilized 0.9% NaCl solution (NSS). Then the last supernatant was decanted. The cells were diluted to an appropriate inoculums with NSS.

2.5.8 Assay Media: Bacto-Folic acid assay medium was used for the assay procedure. Its pH was adjusted up to 6.2 before use.

2.6 Preparation of Standard Folic acid Solution

2.6.1 Stock Standard: 25 mg of USP Folic acid was dissolved in 25% of ethyl alcohol and the pH of this solution was adjusted up to 7.0 by using 0.1N NaOH. The volume was made up to 250 ml with 25% of ethyl alcohol in a volumetric flask where the concentration of folic acid was 100 µg/ml.

2.6.2 Preparation of Phosphate Buffer, pH 7.0: 58.5 ml of Buffer A and 91.5 ml of Buffer B were mixed together and the pH of the solution was adjusted up to 7.0 by using 0.1N NaOH and 0.05N HCl. Then the volume was made up to 300 ml with de-ionized water.

2.6.3 Intermediate Standard I: 1 ml (100 µg/ml) of stock standard was diluted to 100 ml in a volumetric flask so that the concentration of Folic acid in this diluted solution had become 1000 ng/ml.

2.6.4 Intermediate Standard II: 1 ml (1000 ng/ml) of Intermediate Standard I solution was diluted to 100 ml in a volumetric flask so that the concentration of Folic acid in this diluted solution had become 10 ng/ml.

2.6.5 Preparation of Working Standard: 5 ml (10 ng/ml) of Intermediate Standard II solution was diluted to 100 ml in a volumetric flask so that the concentration of Folic acid in this diluted solution had become 0.5 ng/ml. then the pH of this Working Standard was adjusted up to 7.0 by using 0.1N NaOH.

2.7 Preparation of Control

About 100 ml of working buffer, pH 6.1 was taken to prepare 'control' for the samples we used in this assay method. After autoclaving this buffer, it was processed following all the steps which had been performed in case of the samples including enzyme treatment, enzyme inactivation, dilution and filtration. The purpose of preparing this 'control' was to eradicate the traces of folic acid that might come from the enzymes used rather than the sample itself and might be used by the test organism after inoculation and during incubation. After subtracting the amount of Folic acid, present in 'control' from the amount of Folic acid, present in sample, we may be able to get a more accurate value of folic acid content present in our selected food samples.

2.8 Folic Acid Determination Procedure From Food Samples

Food samples (approx 1.0 g) were homogenize in 100 ml buffer, pH 6.1 (with ascorbic acid) and autoclaved at 121°C and 15 lbs pressure for 10 minutes. Then samples were allowed for cooling and incubated with the enzyme, protease at 37°C for 3 hours. Inactivation of this enzyme was performed by boiling in water bath for 5 minutes and cooled and incubated with the enzyme, α-amylase at 37°C for 2 hours. The pH should be maintained at 7.2 Then it was incubated with Chicken pancreas powder and 0.5 ml toluene at 37°C for 16 hours and autoclaved at 121°C and 15 lbs pressure for 10 minutes to inactivate the action of enzymes. Cooling, dilution, filtration and adjustment of pH maintained to 6.2 to a portion of clear filtrate. Further dilution and pipetting in duplicate test tubes were done. Addition of working buffer (pH 6.1) and previously prepared media performed in the same test tubes and autoclaved at 121°C and 15 lbs pressure for 10 min for sterilization. Then Inoculation and incubation at 35-37°C for 18-24 hrs were conducted. Boiling at 100°C was done to stop the growth of bacteria. Finally measurement of bacterial growth was recorded by taking the absorbance at 630 nm using UV Spectrophotometer.

2.9 Folic Acid Determination Procedure From Standard Folic Acid Solution

25 mg of Standard Folic acid was dissolved in 25% Ethyl alcohol and pH adjustment was conducted up to 7.0 with 0.1N NaOH. Stock Standard was formulated by making the volume up to 250 ml with 25% ethyl alcohol. Stock standard was then diluted to prepare working Standard and pipetting was done in triplicate test tubes. After adding working buffer (pH 6.1) and previously prepared media in the same test tubes, theses were autoclaved at 121°C and 15 lbs pressure for 10 min for sterilization. Then Inoculation and incubation were performed at 35-37°C for 18-24 hours. Boiling at 100°C was done to stop the growth of bacteria. Finally measurement of bacterial growth was recorded by taking the absorbance at 630 nm using UV-Spectrophotometer.

3. Results

The present study has been carried out with a view to estimate the folate content of six types of selected food stuffs such as- lentil, bengal gram, spinach, basil, milk and topa boro rice. In this microbiological assay method, the organism, *L.casei* ATCC 7469 was used and it was observed that, the growth of this organism was increased with each additional amount of folic acid contained in sample.

3.1 Construction of Standard Curve for Estimation of Folate Content in Food Samples

A standard curve was generated by turbidity readings of spectrophotometer due to the growth of *L. casei* in the media containing 0.125, 0.25, 0.5, 0.75 and 1.0 ng per tube of triplicate tubes of standard solutions (Figure 1). Absorbance values were taken at 630 nm of wave length.

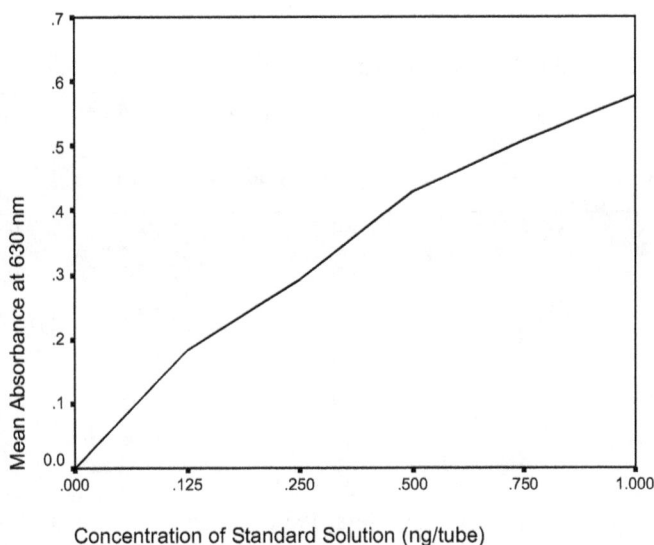

Concentration of Standard Solution (ng/tube)

Figure 1. Standard Curve of Folic acid

3.2 Estimation of Folic Acid in Food Samples

The folic acid of test food samples was determined according to AOAC method (AOAC, 1990). After trienzyme treatment, the free-folates were extracted and diluted with basal medium containing all growth nutrients except folate, and the turbidity of the *L. casei* growth response for the samples was compared quantitatively to that of known standard solutions. Absorbance of control at 630 nm was 0.313 which was essential for determining the folic acid in control after dilution. After dilution, amount of folic acid in control (D.F.=20,000) were 3μg and 3000 ng.

Absorbance for food samples at 630 nm was recorded in Table1 which was used later for determining folic acid concentration (Table 2) from standard curve. The folic acid concentration from different food stuffs after dilution obtained from table 2 expressed in μg was used to estimate actual folic acid content in food samples (Table 3).

Table 1. Absorbance for Food samples at 630 nm

Food Sample	Extract (ml)	Absorbance at 630 nm
Lentil	2.0	0.132
Bengal gram	2.0	0.149
Spinach	2.0	0.215
Basil	2.0	0.174
Milk	2.0	0.118
Topa Boro rice	2.0	0.126

Table 2. Determination of folic acid concentration in Food samples from standard curve

Food samples	Absorbance at 630 nm	Folic acid content in Food sample		After dilution	
		ng/tube	ng/ml	ng	µg
Lentil	0.132	0.092	0.046	3680	3.68
Bengal gram	0.149	0.100	0.050	4000	4.00
Spinach	0.215	0.160	0.080	6400	6.40
Basil	0.174	0.110	0.055	4400	4.40
Milk	0.118	0.080	0.040	3200	3.20
Topa Boro Rice	0.126	0.086	0.043	3440	3.44

Table 3. Determination of Actual Folic acid Content

Food samples	Folic acid, present in Food (µg)	Folic acid, present in Control (µg)	Amount of Folic acid, present only in Food (µg)	Weight of food sample (g/ml)	Actual folic acid content in Food samples	
					µg/g or ml	µg/100 g or 100 ml
Lentil	3.68		0.68	1.084	0.63	63
Bengal gram	4.00		1.00	2.054	0.48	48
Spinach	6.40	3.0	3.40	1.740	1.95	195
Basil	4.40		1.40	1.066	1.31	131
Milk	3.20		0.20	2.000	0.10	10
Topa Boro Rice	3.44		0.44	1.055	0.42	42

From the above tables, it would be seen that the study food samples, namely lentil, bengal gram, spinach, basil, milk and topa boro rice contained approx. 63 µg, 48 µg, 195 µg, 131 µg, 10 µg and 42 µg of folic acid per 100 g of Food samples respectively.These values also compared with reference values in Table 8.

Table 4. Comparison of experimental folic acid value with reference folic acid value of study food samples

Food samples	Experimental folic acid value in study food samples (µg/100g)	Reference folic acid value of study food samples (µg/100g)
Lentil	63	24
Bengal gram	48	34
Spinach	195	194
Basil	131	64
Milk	10	5.6
Topa Boro Rice	42	Not available

Source: USDA (www.USDA.gov).

4. Discussion

Folic acid assay is more challenging than many other micronutrients due to its sensitivity to physical environments and various forms in which it exists. It warrants a good knowledge of folic acid chemistry and appropriate extraction and detection techniques. The extraction techniques may differ with the type, nature, state, origin of foods as well as with the methods of detection. Besides, a careful selection of buffer type, antioxidant, heating condition, conjugate type, incubation conditions and purification methods is needed as each of these steps is likely to affect the final yeild of the vitamin.

Although microbiological assay is the most commonly used method of folic acid assay, it is time-consuming, needs great care and skill and cannot differentiate the individual folic acid. HPLC is a better analytical technique but involves a complex extraction and purification procedure. The more recent methods using the LC/MS/MS techniques offer hope with accurate quantification and better specificity of the folic acid forms. Immunoassay techniques are rapid, easier and much cheaper but less suitable for determination of folic acid in food samples.

The information available on the effect of a particular analytical technique on folic acid content of a particular food is insufficient. Therefore, it is suggested that an optimization of the extraction and detection of folic acid in each food group should be carried out before actual analysis is carried out. It appears that no folic acid analytical method is perfect. The choice of a particular method is largely determined by the purpose of analysis, e.g. food composition, nutritional intervention, regulatory purpose, and to a lesser degree by the resources available, assay time and cost, and analysis themselves.

As summarized in Table 8, the values found in the present study are markedly higher than those published in different articles previously. The reason for these higher values is likely due to the newly developed method of folic acid extraction from foods called tri-enzyme treatment, which generally provides the highest detectable value of food folic acid concentration in a certain food items among the majority of the existing methods (Tamura, 1998).

We are not certain how much of these folic acids are bio-available for absorption and metabolism in humans. There are possibilities that, the previously estimated bioavailability and requirements of food folic acid are grossly underestimated (Tamura, 1998). It is now obvious that, we must evaluate what these higher values actually means in terms of maintaining adequate folic acid nurture for overall health and prevention of diseases. It is evident that, more studies are needed in the areas of food folic acid concentrations and bioavailability as well as appropriate dietary folic acid intake recommendations.

5. Conclusion

Folic acid content of six common Bangladeshi foods was analyzed with the microbiological assay, using the tri-enzyme extraction method. From this study it indicates that the Bangladeshi people have access to foods with high folic acid levels and it is possible to meet the recommended dietary intake for folic acid of 200-300 μg/day by the population. We hope that, these new values will serve as a useful means to calculate dietary intake of folic acid in the general population. It is our hope that, the wide use of an more sophisticated and appropriate procedure of the tri-enzyme extraction method, in combination with a reasonable detection method, help in establishing more accurate and reliable food- folic acid tables and then adequate folic acid nutritional status will be ensured among the mass people through consumption of folic acid -rich food items that will contribute for maintaining good health of people, especially for the adolescent girls and the pregnant and lactating women.

Acknowledgment

This study was conducted in the Institute of Nutrition and Food Science, University of Dhaka, Dhaka-1000, Bangladesh with a partial financial support from UN FAO (United Nation Food and Agricultural Organization) at preliminary stage. Later no fund was available to further continue this research study.

References

AACC. (2000). AACC Method 86-47 Total Folate in Cereal Products - Microbiological Assay Using Trienzyme Extraction. *Approved Methods of the American Association of Cereal Chemists* (10th ed.). St Paul MN, American Association of Cereal Chemists.

Aiso, K., & Tamura, T. (1998). Trienzyme treatment for food folate analysis: optimal pH and incubation time for α-amylase and protease treatment. *Joural of Nutritional Science and Vitaminology, 44*, 361-370. http://dx.doi.org/10.3177/jnsv.44.361

AOAC. (1990b). Vitamin B6 (pyridoxine, pyridoxal, pyridoxamine) in food extracts. Microbiological method. Final action 1975. In K. Helrich (Ed.), *AOAC Official Methods of Analysis* (15th ed.). Association of Official Analytical Chemists, Inc., Arlington, VA, 961.15.

AOAC. (2006a). *AOAC Official Method 992.05 Total Folate (Pteroylglutamic Acid) in Infant Formula.* AOAC International.

Ashok, K., Shrestha, J. A., & Janet, P. (2000). Folate assay of foods by traditional and tri-enzyme treatments using cryoprotected Lactobacillus casei. *Food Chemistry, 71*, 545-552. http://dx.doi.org/10.1016/S0308-8146(00)00210-7

Bird, O. D., Bressler, B., Brown, R. A., Campbell, C. J., & Emmett, A. D. (1945). The microbiological assay of vitamin B conjugate. *J Biol Chem., 159*, 631-635.

Bird, O. D., McGlohon, V. M., & Vaitkus, J. W. (1969). A microbiological assay system for naturally occurring folates. *Can. J. Microbiol., 15*, 465–72. http://dx.doi.org/10.1139/m69-082

Botto, L. D., Moore, C. A., Khoury, M. J., & Erickson, J. D. (1999). Natural Tube Defects. *New England Journal of Medicine, 341*(20), 1509-19. http://dx.doi.org/10.1056/NEJM199911113412006

Cerna, L., & Kas, J. (1983). New conception of folacin assay in starch or glycogen containing food samples. *Nahrung, 10*, 957-964. http://dx.doi.org/10.1002/food.19830271013

Choi, S. W., & Manson, J. B. (2000). Folate and carcinogesis: an integreted scheme. *J Nutr., 130*, 129-132.

Cook, R., Carmel, R., & Jacobsen, D. W. (2001). Folate metabolism. *Homocysteine in Health and Disease* (pp. 13-134). Cambridge, UK. Cambridge University Press.

Dawson, R. M. C, Elliott, D. C, Elliott, W. H., & Jones, K. M. (1986). Data for Biochemical Research (3rd ed.). *Vitamins and coenzymes.* Oxford. Clarendon Press.

De Souza, S., & Eitenmiller, R. (1990). Effects of different enzyme treatments on extraction of total folate from various foods prior to microbiological assay and radioassay. *Journal of Micronutrient Analysis, 7*, 37-57.

Dick, M., Harrison, I. T., & Farreer, K. T. H. (1948). The thermal stability of folic acid. *Australian Journal of Experimental Biology Medicinal Science, 26*, 239-240. http://dx.doi.org/10.1038/icb.1948.25

Fava, M., Borus, J. S., Alpert, J. E., Nierenberg, A. A., Rosenbaum, J. F., & Bottiglieri, T. (1997). Folate, vitamin B_{12}, and homocysteine in major depressive disorder. *The American journal of psychiatry, 154*, 426-428. http://dx.doi.org/10.1176/ajp.154.3.426

Green, R., & Miller, J. W. (1999). Folate deficiency beyond megaloblastic anemia: hyperchomocysteinemia and other manifestations of dysfunctional folate status. *Semin. Hematol., 36*, 47-64.

Hawkes, J. G., & Villota, R. (1989). Folates in foods: Reactivity, stability during processing and nutritional implications. *Critical Reviews in Science and Nutrition, 28*(6), 439-539.

Houston, D. K., Johnson, M. A., Nozza, R. J., Gunter, E. W., Shea, K. J., Cutler, G. M., & Edmonds, J. T. (1999). Age-related hearing loss, vitamin B-12, and folate in elderly women. *The American journal of clinical nutrition, 69*(3), 564-571.

Keagy, P. M. (1986). Computerized semiautomated microbiological assay of folacin. *Journal of Association of Official Analytical Chemists, 69*, 773-777.

Kim, Y. I. (1999). Folate and cancer prevention: a new medical application of folate beyond hyperhomocysteinemia and neural tube defects. *Nutr Rev., 57*, 314-321.

Krumdeick, C. L., Tamura, T., & Ito, I. (1983). Synthesis and analysis of the pteroylpolyglutamates. *Vitamins and Hormones, 40*, 45-104. http://dx.doi.org/10.1016/S0083-6729(08)60432-X

Rimm, E. B., Willett, W. C., Hu, F. B., Sampson, L., Colditz, G. A., Manson, J. E., ... & Stampfer, M. J. (1998). Folate and vitamin B6 from diet and supplements in relation to risk of coronary heart disease among women. *Jama, 279*(5), 359-364. http://dx.doi.org/10.1001/jama.279.5.359

Robinson, K., Arheart, K., Refsum, H., Brattström, L., Boers, G., Ueland, P., ... & Graham, I. (1998). Low circulating folate and vitamin B6 concentrations risk factors for stroke, peripheral vascular disease, and coronary artery disease. *Circulation, 97*(5), 437-443. http://dx.doi.org/10.1161/01.CIR.97.5.437

Snowdon, D. A., Tully, C. L., Smith, C. D., Riley, K. P., & Markesbery, W. R. (2000). Serum folate and the severity of atrophy of the neocortex in Alzheimer disease: findings from the Nun study. *The American journal of clinical nutrition, 71*(4), 993-998.

Stokstad, E. L. R. (1943). Some properties of a growth factor for *Lactobacillus casei. J Biol Chem., 149*, 573-575.

Swaminathan, M. (1985). Essentials of Food and Nutrition (2nd ed.). Vitamins. Bangalore. The Bangalore Print and Publishing Co. Ltd.

Tamura, T. (1998). Determination of food folate. *Nutritional Biochemistry, 9*, 285-293. http://dx.doi.org/10.1016/S0955-2863(98)00013-8

Tamura, T., Mizuno, Y., Johnston, K. E., & Jacob, R. A. (1997). Food folate assay with protease, α-amylase, and folate conjugase treatments. *Journal of Agricultural and Food Chemistry, 45*(1), 135-139. http://dx.doi.org/10.1021/jf960497p

USDA Nutrient Data Laboratory. (ND). In *USDA, United States Department of Agriculture*. Retrieved from www.USDA.gov/ food-composition/usda-nutrient-data-laboratory

19

The Potential of Galangal (*Alpinia galanga* Linn.) Extract against the Pathogens that Cause White Feces Syndrome and Acute Hepatopancreatic Necrosis Disease (AHPND) in Pacific White Shrimp (*Litopenaeus vannamei*)

Tidaporn Chaweepack[1], Boonyee Muenthaisong[1], Surachart Chaweepack[1] & Kaeko Kamei[2]

[1] Chanthaburi Coastal Fisheries Research and Development Center, Department of Fisheries, Muang, Chanthaburi, 22000, Thailand

[2] Department of Biomolecular Engineering, Kyoto Institute of Technology, Matsugasaki, Sakyo-ku, Kyoto 606-8585, Japan

Correspondence: Tidaporn Chaweepack, Chanthaburi Coastal Fisheries Research and Development Center, Department of Fisheries, Muang, Chanthaburi, 22000, Thailand. E-mail: tidaporn2513@gmail.com

Abstract

White feces syndrome and acute hepatopancreatic necrosis disease (AHPND) are serious diseases that have recently been noted in Pacific white shrimp (*Litopenaeus vannamei*). Vibrio bacteria and 6 species of fungi (*Aspergillus flavus, A. ochraceus, A. japonicus, Penicillium* sp., *Fusarium* sp., and *Cladosporium cladosporioides)* were isolated from shrimp naturally infected with white feces syndrome. Antibiotics have been used to treatment the disease for many years, but these have been ineffective and have resulted in drug residue problems for the shrimp industry. In this study, an alternative method was tested for its efficacy in controlling these pathogens. The crude extract of galangal (*Alpinia galanga* Linn.), an herbal medicine, inhibited the growth of 8 vibrio species of the pathogen, *V. parahaemolyticus* (EMS/AHPND) in particular. The results also showed that 0.5 mg/ml of the galangal extract was a concentration that produced the strongest inhibition of the fungi *A. ochraceus*. Naturally infested shrimp *L. vannamei* were fed 2 and 4% (v/w) portions of the herb extract for 12 days and their progress was compared with that of a control group (no herb extract). At the end of the feeding trial, the numbers of total *Vibrio* spp. and the incidence of fungi infestation in the hepatopancreas and intestines of treated shrimp were significantly lower than that in the control group ($P<0.05$). Furthermore, the survival rates for the treatment groups, after injections with *V. parahaemolyticus* (EMS/AHPND), were significantly higher than that of the control group ($P<0.05$). Based on these results, we can report that the galangal extract has antimicrobial properties that are applicable as bio-medicinal agents against white feces syndrome and AHPND. Therefore, in the future this herb should be an alternative to chemotherapeutic agents that are being used in the shrimp industry.

Keywords: galangal, *Alpinia galanga Linn.*, *Vibrio* spp., *V. parahaemolyticus* EMS/AHPND, fungi, *Litopenaeus vannamei*

1. Introduction

Disease outbreaks caused by viruses, bacteria and protozoa are important disease agents within the shrimp aquaculture, a major industry in Southeast Asia, because they can lead to serious economic losses for long periods of time. These bacterial disease pathogens have grown rapidly due to global warming, which has stressed the shrimp population by reducing immunity and enhancing the infection rates. In 2009, the first reports of an emerging disease in *Penaeus* sp. shrimp was initially named Early Mortality Syndrome (EMS) in China. This disease was later referred to as Acute Hepatopancreatic Necrosis Disease (AHPND). In 2011, EMS/AHPND occurred frequently in Vietnam, which led to the isolation of the causative agent, *V. parahaemolyticus* (Tran et al., 2013). In 2012, AHPND had a major impact on shrimp farmers in Thailand and Southeast Asian who sustained heavy losses in production (Flegel, 2012; NACA-FAO, 2011). Outbreaks of disease reported in Thailand during 2010-2011 called White feces syndrome were found mostly in Pacific white shrimp (*L. vannamei*). Infestations of Gregarine protozoa and Vibrio bacteria have been found that caused loose shells, decreased appetite, retarded growth, and finally sporadic mortality of the shrimp in grow out ponds. Typical symptoms of the disease can be noted by white feces floating on the water surface in the rearing pond (Tangtrongpiros, 2010). Several reports have shown that vibrio bacteria including *V. alginolyticus, V. cholerae, V. parahaemolyticus, V. vulnificus, V. mimicus*, and *V. harveyi* are the important pathogens, as these are always isolated from diseased *L. vannamei* (Lighter, 1993; Vandenberghe et al., 1999; Montero & Austin, 1999). Bacterial pathogens can cause problems that range from growth retardation to sporadic mortalities and mass mortalities of shrimp. From this point of view, vibrio bacteria represent the greatest bacterial threat to the shrimp industry. When shrimp are under stressful conditions, such as when they are reared in a high-density setting, their immunity is lowered. Under those conditions, vibrio bacteria easily attack, which leads to the occurrence of disease (Ruangpan et al., 1997; Sung et al., 2001; Karunasagar et al., 2004; Lee et al., 2009). Some aquatic fungi are also opportunistic pathogens. Fusaruosis or black gill disease is caused by *Fusarium* spp., and it may affect all developmental stages of penaeid shrimp (Karunasagar et al., 2004). Shrimp farmers have few choices to control the diseases that threaten their shrimp production. Many types of drugs and chemicals have been used to treat vibrio bacteria, and this has created a residual drug problem that has led to the spread of drug-resistant bacterial strains in the shrimp industry. To solve these problems, medicinal herbs have been introduced as alternative methods for the control and treatment of bacterial diseases, because herbs are known to have anti-parasite and antimicrobial effects. Galangal (*Alpinia galanga* Linn.) is one of the typical herbs that can be used for the treatment of diseases in shrimp, and it is available year-round at a low cost. This herb is rich in phenolic compounds such as flavonoids and phenolic acids (Mayachiew & Devahastin, 2008). The essential oil of the crude galangal extract is responsible for its antimicrobial activity (de Pooter et al., 1985; Turker et al., 2002; Oonmetta-areea et al., 2006; and, Rao et al., 2010). Therefore, we tested the potential efficacy of an ethanol galangal crude extract to inhibit and reduce the number of *Vibrio* spp. bacteria and fungi in natural infestations of white feces syndrome and AHPND in Pacific white shrimp (*L. vannamei*).

2. Materials and Methods

2.1 Preparation of Galangal-Ethanol Extract

Fresh galangal rhizomes were purchased from a local market in Chanthaburi, Thailand. The rhizomes were sliced into thin layers and dried at room temperature in a tray dryer followed by heating at 45 °C for 24 h. After drying, they were ground into powder using an electric blender (Philips, Cucina, Thailand). Ten grams of the galangal powder were suspended in 100 ml of ethanol, and then let stand at room temperature overnight. After filtration through filter paper (No. 1, Whatman International Ltd., Maidstone, UK), the extract was dried in a rotary evaporator (Rotary Evaporator BUCHI R-114, Vacuum pump BUCHI B-169 Switzerland) (Oonmetta-aree et al., 2006).

2.2 Isolation and Identification of Bacteria

Pathogenic *Vibrio* species that included *V. cholerae, V. parahaemolyticus, V. parahaemolyticus* (EMS/AHPND), *V. fluvialis, V. vulnificus, V. alginolyticus, V. mimicus*, and *V. harveyi* were isolated from diseased Pacific white shrimp (*L. vannamei*) obtained from grow-out ponds in Chanthaburi Province, using the Ruangpan & Kitao (1991) method. The isolate was streaked on Thiosulfate Citrate-Bilesalt Sucrose Agar (TCBS Agar, Difco USA) for selective isolation of *Vibrio*, and cultivated at 30 °C for 18-24 h. The *Vibrio* species was identified according to a scheme developed by Colwell (1984) using API 20E kits (ATB System, BioMérieux, France) and the PCR technique for the detection of *V. parahaemolyticus* (EMS/AHPND) (Joshi et al., 2014).

2.3 Bacterial Inoculum

The isolate was streaked on TCBS agar and incubated at 30 °C for 18–24 h. A pure colony was streaked on Trypticase Soy Agar (TSA, Difco, USA), and incubated at 30 °C for 18–24 h. Several colonies of bacteria were inoculated into a 1.5% sterile sodium solution, which was mixed well with a vortex mixer. The bacterial suspension was tested using McFarland Standard No. 0.5 ($\times 10^8$cfu/ml) then diluted to $\times 10^6$ cfu/ml for antibacterial activity testing.

2.4 Disc Diffusion Test

The efficacy of herbs for the growth inhibition of *Vibrio* was done using an agar disc diffusion assay and efficacy testing method developed by Oometta-aree et al. (2006). The *Vibrio* inoculums ($\times 10^6$ cfu/ml) were spread on Mueller Hinton agar (Difco, USA) supplemented with 1.5% NaCl (MHA–1.5 % NaCl), and then kept for about 5 min to allow the surface of the agar to dry. Each herb extract (80 µl) was applied to a paper disc ($\Phi 8$ mm, Advantec, Tokyo, Japan) then both sides were dried in a sterile laminar flow. The paper disc was then placed onto an MHA–1.5% NaCl plate inoculated with *Vibrio*, followed by incubation at 30 °C for 18-24 h. The efficacy of herbs for the growth inhibition of *Vibrio* was estimated by measuring the diameter of the clear zone surrounding the herbal disc.

2.5 Minimum Inhibitory Concentration (MIC) Test

One milliliter of *Vibrio* ($\times 10^6$ cfu/ml) was inoculated into liquid Mueller Hinton broth containing 1.5% NaCl (MHB–1.5 % NaCl). Each herbal extract (0.25 g/ml) was diluted with ethanol using a 2-fold dilution method. In this method, 160 µl of each dilution was dropped onto a paper disc (Φ8 mm, Advantec, Tokyo, Japan), followed by drying in a sterile laminar flow. Then the paper disc was placed into a MHB–1.5% NaCl liquid medium and incubated at 30 °C for 18-24 h. The MIC of ethanol extract was regarded as the lowest concentration of the extract in the liquid medium that would permit no turbidity of the tested microorganism (Oometta-aree et al., 2006).

2.6 Minimum Bactericidal Concentration (MBC) test

The tubes that showed no turbidity of the bacteria and the last tubes showing turbidity from the MIC test were used for further MBC testing. Then, 0.1 ml of culture medium used in the MIC test was spread onto TCBS Agar (Difco, USA), incubated at 30 °C for 18-24 h, and the formations of colonies were counted. The MBC was the lowest concentration of herbal extract that formed less than 20 colonies, corresponding to an inhibition of the bacterial growth at 99.9%, or more (Oometta-aree et al., 2006).

2.7 Efficacy of the Galangal Extract against Pathogenic Fungi

2.7.1 Fungi Isolation and Identification

The samples of naturally infected shrimp were collected from 50 grow-out ponds. Artificial feed was simultaneously collected for use in rearing shrimp samples. Fungi were isolated from the hepatopancreas and intestines of diseased shrimp and artificial feed were cultured on Glucose – Yeast in seawater (GYS) agar then incubated at 25–27 °C for 2-5 days (Khomvilai, 2006). Several isolation samples were streaked on GYS agar until the fungus had formed a pure culture. Then the isolations were identified by the National Center for Genetic Engineering and Biotechnology, Thailand, as described by Klich (2002), Raper & Fennell (1965) and by Samson et al. (1994 & 1995).

2.7.2 Fungi growth Inhibition Test

The galangal crude extract was dissolved with sterilized seawater in 8 dilutions: 0.0005, 0.005, 0.05, 0.5, 1, 2, 4 and 8 mg/ml. Agar plates of pure fungus were cultured for normal growth. Each fungus colony was cut into a 1 cm diameter using a cork border and was immersed in each galangal extract dilution for 3 replications. The treatment group was exposed to each extract dilution for 30 minutes, then incubated at 25 °C, and the control group was immersed in sterilized seawater. After the fungi had been completely immersed, the fungi plugs were cleaned twice in sterilized seawater and placed on GYS agar followed by incubation at 25–27 °C. After incubation for 2 days, the fungi colony diameters from each plate were measured. The efficacy of the crude extract against growth inhibition was established by comparing the diameters of the colony radii of the fungal mycelia of the treatment and control groups, which was calculated as follows:

$$\text{Colony radius growth rate (\%)} = \frac{\text{(colony radius of the treated fungus)}}{\text{(colony radius of the control)}} \, x \, 100$$

The growth inhibition of fungi (%) = 100 – Colony radius growth rate (%)

2.8 Effect of the Galangal Extract against Vibrio spp. Bacteria and Fungi in L. vannamei

2.8.1 Test Diets

The galangal extract was dissolved in ethanol to a concentration of 0.25 g/ml, then was mixed with commercial pellet feed (CP feed, Thailand) at 2 (5 g/kg diet) and 4% (10 g/kg diet). For the control group, no herbal extract was added. The feed was kept at room temperature for 30 min to allow the absorption of the extract and the evaporation of the ethanol. The control diet was also absorbed with ethanol, and then evaporated. Next, the pellets were coated with squid fish oil (Agrithai And Development Co., Ltd., Thailand) at 10 g/kg feed to prevent the dispersion of the galangal extract in water and to reduce the smell of the extract. The feed was then dried at room temperature. These tested diets were prepared for shrimp each day. The feeding rate was 3.0% of the shrimp body weight.

2.8.2 Treatment of Pathogenic Vibrio Bacteria and Fungi

About nine hundred white feces syndrome diseased shrimp, *L. vannamei,* with average body weight 10.7 ± 1.6 g were obtained from a shrimp farm in Chanthaburi Province, Thailand. The samples were acclimatized for 3 days in 3 plastic tanks, each with a capacity of 500 L. Then, the number of *Vibrio* spp. bacteria and fungi in the hepatopancreas and intestines of 15 shrimp were examined and recorded. Experimental aquariums (90 L capacity) were filled with 70 L of chlorine-treated 29 ppt seawater. Flow-through water equipped with an aeration system was arranged. Each group of 25 shrimp was sampled from the stocked tanks and was transferred into the individual experimental aquariums for 5 replications of treatment and the control group. The shrimp were fed the test diet 3 times a day for 12 days. Excess feed and waste was removed before each feeding. The parameters of the water were maintained for optimal quality. Three shrimp were sampled at 1, 3, 5, 7, 10, and 12 days from each aquarium after the feeding trials. The hepatopancreas and intestines of the shrimp were examined for *Vibrio* spp. and fungi via the spread-plate method.

$$\text{Incidence of fungi infestation (\%)} = \frac{\text{The number of infested shrimp}}{\text{Total number of observed shrimp}} \; x \; 100$$

2.9 Effect of the Extract on Disease Resistance Caused by Vibrio spp.

2.9.1 Preparation of the Bacterial Samples

The pathogenic *V. parahaemolyticus* (EMS/AHPND) was isolated from infected shrimp. This bacterium was identified based on a scheme established by Colwell (1984) using API 20E kits (ATB System, BioMérieux, France). Identification of *V. parahaemolyticus* (EMS/AHPND) was accomplished via the PCR technique (Joshi et al. 2014). The isolate was grown on Thiosulfate Citrate Bile Sucrose Agar (TCBS Agar, Difco USA) incubated at 30 °C for 18-24 h. Then a pure colony was streaked on Trypticase Soy Agar (TSA Difco, USA), and incubated at 30 °C for 18-24 h. Heavy streaks of isolated bacteria were inoculated into a 0.85% sterile sodium solution, which was then mixed well via a vortex mixer. The bacterial suspensions were tested using McFarland Standard No. 0.5 (10^8cfu/ml) then diluted to 10^5cfu/ml for the antibacterial activity test.

2.9.2 Resistance Test

The water in the 90 L aquariums was managed according to the method mentioned above. Triplicate aquariums were prepared for each group. Ten shrimp (10.7 ± 0.8 g/shrimp) were transferred to each aquarium for 2 treatment groups (2 and 4% of the extract diet) and 1 control group (0% of the extract diet). The shrimp were fed test diets 3 times a day for 12 days. After the shrimp were fed a treated diet for 12 days, disease resistance tests were conducted via the injection of 100 µl of bacterial suspension, *V. parahaemolyticus* (2.85 ± 0.63 X 10^5 cfu/ml), into the abdominal segment. After injection, shrimp were fed a commercial diet twice a day for 15 days. Disease symptoms and mortality rates were observed and recorded daily; the surviving shrimp were counted, and statistical analysis was calculated after the end of the experiment.

2.10 Statistical Analysis

A multiple comparison (Pair wise Comparison Test: Fisher's LSD) test was used to examine the significant differences ($P<0.05$) among treatments and control groups using SYSTAT VERSION 5.0.

3. Results

3.1 Antibacterial Activities of Galangal Extract

Table 1. Growth Inhibition zones, Minimum Inhibitory Concentration (MIC) and Minimum bactericidal Concentration (MBC) by the galangal extract to 8 species of *Vibrio* spp.

Vibrio spp.	Growth inhibition zone (mm.)		MIC	MBC	Sensitivity of bacteria
	Galangal	Ethanol	(mg/ml)	(mg/ml)	MBC/MIC
V. cholerae	31.0 ± 0.5	0 ± 0	0.15	0.15	1
V. parahaemolyticus	29.8 ± 0.8	0 ± 0	1.25	2.50	2
V. parahaemolyticus (EMS/AHPND)	18.0 ± 2.3	0 ± 0	2.50	5.00	2
V. fluvialis	28.0 ± 0.7	0 ± 0	0.31	0.63	2
V. vulnificus	27.6 ± 0.5	0 ± 0	1.25	1.25	1
V. alginolyticus	27.1 ± 0.8	0 ± 0	0.63	2.50	4
V. mimicus	27.6 ± 0.5	0 ± 0	0.31	0.63	2
V. harveyi	21.3 ± 2.1	0 ± 0	2.50	5.00	2

Note: Growth inhibition zone (mm.); Resistant: ≤ 9 mm; Intermediate: ≥ 10 – 13 mm; Susceptible: ≥ 14 mm (Lorian, 1995 in Oonmetta – aree et.al., 2006), Data in table, Average ± SD.

As reported by Canillac & Mourey (2001), if the MBC/MIC ratio is found to be less than or equal to 4, the strain is considered to be susceptible; on the other hand, if this ratio is greater than 4, the strain is considered to be tolerant.

The results showed a growth inhibition zone surrounding the galangal extract disc of all species' bacteria. The MBC and MIC of 2 bacterial species, *V. cholerae* and *V. vulnificus*, showed the highest degree of sensitivity. The other 5 species, *V. parahaemolyticus, V. parahaemolyticus* (EMS/AHPND), *V. fluvialis, V. mimicus*, and *V. harveyi*, showed intermediate sensitivity, and *V. alginolyticus* showed low sensitivity among the bacteria tested. The results of the antibacterial activity of the galangal extract in this study showed most of the species of *Vibrio* spp. to be sensitive (Table 1).

3.2 Fungi Isolation and Identification

Pathogenic fungi were isolated from *L. vannamei,* feces, and artificial feed. All fungi were identified by the 6 species shown in Table 2.

Table 2. The 6 species of fungi isolated from the hepatopancreas and intestines of shrimp naturally infected with white feces syndrome *L. vannamei*, and from artificial feed

Fungi species	Source of isolated
Aspergillus flavus	Hepatopancreas, intestine, and artificial feed
Aspergillus ochraceus	Hepatopancreas and artificial feed
Aspergillus japonicus	White feces and artificial feed
Penicillium sp.	Hepatopancreas and intestine
Fusarium sp.	Hepatopancreas, intestine, and artificial feed
Cladosporium cladosporioides	Intestine

3.3 Efficacy of the Extract against Pathogenic Fungi

Table 3 shows the efficacy of the galangal extract on 6 species of fungi. The galangal extract at a concentration of 0.5 mg/ml represented the strongest inhibition of *A. ochraceus*. A concentration of 1 mg/ml inhibited the growth of *Penicillium* sp.; *Fusarium* sp., and *C. cladosporioides*. Two species, *A. flavus* and *A. japonicus*, were least susceptible to the galangal extract (8 mg/ml).

Table 3. Efficacy of the galangal extract on the growth inhibition of fungi (%) isolated from an infection of white feces syndrome pathogen *L. vannamei* and from artificial feed

Extract Concentration (mg/ml)	The growth inhibition of fungi (%)					
	A. flavus	*A. ochraceus*	*A. japonicus*	*Penicillium* sp.	*Fusarium* sp.	*C. cladosporioides*
0.0005	6.12 ± 5.77	0.00 ± 0.00	0.00 ± 0.00	0.00 ± 0.00	0.00 ± 0.00	6.67 ± 6.29
0.005	16.33 ± 2.89	6.82 ± 2.50	18.64 ± 9.59	0.00 ± 0.00	4.23 ± 1.48	17.05 ± 11.25
0.05	48.98 ± 14.43	18.18 ± 9.28	15.25 ± 0.00	16.67 ± 0.00	27.95 ± 12.58	40.00 ± 3.14
0.5	56.73 ± 2.89	100.00 ± 0.00	25.42 ± 4.79	56.23 ± 15.89	42.35 ± 19.49	51.11 ± 6.29
1	77.55 ± 31.75	100.00 ± 0.00	37.29 ± 1.15	100.00 ± 0.00	100.00 ± 0.00	100.00 ± 0.00
2	77.55 ± 31.75	100.00 ± 0.00	57.63 ± 11.98	100.00 ± 0.00	100.00 ± 0.00	100.00 ± 0.00
4	82.66 ± 49.06	100.00 ± 0.00	55.93 ± 9.59	100.00 ± 0.00	100.00 ± 0.00	100.00 ± 0.00
8	100.00 ± 0.00	100.00 ± 0.00	100.00 ± 0.00	100.00 ± 0.00	100.00 ± 0.00	100.00 ± 0.00

Note: Data in table, Average ± SD.

3.4 Effect of the Galangal Extract against Vibrio spp. and Fungi

3.4.1 Health Status of the Naturally infected Shrimp

The hepatopancreas and intestines of shrimp from a grow-out pond that were infected with white feces syndrome were observed, and 4 species of fungi were found including *C. cladosporioides, Penicillium* sp., *A. japonicus*, and *Fusarium* spp. In addition, a rather high incidence of *Vibrio* spp. were also isolated from the same samples (100.0 ± 0.0 %). The total counts from the hepatopancreas and intestines were as high as $386.73 \pm 323.73 \times 10^3$ cfu/g and $426.46 \pm 168.21 \times 10^5$ cfu/g, respectively. The dominant numbers of *Vibrio* spp. isolated in the feces were green and yellow colonies.

3.4.2 Effect of the Galangal Extract on the Total Number of *Vibrio* spp.

The totals for the number of *Vibrio* spp. in the hepatopancreas and intestines of infested *L. vannamei* shrimp fed the galangal extract (treated diet) were significantly less than that for the control group ($P < 0.05$). Figure 1 lists the numbers after 3 days in the hepatopancreas (A) and 5 days in the intestines (B) after feeding with the treated diet. In contrast, the total number of *Vibrio* spp. in the hepatopancreas (A) and intestines (B) of the control group had increased through 12 days (Figure 1). No green colonies of *Vibrio* spp. were found in either of the groups fed 2 or 4% treated diets.

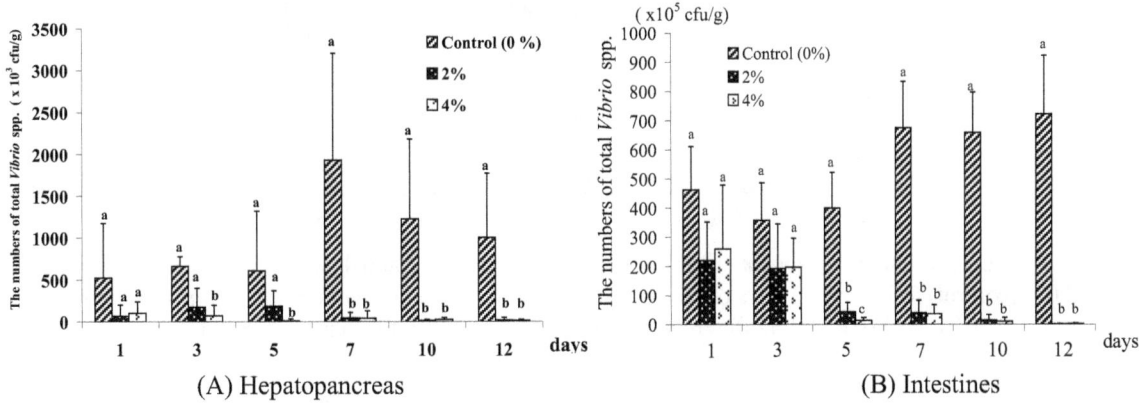

Figure 1. The total number of *Vibrio* spp. in (A) hepatopancreas and (B) intestines of infested *L. vannamei* fed the 2 and 4% galangal extract, as well as the control (0%), for 12 days. Data are the means ± SD; different letters for the time interval indicates a significant difference ($P<0.05$)

3.4.3 Effect of the Galangal Extract on Pathogenic Fungi

The incidence of fungi infestation in the hepatopancreas and intestines of infested *L. vannamei* shrimp fed the crude galangal extract (treated diet) was significantly lower than that in the control group ($P<0.05$). Also, no fungi could be found in the hepatopancreas and intestines of white shrimp after feeding with the treated diet for 10 to 12 days (Figure 2 (A) and (B)).

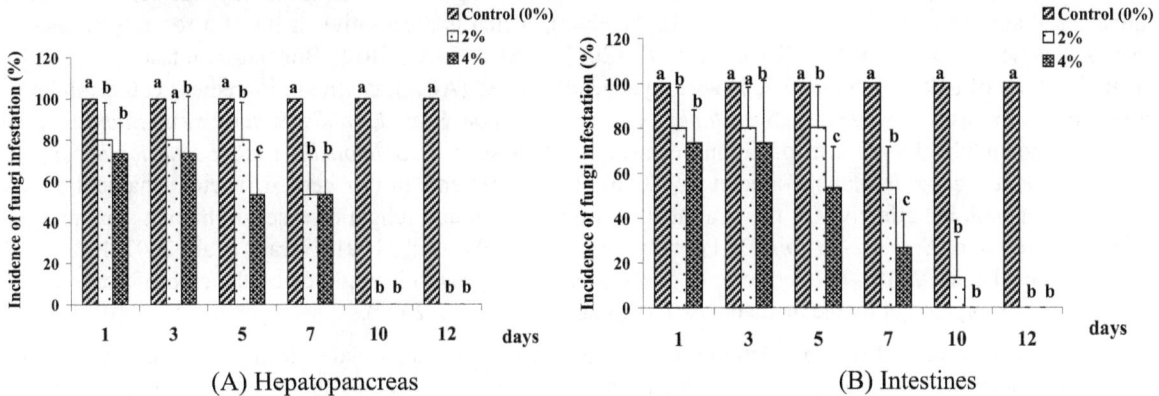

Figure 2. Incidence of fungi infestation (%) in the hepatopancreas (A) and intestines (B) of infested *L. vannamei* fed galangal extract at 2 and 4%, along with that of the controls (0%), for 12 days. Data are the means ± SD; different letters for the time interval indicates a significant difference ($P<0.05$)

3.5 Effect of the Galangal Extract on Disease Resistance Challenged by V. parahaemolyticus (EMS/AHPND)

The survival rates for shrimp fed 2 and 4% galangal extract were 73.3 ± 5.8% and 83.3 ± 5.8%, respectively. Only 16.7 ± 5.8% of the survival shrimp were from the control group (0% galangal extract). The results differed significantly between the treated groups and the control group ($P<0.05$). Obviously, shrimp fed the galangal extract at 4% had the highest survival rate (Figure 3).

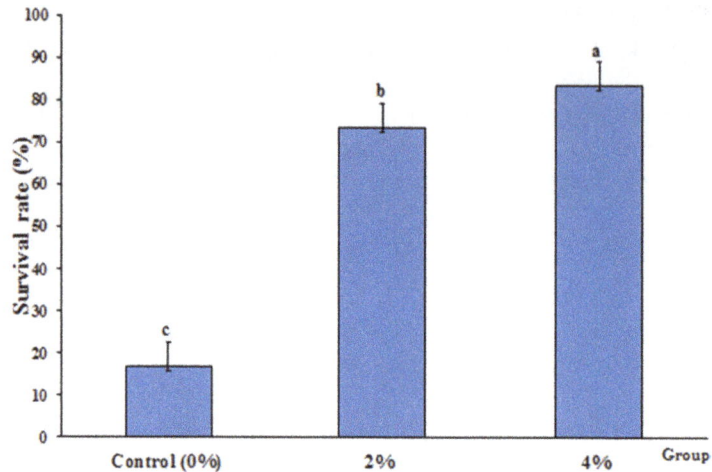

Figure 3. The survival rate of *L. vannamei* shrimp fed diets containing 2 and 4% of galangal extract (treated diet) and 0% (control) following 12 days of feeding trials prior to being challenged with *V. parahaemolyticus* (EMS/AHPND). Different letters indicate a significant difference (*P*<0.05)

4. Discussion

Investigation into the bacteria that are important and common in the aquatic environment of EMS/AHPND has been used to develop a safe and environmentally friendly product in shrimp aquaculture via alternative methods to control and treat vibrios diseases. The essential oil of crude galangal extract is a candidate for such a product since many researchers have documented its antimicrobial activities (de Pooter et al., 1985; Thomas et al., 1996; Turker et al., 2002; Oonmetta-areea et al., 2006; Tachakittirungrod & Chowwanapoonpohn, 2007; Vuddhakul et.al., 2007; Mayachiew & Devahastin, 2008; Latha et al., 2009; Rao et al., 2010). Canillac & Mourey (2001) reported that strains of vibrios with an MBC/MIC ratio of less than or equal to 4 are considered to be susceptible to the extract, and if the ratio is greater than 4 a strain is considered to be tolerant of the extract. Based on our results, the MBC/MIC ratios showed that 8 species of *Vibrio* were sensitive to the extract. The results also revealed that the extract inhibited the growth of *V. harveyi*. Also, the ethanol galangal extract in this study was considered to be antibacterial against *V. parahaemolyticus* (EMS/AHPND), which is a causative agent of a recently recognized serious disease in Southeast Asia (Lightner et al., 2012; NACA FAO, 2011; Bondad-Reantaso et al., 2012). Ethanol extract of galangal containing 1'-acetoxyeugenol acetate (ACA) as a main ingredient could inhibit the growth of Gram-positive bacteria such as *Staphylococcus cerevisiae*, *S. epidermidis*, *S. aureus* and *Bacillus cereus*, but it did not inhibit the growth of Gram-negative bacteria such as *Salmonella spp.*, *Escherichia coli, and Enterobacter aerogenes* (Oonmetta-aree et al., 2006). The ingredients in this herb, 1'-acetoxychavicol acetate, ACA and eugenol, are effective for the treatment of inflammation and fungal diseases on human skin, and they inhibit the growth of *Escherichia coli* (De Pooter et al., 1985; Kubo et al., 1991; Norajit et al., 2007; Natta et al., 2008; Chudiwal et al., 2010). Furthermore, *V. parahaemolyticus* growth is inhibited by the chloroform extract of fresh galangal root, but not by the methanol extract (Vuddhakul et. al., 2007).

These studies were carried out to determine the fungal species that are isolated from the hepatopancreas and intestine of white feces syndrome *L. vannamei* and from artificial feed. Among these species, *Fusarium* sp. were observed more frequently than the others. The study was focused on the degree that galangal crude extract inhibited the growth of these pathogenic fungi. De Silava et al. (2011) isolated 18 fungi species from *L. vannamei,* and among those, *Aspergillus flavus* and 2 species of *A. parasiticus* were able to produce aflatoxin B1. Contamination of *F. moniliforme* (Li et al., 1994) and *A. flavus* (Lighter, 1993) was reported in raw materials used to produce commercial pellet feed. Yulia (2005) previously reported that the ethanol extract of galangal rhizomes was the most effective in reducing spore germination of the fungi. Furthermore, this extract also inhibited the colony areas of *F. moniliforme, A. flavus* and *A. niger* (Handajani & Purwoko, 2008). This study's results also showed the growth inhibition of fungi by crude galangal extract. The study found fungi that produce alflatoxin to be few in number in artificial feed, which may be due to moisture retention during transport or storage on the farm. To prevent fungi in feed, it should be kept in a dry place and it should not be expired.

Our tests on the inhibition of *Vibrio* spp. and fungi *in vitro* and *in vivo* returned similar results. The infected shrimp fed the galangal extract diets had a lower number of *Vibrio* spp. and fungi compared with the control group. It is remarkable that the galangal extract could be used to inhibit growth and reduce the number of *Vibrio* spp. and fungi that cause white feces syndrome and AHPND in shrimp. In the present study, after shrimp were fed a diet mixed with 2 and 4% of the extract with ethanol, from day 1 the numbers of *Vibrio* spp. in the hepatopancreas and intestines were lower than in the control group. Remarkably, galangal extract can be used to inhibit the growth and reduce the number of *Vibrio* spp. and fungi to prevent white feces syndrome. Moreover, resistance in *L. vannamei* against *V. parahaemolyticus* causing AHPND occurred when the shrimp were fed a diet mixed with galangal extract at 2 or 4% for 12 days. Our results also showed a higher survival rate (83.3%) of the treatment group (fed with 4% galangal extract) than the control group (feeding with 0% extract). The results revealed the efficiency of the extract, which has ability to inhibit *V. parahaemolyticus* (EMS/AHPND), and it may stimulate the immune system in the tested shrimp (Chaweepack et al., 2015). Therefore, shrimp fed galangal extract diet are healthier than the control group, and shrimp fed a galangal extract diet could resist infection well. For several years, antibiotics and some hazardous chemicals have been used to treat pathogenic agents in the aquaculture industry, particularly in intensive shrimp-farming systems. The development of resistant bacteria and residue in shrimp caused a major obstacle to trade and had a negative impact on the environment as well as on consumer health. The use of herbal medicines is a viable alternative to the overuse of antibacterial agents. The main advantage of herbal agents is that the crude extract contains a mixture of compounds such as phenols, acids, esters, and aldehydes, to which bacteria are unlikely to develop resistance. A synthetic agent that contains a single compound, however, is easier for bacterial stains to resist (Abraham et al., 1997). Based on the results of the present study, we recommend alternative methods for control of the disease that causes white feces syndrome and AHPND in the shrimp industry by using galangal crude extract, which has proven to be effective and safe for the environment as well as for consumers.

5. Conclusion

This investigation was focused on the efficacy of galangal extract for treatment of the pathogenic organisms that cause white feces syndrome and AHPND in Pacific white shrimp, *L. vannamei*. Ethanol galangal extract exhibited the highest potential for *Vibrio* spp. and fungi reduction. The highest percentage (4%) of ethanol extract was most effective in reducing all species of the pathogens of *V. parahaemolyticus* (EMS/AHPND). Therefore, galangal extract should be used as an antimicrobial for white feces syndrome and AHPND therapeutics in Pacific white shrimp *L. vannamei* and in shrimp cultures. This alternative method could assist in reducing the impact of antibiotic or chemical residue in shrimp products as well as helping to reduce the presence of resistant bacterial strains in the environment.

Acknowledgments

This research was supported by the Department of Fisheries (DOF), Agriculture Research Development Agency (ARDA), Ministry of Agriculture and Cooperatives of Thailand and the Kyoto Institute of Technology (KIT), Japan.

References

Abraham, T., Manley, J. R., Palaniappan, R., & Devendran, K. (1997). Pathogenicity and antibiotic sensitivity of luminous *Vibrio harveyi* isolated from diseased penaeid shrimp. *J Aquac Trop., 12*(1), 1-8.

Bondad-Reantaso, M.G., Subasinghe, R.P., Josupeit, H., Cai, J., & Zhou, X. (2012). The role of crustacean fisheries and aquaculture in global food security: past, present and future. *J Invertebr Pathol., 110*, 158–165. http://dx.doi.org/10.1016/j.jip.2012.03.010

Canillac, N., & Mourey, A. (2001). Antibacterial activity of the essential oil of *Picea excels* on Listeria, *Staphylococcus aureus* and coliform bacteria. *Food Microbiology, 18*, 261-268. http://dx.doi.org/10.1006/fmic.2000.0397

Chaweepack, T., Chaweepack, S., Muenthaisong, B., Ruangpan, L., Nagata, K., & Kamei, K. (2015). Effect of galangal (*Alpinia galanga* Linn.) extract on the expression of immune-related genes and *Vibrio harveyi* resistance in Pacific white shrimp (*Litopenaeus vannamei*). *Aquacult Int., 23*(1), 385-399. http://dx.doi.org/10.1007/s10499-014-9822-2

Chudiwal, A. K., Jain, D. P., & Somani, R. S. (2010). *Alpinia galangal* Wild.-An overview on phyto-pharmacological properties. *Indian Journal of Natural Products and Resources, 1*(2), 143-149.

Colwell, R. R. (1984). Vibrios in the environment. *Wiley*, New York.

de Pooter, H. L., Omar, M. N., Coolaset, B. A., & Schamp, N. M. (1985). The essential oil of greater galanga (*Alpinia galanga*) from Malaysia. *Phytochemistry., 24*, 93-96. http://dx.doi.org/10.1016/S0031-9422(00)80814-6

de Silva, L. R. C., de Souza, O. C., dos Santos Fernandes, M. J., Débora, M. M., Lima, D. M. M., Coelho, R. R. R., & Souza-Motta, C. M. (2011). Culturable Fungal Diversity of shrimp *Litopenaeus vannamei* Boone from Breeding frams in Brazil. *Brazilian Journal of Microbiology, 42*, 49-56.

Flegel, T. W. (2012). Historic emergence, impact and current status of shrimp pathogens in Asia. *Journal of Invertebrate Pathology, 110*(2), 166-173. http://dx.doi.org/10.1016/j.jip.2012.03.004

Handajani, N. S., & Purwoko, T. (2008). The activity of galanga (*Alpinia galanga*) rhizome extract against the growth of filamentous fungi *Aspergillus* spp. that produce aflatoxin and *Fusarium moniliforme*. *Biodiversitas., 9*, 161-164. http://dx.doi.org/10.13057/biodiv/d090301

Joshi, J., Srisala, J., Hong, V. T., Chen, I. T., Nuangsaeng, B., Suthienkul, O., ... Thitamadee, S. (2014). Variation in *Vibrio parahaemolyticus* isolates from a single Thai shrimp farm experiencing an outbreak of acute hepatopancreatic necrosis disease (AHPND). *Aquaculture, 428-429*, 297-302. http://dx.doi.org/10.1016/j.aquaculture.2014.03.030

Karunasagar, I. D., Karunasagar, I. N., & Umesha, R. K. (2004). Microbial Diseases in Shrimp Aquaculture. Marine Microbiology. In N. Ramaiah (Ed.), *National Institute of Oceanography, Goa India. Facets & Opportunities* (pp. 121-134).

Khomvilai, C. (2006). *Application of sodium hypochlorite as an antifungal agent to aquaculture in Japan and Thailand* (Unpublished doctoral dissertation). Mie University, Japan.

Klich, M. A. (2002). Identification of common *Aspergillus* species. CBS. The Netherlands.

Latha, C., Shriram, V. D., Jahagirdar, S. S., Dhakephalkar P. K., & Rojatkar, S. R. (2009). Antiplasmid activity of 1'-acetoxychavicol acetate from *Alpinia galanga* against multi-drug resistant bacteria. *Ethnopharmacology, 123*, 522-525. http://dx.doi.org/10.1016/j.jep.2009.03.028

Lee, S. W., Najiah, M., Wendy, W., & Nadirah, M. (2009). Comparative study on antibiogram of *Vibrio* spp. isolated from diseased postlarval and marketable-sized white leg shrimp (*Litopenaeus vannamei*). *Front Agric China, 3*, 446-451. http://dx.doi.org/10.1007/s11703-009-0068-0

Li, M. H., Raverty, S. A., & Robinson, E. H. (1994). Effect of dietary mycotoxins produced by the mold *Fusarium moniliforme* on channel catfish *Ictalurus punctatus*. *J.World Aquacult. Soc., 25*(4), 512-516. http://dx.doi.org/10.1111/j.1749-7345.1994.tb00820.x

Lightner, D. V. (1993). Diseases of culture Penaeid shrimp. In J. P. McVey (Ed.), *CRC Handbook of Mariculture: Crustacean Aquaculture* (pp. 455-474)(ed.) CRC Press. BocaRaton.

Lightner, D. V., Redman, R. M., Pantoja, C. R., Noble, B. L., & Tran, L. H. (2012). Early mortality syndrome affects shrimp in Asia. *Glob Aquacult Advocate., Jan/Feb 2012*, 40.

Mayachiew, P., & Devahastin, S. (2008). Antimicrobial and antioxidant activities of Indian gooseberry and galangal extracts. *Food Science and Technology, 41*, 1153-1159. http://dx.doi.org/10.1016/j.lwt.2007.07.019

Montero, A. B., & Austin, B. (1999). Characterization of extracellular products from an isolate of *Vibrio harveyi* recovered from diseased post-larval *Penaeus vannamei* (Bonne). *Journal of Fish Diseases, 22*, 377-386. http://dx.doi.org/10.1046/j.1365-2761.1999.00189.x

NACA-FAO (Network of Aquaculture Centers in Asia-Pacific—Food and Agriculture Organization of the United Nations). (2011). Quarterly aquatic animal diseasereport (Asia and Pacific Region), 2011/2, April-June 2011. NACA, Bangkok

Oonmetta-aree, J., Suzuki, T., Gasaluck, P., & Eumkeb., G. (2006). Antimicrobial and action of galangal (*Alpinia galanga* Linn.) on *Staphylococus aureus*. *Food Science and Technology., 39*, 959-965. http://dx.doi.org/10.1016/j.lwt.2005.06.015

Rao, K., Ch, B., Narasu, L. M., & Giri, A. (2010). Antibacterial Activity of *Alpinia galangal* (L) Willd Crude Extracts. *Appl Biochem Biotechnol., 162*, 871-884. http://dx.doi.org/10.1007/s12010-009-8900-9

Raper, K. B., & Fennell, D. I. (1965). The genus *Aspergillus*. The Williams & Wilkins Company, Baltimore.

Ruangpan, L., & Kitao, T. (1991). Vibrio bacteria isolated from black tiger shrimp *Penaeus monodon* Fabricius. *J Fish Dis., 14*, 383-388. http://dx.doi.org/10.1111/j.1365-2761.1991.tb00836.x

Ruangpan, L., Thanasomwang, V., & Sangrungrauge, K. (1997). Bacteria in Black tiger Shrimp pond farming development Systems. *The Proceeding of 35th Kasetsart University Annual Conference. Thailand.*, 3-10.

Samson, R. A., & Hoekstra, E. S. (1994). Common fungi occurring in indoor environments. In R. A. Samson (Ed.), Health Implications of Fungi in indoor environments (pp. 541-587).   Elsevier, Amsterdam.

Samson, R. A., Hoekstra, E. S., Frisvad, J. C., & Filtenborg, O. (1995). Introduction to Food-borne Fungi. *CBS*, the Netherlands.

Sung, H. H., Hsu, S. F., Chen, C. K., Ting, Y. Y., & Chao, W. L. (2001). Relationships between disease outbreaks in cultured tiger shrimp (*Penaeus monodon*) and composition of Vibrio communities in pond water and shrimp hepatopancreas during cultivation. *Aquaculture., 192*, 101-110. http://dx.doi.org/10.1016/S0044 -8486(00)00458-0

Tachakittirungrod, S., & Chowwanapoonpohn, S. (2007). Comparison of Antioxidant and Antimicrobial activities of Essential Oils from *Hyptis suaveolens* and *Alpinia galangal* Growing in Northern Thailand, CMU, *J Nat Sci., 6*, 31-42.

Tangtrongpiros, J. (2010). Chula Researcher Unveils Groundbreaking Study On Disease Killing Off Cultured Shrimp. Preventive approaches proposed to avert business losses in national shrimp farming industry. *The Gazette of Chulalongkorn University, Thailand, 2*, 3.

Thomas, E., Shanmugan, J., & Rafi, M. M. (1996). Antibacterial activity of plants belonging to Zingiberaceace family. *Biomedicine., 16*, 15-20.

Tran L., Nunan, L., Redman, R. M., Mohney, L. L., Pantoja, C.R., Fitzsimmons, K., & Lightner, D. V. (2013). Determination of the infectious nature of the agent of acute hepatopancreatic necrosis syndrome affecting penaeid shrimp. *Dis Aquat Org., 105*, 45-55. http://dx.doi.org/10.3354/dao02621

Turker, A., & Usta, C. (2002). Biological activity of some medicinal plants sold in Turkish Health-food stores. *Biodiversity Ecosyst., 34*(19), 105-113.

Vandenberghe, J., Verdonck, L., Robles-Arozarena, R., Rivera, G., Blland, A., Balladares, M., Gomez-Gil, B., Calderon, J., Sorgeloos, P., & Swings, J. (1999). Vibrios Associated with *Litopenaeus vennamei* Larvae, Postlarvae, Broodstock, and Hatchery Probionts. *Applied and Environmental Microbiology., 65*, 2592-2597.

Vuddhakul, V., Bhoopong, P., Hayeebilan, F., & Subhadhirasakul, S. (2007). Inhibitory activity of Thai condiments on pandemic strain of *Vibrio parahaemolyticus. Food Microbiology., 24*, 413-418. http://dx.doi.org/10.1016/j.fm.2006.04.010

Yulia, E. (2005). Antifungal activity of plant extracts and oils against fungal pathogens of pepper (*Piper nigrum* L.), cinnamon (*Cinnamomum zeylanicum* Blume.), and turmeric (*Curcuma domestica* Val.). Thesis of Master of Science the School of Tropical Biology, James Cook University, Australia.

Pseudomonas aeruginosa – Pathogenesis and Pathogenic Mechanisms

Alaa Alhazmi[1,2]

[1] Department of Biology, Lakehead University, Thunder Bay, Ontario, Canada

[2] Department of Medical Laboratory Technology, Jazan University, Jazan, Kingdom of Saudi Arabia

Correspondence: Alaa Alhazmi, Department of Biology, Lakehead University, 955 Oliver Road, Thunder Bay, ON, P7B 5E1, Canada. Email: aalhazmi@lakeheadu.ca

Abstract

Pseudomonas aeruginosa is a common bacterium, Gram-negative opportunistic pathogen capable of infecting humans with compromised natural defenses and causing severe pulmonary disease. It is one of the leading pathogen associated with nosocomial infections. It has a vast arsenal of pathogenicity factors that are used to interfere with host defenses. Pathogenesis in *P. aeruginosa* facilitates adhesion, modulate or disrupt host cell pathways, and target the extracellular matrix. The propensity of *P. aeruginosa* to form biofilms further protects it from antibiotics and the host immune system. *P. aeruginosa* is intrinsically resistant to a large number of antibiotics and can be acquired resistance to many others, making treatment difficult. *P. aeruginosa* provokes a potent inflammatory response during the infectious process. The majority of mortalities in immunocompromised patients; cystic fibrosis, can be attributed to the progressive decline of lung function resulting from chronic infection by pathogens such as *P. aeruginosa*. Antibiotic treatment of chronic *P. aeruginosa* infections may temporarily suppress symptoms; however, they do not eradicate the pathogen. Lung diseases caused by *P. aeruginosa* are a leading cause of death in immunocompromised individuals as well as in children. Although immunocytes recruitment is critical to augment the host defense, excessive neutrophil accumulation results in life-threatening diseases, such as acute lung injury, as well as acute respiratory distress syndrome. Several virulence factors have been studied for their roles as potential vaccine candidate, although there is currently no clinically accepted vaccine. Understanding host-pathogen interaction is critical for the development of effective therapeutic strategies to control the damage in the lung.

Keywords: *Pseudomonas aeruginosa*, nosocomial infection, virulence factors, antibiotic resistance, cystic fibrosis

P. aeruginosa is a motile, non-fermenting, Gram-negative organism belonging to the family Pseudomonadaceae. In 1850s, Sédillot observed that a blue-green discharge was frequently present and associated with infection in surgical wound dressings (Lyczak, Cannon, & Pier, 2000). The infectious organism has a rod-shaped and blue-green pigmented bacterium (Lartigau, 1898). By 1961, the ability of this organism to cause both severe acute and chronic infections was recognized (Freeman, 1916). In 1960s, *P. aeruginosa* emerged as an important human pathogen (Doggett, 1979). Despite anti-pseudomonas activity being one of the pharmaceutical drug discoveries for several decades. It remains one of the most recalcitrant and difficult to treat organisms. Accordingly, *P. aeruginosa* result has achieved Superbug status.

At 6.3 million base pair (Mbp), *P. aeruginosa* genome is markedly as a large genome (encoding 5567 genes) compared to 4.64 Mbp (4279 genes) in *Escherichia coli* K12, 2.81 Mbp (2594 genes) in *Staphylococcus aureus* N315, and 1.83 Mbp (1714 genes) in *Haemophilus influenza* Rd. Also, the proportion of predicted regulatory genes in *P. aeruginosa* genome is greater than in all other sequenced bacterial genomes (Stover et al., 2000; Lambert, 2002), thus lending to its adaptability to varying environments. *P. aeruginosa* has a broad range of growth substrate, minimal nutrient requirements; non-fastidious microorganisms (Favero, Carson, Bond, & Petersen, 1971). The organism is tolerant of temperatures as high as 50°C and is capable of growing under aerobic conditions, as well as anaerobic conditions (van Hartingsveldt & Stouthamer, 1973). Despite possessing a large number of virulence factors. In spite of this, *P. aeruginosa* is truly a challenging pathogen in the hospital setting as it is resistant to many antibiotics. Also, it is capable of forming hardy biofilms, both within the body and on the surfaces of medical instruments (Consterton, Stewart, & Greenberg, 1999; Hancock, 1998, Moreau-Marquis, Stanton, & O'Toole, 2008). *P. aeruginosa* continues to be problematic from a treatment perspective.

P. aeruginosa is armed with potent virulence factors. Although ubiquitously present in the environment, *P. aeruginosa* never causes disease in an immunocompetent host as the immune system effectively prevents the infection. However, the pathogen causes severe infections in Cystic Fibrosis (CF) patients. In CF, a genetic defect in lung innate immunity underlies the development of persistent infection with *P. aeruginosa* that gradually leads to irreversible tissue damage. Several conserved microbial structures in *P. aeruginosa* are recognized by Toll-like receptors (TLRs) and NOD-like receptors (NLRs); which have been implicated in activating the host innate immune responses to *P. aeruginosa* (DiMango, Zar, Bryan, & Prince, 1995; Skerrett, Wilson, Liggitt, & Hajjar, 2007).

There are a number of clinical diseases associated with *P. aeruginosa* infection. However, *P. aeruginosa* is an opportunistic organism infecting; burn, CF, leukemic, transplant, neutropenic, long-term urinary catheters, and diabetic patients as well intravenous drug abusers.

1. Nosocomial Infection Due to *P. aeruginosa*

Nosocomial (hospital-acquired) infections are those not present or incubating at the time of hospital admission, but usually develop post-admission. The 2006-7 report by National Healthcare Safety Network (NHSN) at the Centers for Disease Control and Prevention (CDC) ranked *P. aeruginosa* as the sixth most common healthcare associated pathogen-causing infection. Also, it is typically found at even higher rank in studies focused on the intensive care unit (ICU). The NHSN reports that in United States in 2006-7, 8% of all hospital-associated infections were due to *P. aeruginosa*, with *P. aeruginosa* causing 3% of central line-associated bloodstream infections, 6% of surgical site infections, 10% of catheter-associated urinary tract infection and 16% of ventilator-associated pneumonia (VAP) infections (Hidron et al., 2008). Mechanical ventilation, antibiotic therapy, surgery, and chemotherapy are the major predisposing factors contributing to the acquisition of a *P. aeruginosa* infection in the hospital (Thuong et al., 2003). It is worth noting however that difficulties in treatment of such infections and the associated morbidity and mortality, have made *P. aeruginosa* on of the most feared hospital pathogens.

1.1 Burn Wound Infections

P. aeruginosa is the leading cause of invasive infections in burn patients; 75% of all deaths in patients with severe burn are related to sepsis from invasive burn wound infection (Baker, Miller, & Trunkey, 1979; Bang, Sharma, Sanyal, & Najjadah 2002; Barrow, Spies, Barrow, & Herndom, 2004). In addition to wounded skin injury, inhalation injury is common in burn patients. This results in edema and sloughing of the respiratory tract mucosa and impairment of the normal mucociliary clearance mechanism, thus making these patients more susceptible to upper respiratory tract infections as well as *P. aeruginosa* pneumonia (Church, Elsayed, Reid, Winston, & Lindsay, 2006).

Although Gram-positive organisms such as *Staphylococcus aureus* and *Streptococcus pyrogens* are typically the first microorganisms to colonize the site of infection, after that other microbes including *P. aeruginosa* being to colonize these wounds (Altoparlak, Erol, Akcay, Celebi, & Kadanali, 2004; MacMillan, 1980). Success with early wound excision practices was shown to contribute to the prevention of invasive infections disseminating from the wound site (Barret & Herndon, 2003). Animal studies of partial-thickness cutaneous burns showed that mature biofilms could develop in 48 to 72 hours, indicating a major potential source of further difficulties in antimicrobial therapy at these sites (Trafny, 1998).

In addition to the *P. aeruginosa* virulence factors that undoubtedly contribute to the success of *P. aeruginosa* as a pathogen in the burn patients, described later, the impairment of host immunity, beyond simple loss of the skin's physical barrier, plays a role in enhancing susceptibility to infection. Recent studies have demonstrated that thermal injury causes impaired production of the host defense peptides β-defensins in the tissues surrounding the wound. These immunomodulatory peptides have been proposed to play an important role in primary defense against *P. aeruginosa* and synthetic β-defensin was recently shown to be protective against *P. aeruginosa* infection in a burned mouse model (Kobayashi et al., 2008).

1.2 Bacteremia

P. aeruginosa is among the five leading causes of nosocomial bacteremia and frequently leads to sepsis. In the 1960s and early 1970s, aminoglycosides and polymyxins were the only options for treatment of *P. aeruginosa* bacteremia but were found to be fairly ineffective for these infections. Mortality of greater than 50% was reported when mortality was used as the end point (Fishman & Armstrong, 1972; Whitecar, Luna, & Bodey, 1970), and was as high as 70% in febrile neutropenic patients (Bodey, Jadeja, & Elting, 1985). Despite the introduction of effective anti-pseudomonal β-lactams and the associated reduction in mortality rates, *P. aeruginosa* bacteremia is still one of the most feared nosocomial infections. These infections are generally associated with higher mortality than with other infecting pathogens, and persistence, particularly related to device-related bacteremia, continues to plague patients (Rello, Ricart, Mirelis, Quintana, Gurgui, Net, & Prats, 1994).

The main distinguishing feature of *P. aeruginosa* sepsis is the presence of ecthyma gangrenous, and these infarcted skin lesions occur only in markedly neutropenic patients (Pier & Ramphal, 2005). When *P. aeruginosa* disseminates from a site of local infection, it gains access to the bloodstream by breaking down epithelial and endothelial tissue barriers (Kurahashi et al., 1999). To evade the bactericidal activity of the serum complement, *P. aeruginosa* produces a smooth lipopolysaccharide (LPS) (Hancock et al., 1983; Pier & Ames, 1984); full-length O side-chain of the bacteria.

1.3 Hospital-Associated Pneumonia

The human respiratory tract presents a favorable environment to which *P. aeruginosa* has become particularly well adapted. *P. aeruginosa* has the formidable ability to cause both chronic infections in the lung of CF patients and acute nosocomial pneumonia (Blondel-Hill and Fryters, 2006). Animal model studies of *P. aeruginosa* pneumonia have demonstrated the involvement in virulence of proteases, flagella, pili and LPS O side chains as well as the delivery of the extracellular toxins ExoS, ExoT and ExoU via a type III secretion system (T3SS). For example, administration of anti-pcrV antibodies blocking the T3SS has been shown to offer protection against acute *P. aeruginosa* pneumonia when tested in animal studies (Faure et al., 2003; Shime et al., 2001).

1.4 Ventilator-Associated Pneumonia

P. aeruginosa is commonly found to be the first or second major pathogen causing VAP (Hidron et. al., 2008). It is the most common multidrug resistance pathogen involved in this disease and recovery rate of *P. aeruginosa* is increased with increased duration of mechanical ventilation. In addition to being amongst the most common pathogens causing VAP, *P. aeruginosa* is also amongst the most lethal pathogens, since reports suggest up to 70-80% mortality when the organism remains confined to the lung (Chastre & Fagon, 2002), with directly attributable mortality rates reaching 38% (Fagon et al., 1993).

2. *P. aeruginosa* Infections in Cystic Fibrosis

Cystic fibrosis (CF) is; an autosomal receive genetic disorder, the most common fatal genetic disease. CF is caused by a mutation in a gene on chromosome 7 known as CFTR (cystic fibrosis transmembrane conductance regulator). Most common mutant allele is ΔF508 (or F508del) mutation, which is a three-nucleotide deletion of a phenylalanine residue and subsequent defective intracellular processing of the CFTR protein that is an important chloride channel (Bobadilla, Fine, & Farrell, 2002). CF is affecting 1:2,500 in the Caucasian population (Ratjen & Doring, 2003). CF is multi-system disease, which affects mainly lung and digestive system. Most CF-related deaths are due to lung disease (Bobadilla, Fine, & Farrell, 2002).

Mortality in this afflicted population is mainly attributed to chronic respiratory infections and the associated gradual deterioration of lung function. There are several pathogens known to play a role in CF lung infection, with *Staphylococcus aureus* and *Haemophilus influenzae* being the predominant pathogens colonizing in infancy and early childhood, and eventual replacement by *P. aeruginosa*. However, *P. aeruginosa* is often isolated from patients less than 2 years of age and is the most predominant concern in adults (Gibson, Burns, & Ramsey, 2003; Hudson, Wielinski, & Regelmann, 1993). Up to 90% of individuals suffering from CF become infected with *P. aeruginosa* during their lifetime, and this organism is the leading cause of morbidity and mortality among those patients. It is dominant pathogen in chronic lung infection in CF. In the majority of cases, colonization of the CF airway by *P. aeruginosa* leads to a chronic infection that is resistant to antimicrobial therapy (MacEachran et al., 2007; Döring et al., 2014). Chronic colonization and infection with *P. aeruginosa* is an inevitable reality for the majority of adults with CF, as over 80% of adults over the age of 18 years return positive cultures for *P. aeruginosa* (Hodson, 2000). The nature of this disease is critical in understanding why *P. aeruginosa* dominates as the primary pathogen in CF patients and so host pathology is addressed below.

The defective gene involved in CF encodes for CFTR resulting in pathological changes in organs that express CFTR, including lungs. In a normal airway epithelial cell, the gene encoding for CFTR, regulates the transport of chloride, sodium, and water. Abnormalities of the CFTR gene product lead to a thick and dehydrated mucous secretion that impairs mucociliary clearance of bacterial pathogens (Collins, 1992). In the normal lung, the mucus layers function in binding and clearance of inhaled pathogen, and although the bacterial load can be quite high in the upper airways, the lower airways remain free of bacteria (Bals, Weiner, & Wilson, 1999). Due to the characteristic thickened mucus associated with CF lung disease, and resulting an inability of ciliary beating to remove the mucus, invading pathogens become trapped in the mucus layer. As a result, a constant presence of bacteria with expression pathogen-associated molecular patterns (PAMPs) leads to chronic inflammation, consequently damaging the epithelial surface (Donaldson et al., 2006; Greene et al., 2005; Kolberg, Mossberg, Afzelius, Philipson, & Camner, 1978).

A novel concept of host susceptibility emerged that the epithelial cells use CFTR as a receptor for internalization of *P. aeruginosa* and subsequent removal of bacteria from the airway surface (Pier, Grout, & Zaidi, 1997; Pier et al., 1996). Accordingly, CFTR is considered as a pattern recognition molecule that extracts *P. aeruginosa* LPS from the organism's surface into epithelial cells (Schroeder et al., 2002). The prevention of CFTR-*P. aeruginosa* interactions led to decrease bacterial clearance and increased bacterial burdens in the lungs.

2.1 Adaptation Occurring During Chronic Infection

CF patients frequently become colonized in the upper airways by unrelated environmental isolates of *P. aeruginosa* (Salunkhe et al., 2005; Speert et al., 2002). During the process of infection, a number of adaptations occur leading to the characteristic persistence and antibiotic resistance of isolates found from chronic infection. Amongst the most common adaptations of *P. aeruginosa* found in CF isolates, is the conversion to mucoid phenotype due to overexpression of alginate (Govan & Deretic, 1996). Environmental isolates usually present a non-mucoid phenotype, as *P. aeruginosa* penetrates the thickened mucus lining of the airways, travelling down the oxygen gradient, increased expression of alginate and a switch to a mucoid phenotype occur (Grimwood, 1992; Starner & McCray, 2005). This phenotype often occurs coincidently with the establishment of chronic infection and becomes stabilized by regulatory mutations as described earlier. The mucoid form of *P. aeruginosa* is associated with 90% of *P. aeruginosa* CF infections compared to only 2% of *P. aeruginosa* non-CF infections (Doggett, 1969; Doggett, Harrison, & Carter, 1971). This phenotype is often coordinately regulated with a loss of flagella by the alternative sigma factor AlgT (Tart, Wolfgang, & Wozniak, 2005). The loss of flagella causes not only loss of motility, but also decreased activation of host inflammatory mediators (Cobb, Mychaleckyj, Wozniak, & Lopez-Boado, 2004).

Other easily identified morphological adaptations of *P. aeruginosa* include the switch from smooth to rough colony morphology and the appearance of small colony variants. The rough colony morphology is representative of strains that have lost the LPS O-antigen (Hancock et al., 1983). As the O-antigen is the immunodominant portion of the LPS, this adaptation leads to a less virulent phenotype. It also makes rough isolates more susceptible to complement killing and perhaps explains in part why these organisms virtually never cause invasive infections. Modifications to the lipid A moiety of the LPS are also observed. These include the addition of palmitate, aminoarabinose and the retention of 3-hydroxydecanoate (Ernst et al., 2007). The small colony phenotype is less well understood but is of considerable interest as these isolates exhibit increased antibiotic resistance. Isolates exhibiting this phenotype have been found to be hyperpiliated with increased abilities in twitching and biofilm formation, and with decreased ability for swimming (Haussler et al., 2003).

Another phenotype of relevance to antimicrobial therapy and resistance is the hypermutator phenotype, which is frequently observed in CF isolates, but less commonly in nosocomial isolates of *P. aeruginosa* (Oliver, Canton, Campo, Baquero, & Blazquez, 2000). This phenotype, characterized by an up to 1000-fold increased mutation frequency, has been attributed to mutations in genes encoding DNA replication and repair mechanisms, such as mutS, mutL, and mutY. Most importantly, these hypermutator isolates can develop mutational resistance more readily during a course of antimicrobial therapy than do non-mutator isolates. The hypermutator phenotype can give rise to a variety of mixed morphologies within the lung, including those described above (Hogardt et al., 2007). These diverse populations can colonize or infect different compartments within the lung and often have variable antimicrobial susceptibilities with virulence properties (Foweraker, Laughton, Brown, & Bilton, 2005; Irvin, Govan, Fyfe, & Costerton, 1981; MacLeod et al., 2000; Pai & Nahata, 2001).

Comparison of *P. aeruginosa* isolates from the CF lung to strains from non-CF patients showed clearly that CF isolates tend to demonstrate an overproduction of β-lactamase, loss of OprD and an overproduction of MexXY. This efflux pump overproduction leads to high-level aminoglycoside resistance and the overproduction of this and other efflux systems also lead to quinolone resistance, amongst which MexCD-OprJ was the most frequent (Henrichfreise, Wiegand, Pfister, & Wiedemann, 2007).

2.2 Antimicrobial Therapy for Treatment of P. aeruginosa

P. aeruginosa isolates from CF patients frequently develop multi-drug resistance. Combination therapy can be used to avoid resistance development and to exploit the synergistic effects of the bactericidal antibiotics. The use of aerosols allows for drugs to be delivered directly to the lung in CF patients and a number of antibiotics including gentamicin, tobramycin, colistin, ceftazidime, carbenicillin aztreonam and amikacin have been administered as aerosols to CF patients, although approved formulation and adequate controlled studies have not been performed on most of these (Hodson, Penketh, & Batten, 1981; Stead, Hodson, & Batten, 1987).

2.2.1 Antimicrobial Therapy for Colonization and Initial Infection

Eradication of *P. aeruginosa* from the CF lung is possible only in the early stages of colonization. At this point, the bacterial load tends to be low, and the organism is non-mucoid and has not begun to undergo significant morphological changes. Aggressive antimicrobial treatment upon first isolation of *P. aeruginosa* has been demonstrated in most cases to delay and occasionally prevent the onset of chronic infections resulting in a better quality of life and a greater life expectancy (Nixon et al., 2001; Rosenfeld, Ramsey, & Gibson, 2003). Successful eradicating is judged by the observation of at least three consecutive negative cultures at intervals of at least month. After one year of negative cultures following the onset of antimicrobial therapy, any isolation of *P. aeruginosa* is considered to represent a new isolate (Gibson, Burns, & Ramsey, 2003). Aggressive antimicrobial use at the early stage has proven in certain cases to be successful, with a number of patients having remained culture negative for *P. aeruginosa* for several years after treatment (Frederiksen & Hoiby, 1997; Ratjen, Doring, & Nikolaizik, 2001).

2.2.2 Antimicrobial Therapy for Chronic Infections

Once chronic infection has been established by *P. aeruginosa*, the high bacterial load present in the lung, as well as the phenotypic changes occurring in the pathogen complicates antimicrobial therapy. The high bacterial load and thickened mucus are barriers to the attainment of sufficient exposure of the entire bacterial population to bactericidal concentrations of antibiotics (Mendelman et al., 1985). Administration of insufficient concentrations of antibiotics adds increased selective pressure for resistant phenotypes, thereby enhancing the diversity of the population, lending further difficulties to effective treatment (MacLeod et al., 2000; Pai & Nahata, 2001).

Antimicrobial therapy is used during chronic infections in CF for two main purposes: maintenance therapy and treatment of acute exacerbations of infection (Doring et al., 2000). Maintenance therapy is recommended for CF patients with chronic *P. aeruginosa* infections in order to reduce bacterial load and maintain overall lung function. Unfortunately, a number of side effects are associated with long-term antimicrobial use including loss of hearing, increased cough, alterations of the voice, and the appearance of antibiotic resistant strains. The use of on/off cycles of intermittent drug administration led to the reduced occurrence of these side effects (Ramsey et al., 1999).

3. Pathogenesis and Major Virulence Factors

Pathogenesis in *P. aeruginosa* is mediated by multiple bacterial virulence factors that facilitate adhesion and/or disrupt host cell signaling pathways while targeting the extracellular matrix (Figure 1). *P. aeruginosa* stands out as a unique and threating organism as it is capable of causing severe invasive disease and of evading immune defenses causing persisting infections that are nearly impossible to eradicate (Pier & Ramphal, 2005). The subsequent tissue damage, invasion, and dissemination of *P. aeruginosa* are likely attributed to the many virulence factors it produces. These virulence factors play an initial role in motility and adhesion to the epithelium. These virulence factors are thought to be critical for maximum virulence of *P. aeruginosa*; however, based on observations of diverse plant and animal models, the relative contribution of any given factor may vary with the type of infection (Preston et al., 1995; Rahme et al., 1997; Tamura, Suzuki, & Sawada, 1992; Tamura, Suzuki, Kijima, Takahashi, & Nakamura, 1992; Tang et al., 1996). Several of these virulence factors have also been studied for their roles as potential vaccine candidate although there is currently no general accepted vaccine. The following section briefly outlines several prominent putative virulence factors produce by *P. aeruginosa* and their proposed roles in contributing to disease.

3.1 Lipopolysaccharide

The LPS is a predominant component of the outer membrane of *P. aeruginosa*. Bacterial LPS typically consist of a hydrophobic domain known as lipid A (or endotoxin), a non-repeating core oligosaccharide, and a distal polysaccharide (or O-antigen) (Raetz & Whitfield, 2002). The composition of O-antigen determines the serotypes of the *P. aeruginosa* isolate and currently there are 20 serotypes based on serological reactivity of the O-antigen (Liu & Wang, 1990). LPS plays a prominent role in activation the host's innate (TLR4, NLRP1, NLRP2, and NLRP3) and adaptive (or acquired) immune responses; and, eventually causes dysregulated inflammation responses that contribute to morbidity and mortality (Heine, Rietschel, & Ulmer, 2001).

Recognition of LPS occurs largely by TLR4–MD2–CD14 complex, which is present on many cell types including macrophages and dendritic cells. Recognition of lipid A also requires an accessory protein, LPS-binding protein (LBP), which converts oligomeric micelles of LPS to a monomer for delivery to CD14, which is high-affinity membrane protein that can also circulate in a soluble form (Poltorak et al., 1998; Wright, Ramos, Tobias, Ulevitch, & Mathison, 1990; Shimazu et al., 1999; Qureshi et al., 1999; Hoshino et al., 1999). In addition, NLRs regulate both inflammation and pyroptosis. The activation of NLRs results in an assembly of complex structures called inflammasomes (Franchi, Muñoz-Planillo, & Núñez, 2012). The NLRP1 inflammasome was first described in 2002 in human monocytes as a molecular compound that responds to LPS (Martinon, Burns, & Tschopp, 2002). Many stimuli that trigger assembly of the inflammasomes have been described. LPS also reported to activate NLRP3 when administered in the presence of ATP (Stutz, Golenbock, & Latz, 2009), as well as NLRP2 (Lamkanfi, & Dixit, 2009). A number of LPS vaccines have been investigated for use in CF patients in phase II and III clinical trials; however, these have not been successful (Doring & Pier, 2008; Hanessian, Regan, Watson, & Haskell, 1971; Langford & Hiller, 1984; Pennington, 1979; Pennington & Miler, 1979). The LPS based vaccines provided little immunity and did not appear to protect the patients from infection with *P. aeruginosa* (Pier, 2003).

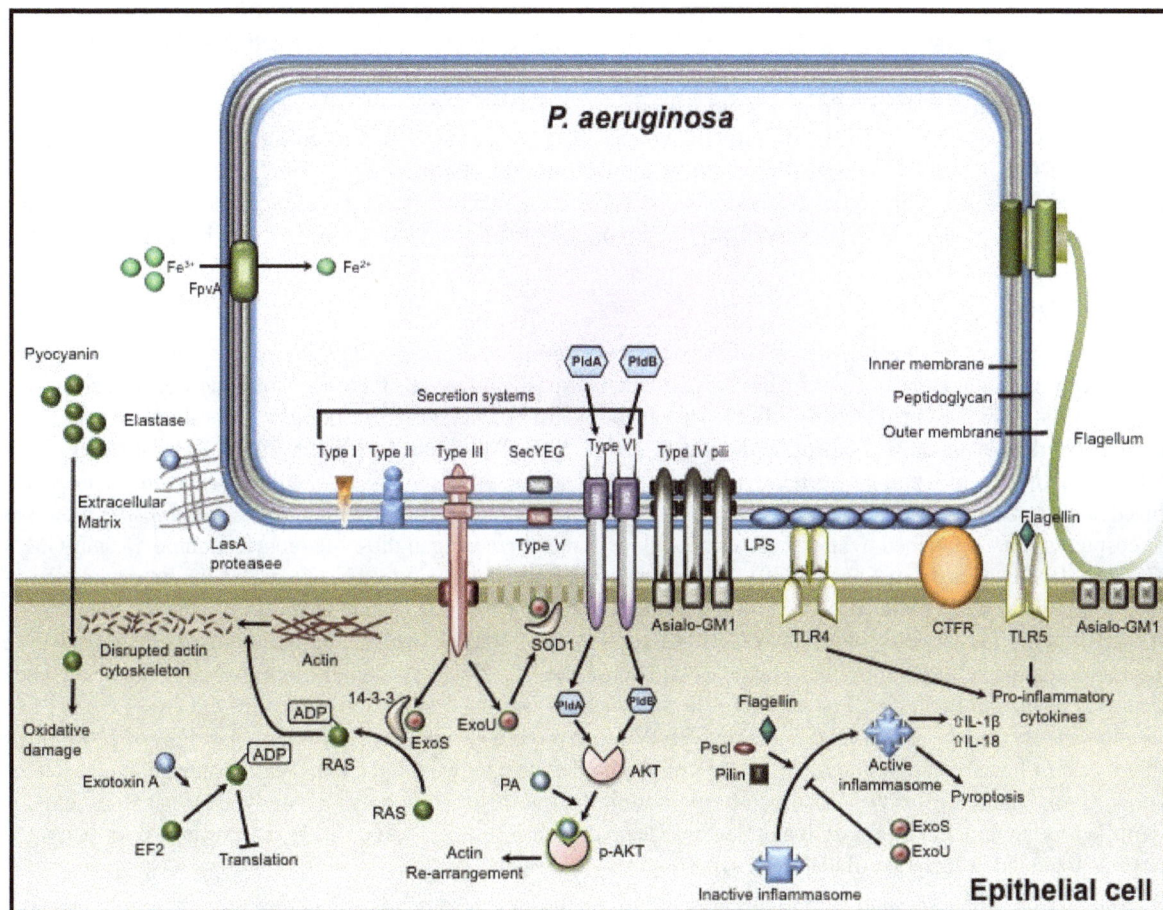

Figure 1. *P. aeruginosa* Pathogenesis and Major Virulence Factors

Pathogenesis in *P. aeruginosa* is mediated by various adhesions and secreted toxins, proteases, effector proteins and pigments that facilitate adhesion, modulate or disrupt host cell pathways, and, target the extracellular matrix. Figure has been recreated from Hauser and Ozer (2011). Abbreviations: **ADP**, adenosine diphosphate; **Asialo-GM1**, asialo-gangliotetraocyl ceramide 1; **EF2**, elongation factor 2; **FpvA**, feeric pyoverdine receptor; **PA,** phosphatidic acid; **RAS**, ribosyltransferase; **SOD1**, superoxide dismutase 1; **14-3-3**, 14-3-3 protein family.

3.2 Flagellum

The single unsheathed polar flagellum of *P. aeruginosa* is responsible for the swimming motility of this organism (Kohler, Curty, Barja, van Delden, &Pechere, 2000). Nonetheless, its role in virulence goes beyond simple motility. Flagellar proteins have been shown to play critical roles in attachment, invasion, biofilm formation and the mediation of inflammatory responses. Flagellar protein synthesis, assembly and regulation involves more than 40 genes and is intricately controlled through transcriptional and post-translational events by the four primary regulators RpoN, FleQ, FleR and FliA (Dasgupta et al., 2003).

Non-flagellated mutants are often isolated from chronic infections in CF patients (Mahenthiralingam, Campbell, & Speert, 1994) due to the repressor activity of AlgT, which acts on the FleQ regulator (Tart, Wolfgang, & Wozniak, 2005). The loss of flagella in these isolates is believed to be useful for the invasion of the host immune system. Flagellin mediators the inflammatory response via the innate immune system, through its specific interaction with a number of pattern recognition receptors (PRRs) of the host (Verma, Arora, Kuravi, & Ramphal, 2005). Flagellin is recognized by both TLR5 (Hayashi et al., 2001) and NLRC4 (Franchi et al., 2007; Cohen, & Prince, 2013); as well NLRP3 (Kim & Jo, 2013). However, different amino acid residues of flagellin were critical for sensing by NLRC4 and TLR5 (Franchi et al., 2007). Moreover, cytosolic delivery of *P. aeruginosa* flagellin is required for the activation of NLRC4 (Miao, Ernst, Dors, Mao, & Aderem, 2008). Surprisingly, the NLRC4 inflammasome can be activated independently of flagellin. The flagellin-deficient strains of *P. aeruginosa* can efficiently activate caspase-1 in an NLRC4-dependent manner. This discrepancy as to the requirement for flagellin in NLRC4 inflammasome activation was recently explained in an elegant study by Maio and colleagues (Miao et al., 2010). They found the NLRC4 inflammasome was activated in response to the basal body rod component of the T3SS apparatus from *P. aeruginosa* (Pscl), as well as, other microorganisms; such as *S. typhimurium* (PrgJ), *Burkholderia pseudomallei* (BsaK), *Escherichia coli* (EprJ and EscI), *S. flexneri* (MxiL) (Miao et al., 2010). These rod proteins contain a sequence motif that resembles one found in flagellin; hence, NLRC4 is activated by both of these similar stimuli. Furthermore, flagellar vaccines have been investigated in pre-clinical studies in mouse models and have reached phase III clinical trials for CF patients; however, limited protection was observed with a monovalent vaccine and development of a bivalent vaccine has been terminated (Doring & Pier, 2008).

3.3 Type IV Pili

The type IV pili of *P. aeruginosa* have a role in adhesion to many cell types and this is likely important in such phenomena as tissue tropism (attachment to particular tissues), initiation of biofilm formation and non-opsonic phagocytosis, which is mediated by phagocyte receptors that recognize corresponding adhesions on microbial surfaces (Barken et al., 2008; Mahenthiralingam & Speert, 1995; Punsalang & Sawyer, 1973). Several studies have indicated a direct correlation between the presence of glycosphingolipid on host cells and *P. aeruginosa* adherence, thus demonstrating the role of this glycosphingolipid as a bacterial receptor. In particular, *P. aeruginosa* pili bind to glycosphingolipid contained within host epithelial cell membranes; ganglio-N-tetraosylceramide (asialo-GM1) (Comolli, Waite, Mostov, & Engel, 1999). The interaction of pili and asialo-GM1 is mediated the internalization of *P. aeruginosa* in host epithelial cells. In addition, these pili also mediate twitching motility, a factor found to be important in the formation of biofilms *in vitro* (Klausen et. al., 2003); as well as, in the initiation of dissemination from an initial point of colonization (Hahn, 1997, Klausen et. al., 2003). Although more than 50 genes have been identified to play either a direct or indirect role in the synthesis, functioning and control of the type IV pili of *P. aeruginosa*, the pili are composed of a single type IV pilim protein encoded by *pilA* (Hansen & Forest, 2006). Five alleles of *pilA* have been identified with group I pili being the most prevalent in CF and environmental isolates (Kus, Tullis, Cvitkovitch, & Burrows, 2004). *P. aeruginosa* pilin, the major component of the type IV bacterial pilus, is identified as an inflammasome-activator factor (Cecilia, Arlehamn, & Evans, 2011); as purified pilin activated caspase-1 and led to secretion of mature IL-1β.

3.4 Type III Secretion System

P. aeruginosa has a variety of secretion systems of which at least four likely play a role in virulence (Type I, II, III, and IV). One of the most intriguing is T3SS that involves a flagellum-basal-body related system for delivering proteins directly from the cytoplasm of *P. aeruginosa* into the cytosol of host cells. A functional T3SS contributes to the successful evasion of phagocytosis by *P. aeruginosa* as well as damage to host tissues, promotion of immune avoidance and bacterial dissemination. The T3SS of *P. aeruginosa* delivers up to four cytotoxins ExoS, ExoT, ExoU and ExoY, directly to host cells (Frank, 1997; Goranson, Hovey, & Frank, 1997; Yahr, Mende-Mueller, Friese, & Frank, 1997).

ExoS and ExoT are bifunctional cytotoxins that possess both Rho GTPase-activating protein and ADP ribosyltransferase activities. These molecules can inhibit phagocytosis by disrupting actin cytoskeletal rearrangement, focal adhesions and signal transduction (Barbieri & Sun, 2004). Moreover, ExoU is a phospholipase that contributes directly to acute cytotoxicity towards epithelial cells and macrophages; while ExoY is an adenylate cyclase that affects intracellular cAMP levels and cytoskeleton reorganization (Sato & Frank, 2004; Yahr, Vallis, Hancock, Barbieri, &Frank, 1998). Recent evidence has implicated a role for the

T3SS in virulence in humans. The presence of large amount of T3SS products, particularly ExoU, in *P. aeruginosa* cultures from intubated patients was linked to increased mortality regardless of whether these patients had symptoms or confirmation of VAP (Zhuo et al., 2008). Also, *P. aeruginosa* T3SS activates NLR inflammasome. However, in the absence of any of the known effector proteins, *P. aeruginosa* T3SS apparatus is sufficient to trigger the activation of caspase-1 by the inflammasome via NLRC4 (Hauser, 2009). In another words, a functional T3SS is critical for the induction of caspase-1 activity, IL-1β secretion and cell death, whereas the effectors ExoS, ExoT and ExoY are dispensable (Franchi et al., 2007).

3.5 Exotoxin A

There are several critical virulence factors that are secret through Type II secretion mechanism, which use a pilus-like apparatus to secrete proteins into the extracellular environment, including exotoxin A, lipase, phospholipase, alkaline phosphatase, and protease; animal experiments have indicated the significant role of these factors in model infection (Passador & Iglewski, 1994). For example, exotoxin A has been demonstrated to be involved in local tissue damage and invasion. This cytotoxin is encoded by the gene toxA and has been found to be present in most clinical isolates of *P. aeruginosa*, although its role in virulence is poorly understood (Passador & Iglewski, 1994). Besides, exotoxin A enters host cells by receptor-mediated endocytosis and catalyzes the ADP-ribosylation of eukaryotic elongation factor-2 (EF-2) (Iglewski, Liu, & Kabat, 1977). EF-2 inhibits protein synthesis, which ultimately leading to cellular death.

3.6 Proteases

P. aeruginosa produces several secreted proteases including the zinc metalloprotease (elastase) LasB, the metalloendopeptidase LasA, and alkaline protease. These proteases work in a concerted fashion to destroy host tissue; thus, play a significant role in both acute lung infections and in burn wound infections (Bielecki, Glik, Kawecki, & Martins dos Santos, 2008; Galloway, 1991; Rumbaugh, Griswold, & Hamood, 2000). A definite role of these destructive proteases in acute infections has been established. LasA and LasB elastases have also been found in the sputum of CF patients suffering from pulmonary exacerbations of infection (Hollsing, Granstrom, Vasil, Wretlind, & Strandvik, 1987; Jagger et al., 1982); yet, their role in chronic infection is not well understood.

3.7 Alginate

P. aeruginosa can produce a mucoid exopolysaccharide capsule, comprised of alginate, an acetylated random co-polymer of β 1-4 linked D-mannuronic acid (poly-M) and L-guluronic acid (Gacesa & Wusteman, 1990). The overproduction of alginate is believed to play a role in cell adherence within the CF lung and is also thought to be involved in resistance to host defense by reducing susceptibility to phagocytosis (Pier, Coleman, Grout, Franklin, & Ohman, 2001), also in resistance to antibiotics. The small minorities of CF patients, who are carrying only nonmucoid *P. aeruginosa*, have significantly better lung function over time compared to those patients infected with mucoid *P. aeruginosa* (Parad, Gerard, Zurakowski, Nichols, & Pier, 1999). The tendency of *P. aeruginosa* to change to a mucoid phenotype is one of the most striking and clinically relevant features of infection by this bacterium. Additionally, poly-M shares with LPS the ability to stimulate human monocytes to cytokines production; in a CD-14-dependent manner (Otterlei et al., 1993). Involvement of TLR2 and TLR4 in cell activation by poly-M has been studied in primary murine macrophages (Flo et al., 2002).

3.8 Quorum Sensing

Quorum sensing is a mechanism of bacterial "cell-to-cell" communication via diffusible chemical compounds. A critical number of bacteria (the quorum) are required to produce a sufficient amount of a secreted signal molecule (termed an autoinducer) to trigger expression of a large regulon (Diggle, Cornelis, Williams, & Camara, 2006; Heurlier et al., 2004; Smith, Harris, Phipps, & Iglewski, 2002). Quorum sensing and biofilm development are two social phenomena exhibited by bacteria. The connection between quorum sensing and biofilms has been named sociomicrobiology (Bjarnsholt et al., 2005; Parsek & Greenberg, 2005). In addition, *P. aeruginosa* is regarded as a "model organism" in the quorum sensing field, which studied in the most detail. Quorum sensing is known to control a number of bacterial genes. More than 300 genes are regulated during quorum sensing in *P. aeruginosa* (Schuster et al., 2003).

The most common class of autoinducer used by Gram-negative bacteria is acyl-homoserine lactones (AHL), which diffuse freely across bacterial membranes. AHL signals produced by *P. aeruginosa* are oxohexanoyl-homoserine lactone and butanoyl-homoserine lactone (Pearson et al., 1994; Pearson et al., 1995). AHL signals produce by AHL synthase (LasI/RhlI), which diffuse into the environment. Increasing in bacterial density during infection leads to an increase in autoinducer concentration. When autoinducer reaches a particular threshold, it subsequently binds to transcriptional activator (LasR/RhlR) forming a complex that activates genes involved in biofilm formation and coding virulence factors (Kipnis, Sawa, & Wiener-Kronish, 2006; Davies et al., 1998; Hirakawa & Tomita, 2013). The production of virulence factors, such as extracellular enzymes and cellular lysins (e.g. rhamnolipid) are important for the pathogenesis of infections as a protective shield against phagocytes (Jensen et al., 2007; Van Gennip et al., 2009; Alhede et al., 2009). Quorum sensing has been shown to determine the tolerance of P. aeruginosa biofilms to antibiotic therapy (Bjarnsholt et al., 2005).

Recent advances in the understanding of quorum sensing in *P. aeruginosa* have generated interest in using quorum sensing as a target for therapeutics. The macrolide antibiotic, azithromycin, has been a promising candidate in this regard as it has been demonstrated to be capable of both penetrating biofilms and interfering with quorum sensing (Hoffmann et al., 2007).

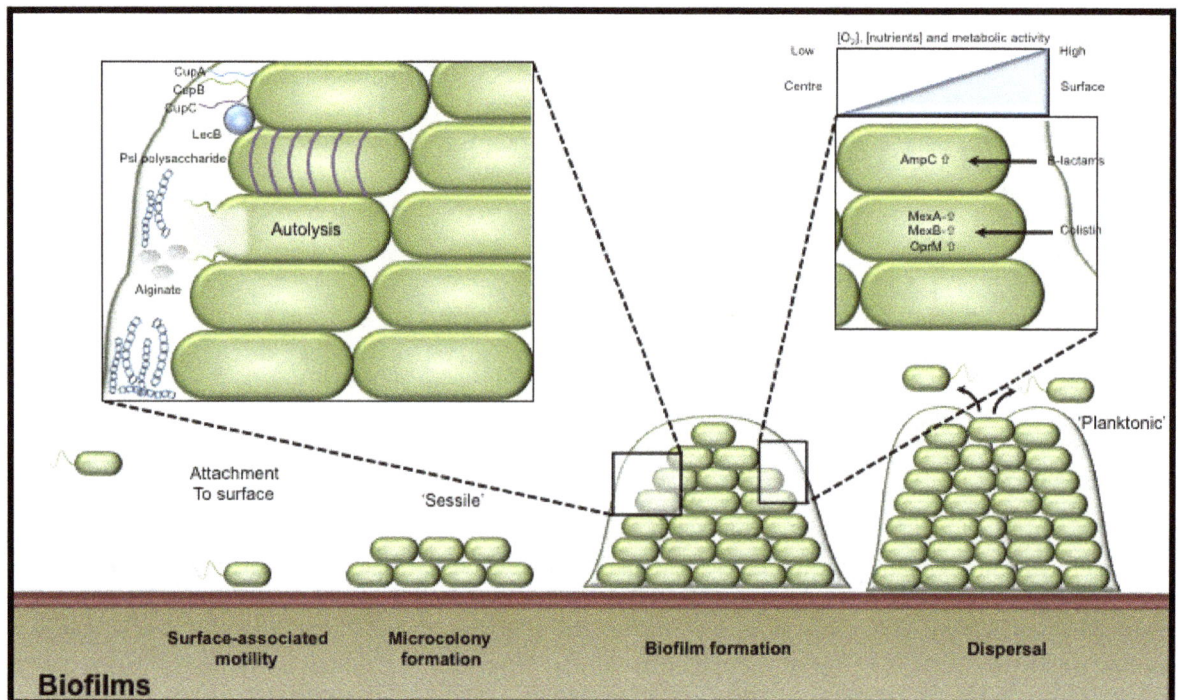

Figure 2. Development of a *P. aeruginosa* biofilm

Biofilm formation starts with the attachment of free-swimming bacteria (planktonic) to a surface via their type IV pili and flagellum, followed by twitching motility and the formation of microcolonies; then quorum sensing signals begin to accumulate. Once a critical threshold of quorum sensing signals is reached, microcolonies become encased in an extracellular matrix. Cells enter a sessile phase of growth and become highly resistant to antimicrobials, which evolve into mature biofilms. Biofilm architecture depends on the production of the biofilm matrix, which consists of the polysaccharides Pel (synthesized by PelA–PelG), Psl (arranged in a helical pattern around cells) and alginate, extracellular DNA (eDNA), and proteins, including the CupA, CupB and CupC fimbriae, which mediate bacterial attachment during initial biofilm formation, and the lectin LecB. The extracellular polymeric matrix delays diffusion of some antibiotics into the biofilm. A gradient of oxygen and nutrients induces the formation of distinct bacterial subpopulations that vary in their susceptibility to antibiotics; exposure to β-lactams or colistin can cause the production of resistance factors (AmpC β-lactamase and MexA–MexB–OprM efflux pumps). Rhamnolipids on bacteria at the surface induce necrosis of neutrophils. Finally, planktonic bacteria are released from parts of a mature biofilm. Individual cells and small microcolonies slough from the mature biofilm initiating further biofilm development. Figure has been recreated from Hauser and Ozer (2011).

3.9 Biofilm Formation

P. aeruginosa is capable of forming complex structures called biofilms. Resistance to antimicrobial agents is the most important features of biofilm infections. Biofilm development is a complex process and partly controlled by quorum sensing signals (Figure 2). Furthermore, a variety of components play a role in the initial attachment of cells to the surface and development of biofilm matrix including extracellular DNA (eDNA)(Whitchurch, Tolker-Nielsen, Ragas, and Mattick. 2002), exopolysaccharide (Psl, Pel, and alginate) (Ma et al., 2009; Karatan & Watnick, 2009), iron siderophore pyoverdine, biosurfactant rhamnolipid (Harmsen, Yang, Pamp, Tolker-Nielsen, 2010), and proteinaceous surface appendages such as type IV pili, flagella (Klausen, Aaes-Jorgensen, Molin, & Tolker-Nielsen, 2003), Cup fimbria (Vallet et al., 2001). Nevertheless, there are still numerous factors that are involved in biofilm formation process and dispersion, which are related to signals, regulatory networks, and materials, reviewed elsewhere (Harmsen, Yang, Pamp, Tolker-Nielsen, 2010; Wei & Ma, 2013; Karatan & Watnick, 2009).

During biofilm formation, cell differentiation occurs, and oxygen and water-filled channels are formed to provide nutrition to the deep-rooted cells of the mature biofilm (Davies, 2002; Friedman and Kolter, 2004; Kirisits, Prost, Starkey, & Parsek, 2005; Ryder, Byrd, & Wozniak, 2007). *P. aeruginosa* has been demonstrated to form biofilms on a variety of indwelling medical devices (Choong & Whitfield, 2000; Khoury, Lam, Ellis, & Costerton, 1992). It is particularly problematic for patients requiring mechanical ventilation and catheterization, as the surfaces of medical devices can readily develop *P. aeruginosa* biofilms that are difficult to remove. Also, *P. aeruginosa* has been demonstrated to grow as a biofilm within the body particularly at the site of burn wounds. It has been proposed that *P. aeruginosa* exists as a biofilm in the CF lung (Consterton, Stewart, & Greenberg, 1999; Singh et. al., 2000) and this has been observed in a mouse model of CF lung infection (Hoffmann et al., 2005).

In addition to evasion of the host immune system, the highly antibiotic resistant nature of biofilms to killing by bactericidal antibiotics contributes to bacterial persistence in chronic infections (Mah, Pitts, Pellock, Walker, Stewart, & O'Toole, 2003). It has been demonstrated that cells growing in a biofilm can be up to 1000 fold more resistant to antibiotics than free-swimming, planktonic cells (Hoyle & Costerton, 1991). Biofilms present not only a diffusion barrier to antibiotics, but also the cells in a biofilm have been demonstrated to have significantly different expression patterns compared to their planktonic counterparts (Sauer, Camper, Ehrlich, Costerton, & Davies, 2002).

3.10 Type VI Secretion Systems

Bacterial pathogens frequently possess number of secretion systems that function to translocate protein secretion. The T6SS represents one of the most recently recognized examples of these secretion systems. An interest in T6SS has led to its rapid study in *P. aeruginosa* in term of structure, mechanical function, assembly, and regulation of secretion (Mougous et al., 2006; Ho, Dong, & Mekalanos, 2014). *P. aeruginosa* T6SS provides defense against other bacteria in the environment (Ho, Dong, & Mekalanos, 2014; Russell, Peterson, & Mougous, 2014) and facilitates interactions with other eukaryotic (Jiang, Waterfield, Yang, Yang, & Jin, 2014). *P. aeruginosa* encodes three distinct T6SS, which are known as H1-, H2-, and H3-T6SS, each involved in bacterium's interaction with other organisms. The H1-T6SS delivers at least six toxic effectors into host bacteria and is a model for studying physiological function of T6SS antimicrobial activity (Ho, Dong, & Mekalanos, 2014; Russell, Peterson, & Mougous, 2014; Whitney et al., 2014). H2- and H3-T6SS have a dual role allowing interaction with both eukaryotic and prokaryotic target cells. The antibacterial activities mediate through H2-T6SS-dependent phospholipase D (PLD) PldA and H3-T6SS-dependent PldB. Both T6SS effectors, PldA and PldB, can degrade membrane phospholipids, resulting in antibacterial activity. T6SSs, H2-T6SS-dependent PldA and H3-T6SS-dependent PldB, have also been linked to *P. aeruginosa* immune evasion by promoting internalization into human epithelial cells (Jiang, Waterfield, Yang, Yang, & Jin, 2014). Interestingly, mutations in the catalytic domains of both PldA and PldB reduced *P. aeruginosa* internalization into epithelial cells, which shows that phospholipase activity is essential for invasion of the mammalian epithelium by *P. aeruginosa* (Jiang, Waterfield, Yang, Yang, & Jin, 2014).

Previous works have shown that the internalization of *P. aeruginosa* is dependent on the activation of the eukaryotic phosphoinositide 3-kinase (PI3K), which results in AKT phosphorylation in presence of phosphatidic acid subsequent actin rearrangement and protrusion formation (Jiang, Waterfield, Yang, Yang, & Jin, 2014; Sana et al., 2012; Kierbel, Gassama-Diagne, Mostov, & Engel, 2005). In addition, PI3K/AKT signaling pathway is crucial to a range of cellular processes including cell growth, proliferation, and programmed cell death (Krachler, Woolery, & Orth, 2011). Notably, epithelial cells that were infected with *P. aeruginosa* mutants (PldA, PldB, H2-T6SS, or H3-T6SS deficient) displayed reduced levels of AKT phosphorylation compared with the wild type strain. Furthermore, PldA and PldB were shown to bind to AKT, and both PldA-AKT and PldB-AKT complexes localized close to the epithelial cell plasma membrane (Jiang, Waterfield, Yang, Yang, & Jin, 2014). These data suggest that PldA and PldB have a central role in the activation of the PI3K-AKT signaling pathway to promote the invasion of epithelial cells by *P. aeruginosa*. Whereas *P. aeruginosa* is known to colonize the lungs of CF patients, previous studies have indicated that PldB and H3-T6SS loci are both up-regulated under low-oxygen conditions (Alvarez-Ortega & Harwood, 2007) and also during biofilm formation (Dötsch et al., 2012). For this reason, induction of both PldB and H3-T6SS in these conditions may allow better colonization of the lung epithelium through invasion. These findings reveal a function for the *P. aeruginosa* H3-T6SS effector PldB and show that PldB and PldA can influence both bacterial competition and interaction with mammalian hosts. T6SSs are promising targets for the development of new approaches to diagnosis, vaccine development and antimicrobial drug design (Yahr, 2006; Baron & Coombes, 2007).

3.11. Oxidant Generation in the Airspace

Oxidative stress refers to an imbalance in the redox status of the cell favoring an oxidizing environment. Extensive reactive oxygen species (ROS) production leads to the depletion of antioxidants and results in cellular damage. In particular, ROS can damage DNA strands by reacting with base pairs and the deoxyribose phosphate backbone of DNA, the primary target of radical damage (Gram, 1997). Without the protection of antioxidants, ROS can also initiate lipid peroxidation of polyunsaturated fatty acid components of cell membrane phospholipids, affecting cellular integrity (Rahman & Adcock, 2006). Amino acids can also be damaged by ROS, leading to protein denaturation and enzyme deactivation (Gram, 1997; Rahman & Adcock, 2006). When left unmanaged, oxidative stress can eventually lead to cell death.

During the infectious process, *P. aeruginosa* induces ROS production within epithelial cells in a few ways. Following its secretion into the microenvironment, pyocyanin permeates the epithelial cell membrane and directly oxidizes intracellular pools of NADPH and glutathione, producing superoxide and downstream ROS (Rada, Gardina, Myers, & Leto, 2011). Recognition of *P. aeruginosa* LPS by the epithelial cells leads to ROS production through protein kinase C (PKC)-NADPH oxidase signaling pathway in human epithelial cells (Yan, Li, Jono, Li, Zhang, Li, & Shen, 2008). Other potential sources of ROS are derived from the activated epithelium via induction of the mitochondrial electron transport chain, cytochrome p450, and xanthine oxidase. In the case of mechanical ventilation, the introduction of excess oxygen can also fuel the production of ROS (Chow, Herrera, Suzuki, & Downey, 2003). In acute lung injury, however, stimulated phagocytes produce the majority of ROS (Ward, 2010). Overwhelming oxidant injury may lead to alveolar collapse and extensive fibrotic scarring, impairing gas exchange between the affected airways and the capillary system (Ward, 2010).

4. Antimicrobial Resistance

P. aeruginosa can be an especially challenging organism to treat once infection has been established as it is intrinsically resistant to many of the available antibiotics. Three mechanisms have been studied by which *P. aeruginosa* resist the action of antibiotics. The outer membrane of *P. aeruginosa* is restricted the penetration of antibiotics and the efficient removal of antibiotics molecules by efflux pumps before acting on their targets. *P. aeruginosa* has the genetic capacity to inactivate and modify of antibiotics. This bacterium can become resistant through mutational changes in antibiotic's targets (Lambert, 2002). Consequently, *P. aeruginosa* has now achieved the status superbug. This section will provide an overview of the main mechanisms of resistance present in clinical isolates of *P. aeruginosa*.

4.1 Intrinsic Resistance

P. aeruginosa exhibits intrinsic resistance to almost all of the available antibiotics, indicating that the wild-type strain possesses number of genetic mechanisms that contribute to reducing susceptibility of the organisms. One of the major factors contributing to this intrinsic resistance is the low permeability of its outer membrane. The outer membrane is essential for passively determining the rate of uptake of antibiotics and small molecules (Hancock, 1998). However, by itself this is insufficient to mediate significant resistance, and antibiotics will equilibrate across the outer membrane. Thus, intrinsic resistance arises from the combination of slow uptake and secondary mechanisms that benefit from this slow uptake including degradative enzymes such as periplasmic β-lactamase and particularly multidrug efflux systems. There are at least four antibiotic efflux systems have been described including MexAB-OprM and MexXY-OprM, MexCD-OprI, and MexEF-oprN elsewhere (Hancock, 1998; Poole, 2001; Lamber, 2002; Lister, Wolter, & Hanson, 2009).

The *P. aeruginosa* outer membrane is an asymmetric membrane composed of an inner leaflet of phospholipid, predominantly phosphatidylethanolamine, an outer layer of polyanionic LPS. The latter presents a negatively charged surface, which, together with the divalent cations bridging the individual LPS molecules, forms a matrix around the cell that is relatively impermeable to polar compounds except polycations (Hancock, 1997). Multidrug efflux pumps also mediated resistance to many classes of antibiotics. The *P. aeruginosa* genome contains a large number of drug efflux systems (Stover et al., 2000), which have been categorized into five superfamilies include: the small multidrug resistance family, the ATP-binding cassette family, the multidrug and toxic compound extrusion family, the resistance-nodulation-cell division family, and major facilitator superfamily (Gotoh, Itoh, Tsujimoto, Yamagishi, Oyamada, & Nishino, 1994; Kohler, Michea-Hamzehpour, Plesiat, Kahr, & Pechere, 1997; Masuda, Sakagawa, Ohya, Gotoh, Tsujimoto, & Nishino, 2000; Masuda, Sakagawa, Ohya, Gotoh, Tsujimoto, & Nishino, 2000; Lister, Wolter, & Hanson, 2009).

P. aeruginosa also expresses periplasmic β-lactamases to degrade β-lactan antibiotics. β-lactamases are hydrolyzing enzymes that cleave the lactam ring of penicillins, carbapenems, cephalosporins and monobactams, thus leading to inactivation of the antibiotic (Richmond and Sykes, 1973; Sanders, 1992; Sykes and Matthew, 1976). In *P. aeruginosa*, this activity is due to a chromosomally encoded AmpC β-lactamase. The AmpC β-lactamase of *P. aeruginosa* can degrade and contribute to intrinsic resistance to ticarcillin, piperacillin and the third-generation cephalosporins. It is strongly induced by carbapenems, particularly imipenem, although these inducing carbapenems are stable against its hydrolytic activity (Balasubramanian et al., 2012).

4.2 Adaptive Resistance

Discrepancies between in *in vitro* susceptibility of *P. aeruginosa* isolates and treatment outcomes in CF patients have been observed, and, can be attributed to the phenomenon of adaptive resistance. Adaptive resistance occurs when cell populations are exposed to non-lethal concentrations of antibiotic and undergo specific changes in gene expression that result in reduced susceptibility. It is a form of inducible resistance that does not require the presence of mutations; it has been demonstrated in in vitro using CF isolates and in mouse models, when isolates were pre-incubated with subinhibitory concentrations of antibiotics (Barclay, Begg, Chambers, Thornley, Pattemore, & Grimwood, 1996; Brazas and Hancock, 2005; Gilleland, Gilleland, Gibson, & Champlin, 1989). However, the concern arises that this induction may allow small population to survive and acquire stably resistant mutations. Similarly, polymyxin susceptibility in *P. aeruginosa* is associated with the LPS structure, which is determined by arnBCADTEF and modulated by PhoPQ and PmrAB (Barrow and Kwon, 2009). Adaptive resistance to polymyxins and antimicrobial peptides has been shown to occur through altered expression of the PhoPQ and PmrAB systems in response to these agents. This also leads to modulation of Lipid A fatty acid composition (Moskowitz, Ernst, & Miller, 2004), which ultimately affects resistance to not only polymyxin and antimicrobial peptides, but also the aminoglycosides, which rely on the LPS binding for self-promoted uptake.

4.3 P. aeruginosa as a Superbug

The accumulation of multiple resistance mechanisms in clinical isolates of *P. aeruginosa* has resulted in strains that are resistance to all available antibiotics. This pandrug resistance, which means resistant to all antimicrobial agents, together with high attributable mortality, has thrust *P. aeruginosa* into the spotlight as an emerging superbug. According to report by the National Nosocomial Infections Surveillance (NNIS) System, which focused on nosocomial infections in ICU, not only were resistance rates increasing, but the incidence of occurrence of most infection types was also increasing (Gaynes and Edwards, 2005). In 2003, the NNIS reported a 9% increase in resistance to the third generation cephalosporins, a 15% increase in ciprofloxacin resistance, and most alarming, a 47% increase in imipenem resistance over a five year period. According to the European Antimicrobial Resistance Surveillance System 18% of *P. aeruginosa* isolates were multidrug resistance with 6% of all isolates being resistant to piperacillin, ceftazidime, fluoroquinolones, aminoglycosides and carbapenems (Souli, Galani, & Giamarellou, 2008). The drug of last resort for infections with multidrug resistance *P. aeruginosa* is colistin (polymyxin E), and while resistance rates remain low (approximately 1% in most countries), mortality of 80% has been observed for infections caused by colistin resistance Gram-negative bacilli (Beno, Krcmery, & Demitrovicova, 2006).

5. Conclusion

In conclusion, it would be impossible to remove *P. aeruginosa* from the environment, even the internal environment of the hospital because it is so hardy and metabolically versatile. However, the last two decades have seen a remarkable addition to active medication and therapy to the regimen for treating CF lung infection. These therapies have enhanced the overall health of patients with CF and, they are apparently part of the reason that demanded survival has increased. However, these therapies do not offer a cure and they primarily treat downstream complications of the pathophysiology of CF lung infection, meaning that patients persist to undergo the morbidity associated with chronic airway infection; predicted survival still lags well below what is normal. However, innate immunity is critical in protecting the host from bacterial invasion, but at the same time it can directly & indirectly damage tissues. In addition, these therapies add to a considerable treatment burden and are thus also associated with poor adherence. What is more, it appears likely that the antibiotic resistance will continue to be a problem in dealing with *P. aeruginosa* infections. The fundamental issues underlying this problem are the condition of the patients that are prone to such infections, and, the high intrinsic resistance of this bacterium, which have remained constant. Chronic *P. aeruginosa* airway infection and the accompanying inflammatory response are clearly the significant clinical problems for CF patients today. Instead, we need an urgent alternative therapeutic strategy.

Acknowledgment

This work was supported by scholarship award granted to A. Alhazmi, from Jazan University, through Saudi Arabian Cultural Bureau in Canada.

References

Alhede, M., Bjarnsholt, T., Jensen, P. Ø., Phipps, R. K., Moser, C., Christophersen, L., ... & Givskov, M. (2009). Pseudomonas aeruginosa recognizes and responds aggressively to the presence of polymorphonuclear leukocytes. *Microbiology, 155*(11), 3500-3508. http://dx.doi.org/10.1099/mic.0.031443-0

Altoparlak, U., Erol, S., Akcay, M. N., Celebi, F., & Kadanali, A. (2004). The timerelated changes of antimicrobial resistance patterns and predominant bacterial profiles of burn wounds and body flora of burned patients. *Burns, 30*, 660-664. http://dx.doi.org/10.1016/j.burns.2004.03.005

Alvarez ‑ Ortega, C., & Harwood, C. S. (2007). Responses of Pseudomonas aeruginosa to low oxygen indicate that growth in the cystic fibrosis lung is by aerobic respiration. *Molecular microbiology, 65*(1), 153-165. http://dx.doi.org/10.1111/j.1365-2958.2007.05772.x

Baker, C. C., Miller, C. L., & Trunkey, D. D. (1979). Predicting fatal sepsis in burn patients. *J Trauma, 19*, 641-648. http://dx.doi.org/0022-5282/79/1909-0641$02.00/0

Balasubramanian, D, Schneper, L, Merighi, M, Smith, R, & Narasimhan, G. (2012). The Regulatory Repertoire of Pseudomonas aeruginosa AmpC ß-Lactamase Regulator AmpR Includes Virulence Genes. *PLoS ONE, 7*(3), e34067. http://dx.doi.org/10.1371/journal.pone.0034067

Bals, R., Weiner, D. J., & Wilson, J. M. (1999). The innate immune system in cystic fibrosis lung disease. *J Clin Invest, 103*, 303-7. http://dx.doi.org/10.1172/JCI6277

Bang, R. L., Sharma, P. N., Sanyal, S. C., & Al Najjadah, I. (2002). Septicaemia after burn injury: a comparative study. *Burns, 28*, 746-51. http://dx.doi.org/10.1016/S0305-4179(02)00183-3

Barbieri, J. T., & Sun, J. 2004. *Pseudomonas aeruginosa* ExoS and ExoT. Rev Physiol Biochem Pharmacol 152:79-92. http://dx.doi.org/10.1007/s10254-004-0031-7

Barclay, M. L., Begg, E. J., Chambers, S. T., Thornley, P. E., Pattemore, P. K., & Grimwood, K. (1996). Adaptive resistance to tobramycin in *Pseudomonas aeruginosa* lung infection in cystic fibrosis. *J Antimicrob Chemother, 37*, 1155-64. http://dx.doi.org/10.1093/jac/37.6.1155

Barken, K. B., Pamp, S. J., Yang, L., Gjermansen, M., Bertrand, J. J., Klausen, M., ... & Tolker ‑ Nielsen, T. (2008). Roles of type IV pili, flagellum ‑ mediated motility and extracellular DNA in the formation of mature multicellular structures in Pseudomonas aeruginosa biofilms. *Environmental microbiology, 10*(9), 2331-2343. http://dx.doi.org/10.1111/j.1462-2920.2008.01658.x

Baron, C., & Coombes, B. (2007). Targeting bacterial secretion systems: benefits of disarmament in the microcosm. *Infectious Disorders-Drug Targets (Formerly Current Drug Targets-Infectious Disorders), 7*(1), 19-27. http://dx.doi.org/10.2174/187152607780090685

Barret, J. P., & Herndon, D. N. (2003). Effects of burn wound excision on bacterial colonization and invasion. *Plastic and reconstructive surgery, 111*(2), 744-50. http://dx.doi.org/10.1097/01.PRS.0000041445.76730.23

Barrow, K., & Kwon, D. H. (2009). Alterations in two-component regulatory systems of phoPQ and pmrAB are associated with polymyxin B resistance in clinical isolates of Pseudomonas aeruginosa. *Antimicrob Agents Chemother, 53*(12), 5150-5154. http://dx.doi.org/10.1128/AAC.00893-09

Barrow, R. E., Spies, M., Barrow, L. N., & Herndon, D. N. (2004). Influence of demographics and inhalation injury on burn mortality in children. *Burns, 30*, 72-7. http://dx.doi.org/10.1016/j.burns.2003.07.003

Beno, P., Krcmery, V., & Demitrovicova, A. (2006). Bacteraemia in cancer patients caused by colistin-resistant Gram-negative bacilli after previous exposure to ciprofloxacin and/or colistin. *Clin Microbiol Infect, 12*, 497-8. http://dx.doi.org/10.1111/j.1469-0691.2006.01364.x

Bielecki, P., Glik, J., Kawecki, M., & dos Santos, V. A. M. (2008). Towards understanding Pseudomonas aeruginosa burn wound infections by profiling gene expression. *Biotechnology letters, 30*(5), 777-790. http://dx.doi.org/10.1007/s10529-007-9620-2

Bjarnsholt, T., Jensen, P. Ø., Burmølle, M., Hentzer, M., Haagensen, J. A., Hougen, H. P., ... & Givskov, M. (2005). Pseudomonas aeruginosa tolerance to tobramycin, hydrogen peroxide and polymorphonuclear leukocytes is quorum-sensing dependent. *Microbiology, 151*(2), 373-383. http://dx.doi.org/10.1099/mic.0. 27463-0

Bjarnsholt, T., Jensen, P. Ø., Burmølle, M., Hentzer, M., Haagensen, J. A., Hougen, H. P., ... & Givskov, M. (2005). Pseudomonas aeruginosa tolerance to tobramycin, hydrogen peroxide and polymorphonuclear leukocytes is quorum-sensing dependent. *Microbiology, 151*(2), 373-383. http://dx.doi.org/10.1099/mic.0. 27463-0

Blondel-Hill, E., & Fryters, S. (2006). Bugs and Drugs, vol. Capital Health, Edmonton.

Bobadilla J, Jr M, Fine, J., & Farrell, P. (2002). Cystic Fibrosis: A Worldwide Analysis of *CFTR* Mutations –Correlation With Incidence Data and Application to Screening. *Human Mutation, 19*, 575-606. http://dx.doi.org/10.1002/humu.10041

Bodey, G. P., Jadeja, L., & Elting, L. (1985). Pseudomonas bacteremia: retrospective analysis of 410 episodes. *Archives of Internal Medicine, 145*(9), 1621-1629. http://dx.doi.org/10.1001/archinte.1985.00360090089015

Brazas, M. D., & Hancock, R. E. W. (2005). Ciprofloxacin induction of a susceptibility determinant in *Pseudomonas aeruginosa. Antimicrob Agents Chemother, 49*, 3222-7. http://dx.doi.org/10.1128/AAC.49.8. 3222-3227.2005

Chastre, J., & Fagon, J. Y. (2002). Ventilator-associated pneumonia. *Am J Respir Crit Care Med, 165*, 867-903. http://dx.doi.org/10.1164/ajrccm.165.7.2105078

Choong, S., & Whitfield, H. (2000). Biofilms and their role in infections in urology. *BJU Int, 86*, 935-41. http://dx.doi.org/10.1046/j.1464-410x.2000.00949.x

Church, D., Elsayed, S., Reid, O., Winston, B., & Lindsay, R. (2006). Burn wound infections. *Clin Microbiol Rev, 19*, 403-34. http://dx.doi.org/10.1128/CMR.19.2.403-434.2006

Cobb, L. M., Mychaleckyj, J. C., Wozniak, D. J., & Lopez-Boado, Y. S. (2004). *Pseudomonas aeruginosa* flagellin and alginate elicit very distinct gene expression patterns in airway epithelial cells: implications for cystic fibrosis disease. *J Immunol, 173*, 5659-70. http://dx.doi.org/10.4049/jimmunol.173.9.5659

Cohen, T. S., Prince, A. S. (2013). Activation of inflammasome signaling mediates pathology of acute P: aeruginosa pneumonia. *J Clin Invest., 123*, 1630–1637. http://dx.doi.org/10.1172/JCI66142

Collins, F. S. (1992). Cystic fibrosis: molecular biology and therapeutic implications. *Science, 256*, 774-779. http://dx.doi.org/10.1126/science.256.5058.774

Comolli, J. C., Waite, L. L., Mostov, K. E., & Engel, J. N. (1999). Pili binding to asialo-GM1 on epithelial cells can mediate cytotoxicity or bacterial internalization byPseudomonas aeruginosa. *Infection and immunity, 67*(7), 3207-3214.

Costerton, J. W., Stewart, P. S., & Greenberg, E. P. (1999). Bacterial biofilms: a common cause of persistent infections. *Science, 284*, 1318-22. http://dx.doi.org/10.1126/science.284.5418.1318

Dasgupta, N., Wolfgang, M. C., Goodman, A. L., Arora, S. K., Jyot, J., Lory, S., & Ramphal, R. (2003). A four - tiered transcriptional regulatory circuit controls flagellar biogenesis in Pseudomonas aeruginosa. *Molecular microbiology, 50*(3), 809-824. http://dx.doi.org/10.1046/j.1365-2958.2003.03740.x

Davies, D. G., Parsek, M. R., Pearson, J. P., Iglewski, B. H., Costerton, J. W., & Greenberg, E. P. (1998). The involvement of cell-to-cell signals in the development of a bacterial biofilm. *Science, 280*(5361), 295-298. http://dx.doi.org/10.1126/science.280.5361.295

Davies, J. C. (2002). *Pseudomonas aeruginosa* in cystic fibrosis: pathogenesis and persistence. *Paediatr Respir Rev, 3*, 128-34. http://dx.doi.org/10.1016/S1526-0550(02)00003-3

Diggle, S. P., Cornelis, P., Williams, P., & Camara, M. (2006). 4-quinolone signalling in *Pseudomonas aeruginosa*: old molecules, new perspectives. *Int J Med Microbiol, 296*, 83-91. http://dx.doi.org/10.1016/j.ijmm.2006.01.038

DiMango, E., Zar, H. J., Bryan, R., Prince, A. (1995). Diverse *Pseudomonas aeruginosa* products stimulate respiratory epithelial cells to produce interleukin-8. *J Clin Invest., 96*, 2204-10. http://dx.doi.org/10.1172/JCI118275

Doggett, R. G. (1969). Incidence of mucoid *Pseudomonas aeruginosa* from clinical sources. *Appl Microbiol, 18*, 936-7.

Doggett, R. G. (1979). Microbiology of *Pseudomonas aeruginosa*. In R. G. Doggett (Ed.), *Pseudomonas aeruginosa: Clinical Manifestations of Infection and Current Therapy* (pp. 1-8). New York: Academic Press.

Doggett, R. G., Harrison, G. M., & Carter, R. E. (1971). Mucoid *Pseudomonas aeruginosa* in patients with chronic illnesses. *Lancet, 1*, 236-7. http://dx.doi.org/10.1016/S0140-6736(71)90973-1

Donaldson, S. H., Bennett, W. D., Zeman, K. L., Knowles, M. R., Tarran, R., & Boucher, R. C. (2006). Mucus clearance and lung function in cystic fibrosis with hypertonic saline. *N Engl J Med, 354*, 241-50. http://dx.doi.org/10.1056/NEJMoa043891

Doring, G., & Pier, G. B. (2008). Vaccines and immunotherapy against *Pseudomonas aeruginosa*. *Vaccine, 26*, 1011-24. http://dx.doi.org/10.1016/j.vaccine.2007.12.007

Döring, G., Bragonzi, A., Paroni, M., Akturk, F. F., Cigana, C., Schmidt, A., ... Ulrich, M. (2014). BIIL 284 reduces neutrophil numbers but increases P. aeruginosa bacteremia and inflammation in mouse lungs. *J Cyst Fibros, 13*(2), 156-163. http://dx.doi.org/10.1016/j.jcf.2013.10.007

Döring, G., Conway, S. P., Heijerman, H. G. M., Hodson, M. E., Høiby, N., Smyth, A., ... & Consensus Committee. (2000). Antibiotic therapy against Pseudomonas aeruginosa in cystic fibrosis: a European consensus. *European Respiratory Journal, 16*(4), 749-767. http://dx.doi.org/10.1034/j.1399-3003.2000. 16d30.x

Dötsch, A., Eckweiler, D., Schniederjans, M., Zimmermann, A., Jensen, V., Scharfe, M., ... & Häussler, S. (2012). The Pseudomonas Aeruginosa Transcriptome in Planktonic Cultures and Static Biofilms Using Rna Sequencing. *Plos One, 7*(2), E31092. http://dx.doi.org/10.1371/journal.pone.0031092

Ernst, R. K., Moskowitz, S. M., Emerson, J. C., Kraig, G. M., Adams, K. N., Harvey, M. D., ... & Miller, S. I. (2007). Unique lipid A modifications in Pseudomonas aeruginosa isolated from the airways of patients with cystic fibrosis. *Journal of Infectious Diseases, 196*(7), 1088-1092. http://dx.doi.org/10.1086/521367

Fagon, J. Y., Chastre, J., Hance, A. J., Montravers, P., Novara, A., & Gibert, C. (1993). Nosocomial pneumonia in ventilated patients: a cohort study evaluating attributable mortality and hospital stay. *The American journal of medicine, 94*(3), 281-288. http://dx.doi.org/10.1016/0002-9343(93)90060-3

Faure, K., Fujimoto, J., Shimabukuro, D. W., Ajayi, T., Shime, N., Moriyama, K., ... & Sawa, T. (2003). Effects of monoclonal anti-PcrV antibody on Pseudomonas aeruginosa-induced acute lung injury in a rat model. *Journal of immune based therapies and vaccines, 1*(1), 2. http://dx.doi.org/10.1186/1476-8518-1-2

Favero, M. S., Carson, L. A., Bond, W. W., & Petersen, N. J. (1971). Pseudomonas aeruginosa: growth in distilled water from hospitals. *Science, 173*(3999), 836-838. http://dx.doi.org/10.1126/science.173.3999.836

Fishman, L. S., & Armstrong, D. (1972). *Pseudomonas aeruginosa* bacteremia in patients with neoplastic disease. *Cancer, 30*, 764-73. http://dx.doi.org/10.1002/1097-0142(197209)30:3<764::AID-CNCR2820300326>3.0.CO; 2-G

Flo, T. H., Ryan, L., Latz, E., Takeuchi, O., Monks, B. G., Lien, E., ... & Espevik, T. (2002). Involvement of toll-like receptor (TLR) 2 and TLR4 in cell activation by mannuronic acid polymers. *Journal of Biological Chemistry, 277*(38), 35489-35495. http://dx.doi.org/10.1074/jbc.M201366200

Foweraker, J. E., Laughton, C. R., Brown, D. F. J., & Bilton, D. (2005). Phenotypic variability of Pseudomonas aeruginosa in sputa from patients with acute infective exacerbation of cystic fibrosis and its impact on the validity of antimicrobial susceptibility testing. *Journal of Antimicrobial Chemotherapy, 55*(6), 921-927. http://dx.doi.org/10.1093/jac/dki146

Franchi, L., Muñoz-Planillo, R., & Núñez, G. (2012). Sensing and reacting to microbes through the inflammasomes. *Nat Immunol., 13*, 325-32. http://dx.doi.org/10.1038/ni.2231

Franchi, L., Stoolman, J., Kanneganti, T. D., Verma, A., Ramphal, R., & Núñez, G. (2007). Critical role for Ipaf in Pseudomonas aeruginosa‐induced caspase‐1 activation. *European journal of immunology, 37*(11), 3030-3039. http://dx.doi.org/10.1002/eji.200737532

Frank, D. W. (1997). The exoenzyme S regulon of *Pseudomonas aeruginosa*. *Mol Microbiol, 26*, 621-9. http://dx.doi.org/10.1046/j.1365-2958.1997.6251991.x

Frederiksen, B., Koch, C., & Hoiby, N. (1997). Antibiotic treatment of initial colonization with *Pseudomonas aeruginosa* postpones chronic infection and prevents deterioration of pulmonary function in cystic fibrosis. *Pediatr Pulmonol, 23*, 330-335. http://dx.doi.org/10.1002/(SICI)1099-0496(199705)23:5<330::AID-PPUL4 >3.0.CO;2-O

Freeman, L. (1916). Chronic General Infection with the Bacillus Pyocyaneus. *Ann Surg, 64*,195-202.

Friedman, L., & Kolter, R. (2004). Genes involved in matrix formation in *Pseudomonas aeruginosa* PA14 biofilms. *Mol Microbiol,* *51,* 675-90. http://dx.doi.org/10.1002/(SICI)1099-0496(199705)23:5<330:: AID-PPUL4>3.0.CO;2-O

Fuqua, C., & Greenberg, E. P. (2002). Listening in on bacteria: acyl-homoserine lactone signalling. *Nat Rev Mol Cell Biol, 3,* 685-95. http://dx.doi.org/10.1038/nrm907

Gacesa, P., & Wusteman, F. S. (1990). Plate assay for simultaneous detection of alginate lyases and determination of substrate specificity. *Appl Environ Microbiol, 56,* 2265-2267.

Galloway, D. R. (1991). *Pseudomonas aeruginosa* elastase and elastolysis revisited: recent developments. *Mol Microbiol, 5,* 2315-21. http://dx.doi.org/10.1111/j.1365-2958.1991.tb02076.x

Gaynes, R., & Edwards, J. R. (2005). Overview of nosocomial infections caused by gram-negative bacilli. *Clin Infect Dis, 41,* 848-54. http://dx.doi.org/10.1086/432803

Gibson, R. L., Burns, J. L., & Ramsey, B. W. (2003). Pathophysiology and management of pulmonary infections in cystic fibrosis. *Am J Respir Crit Care Med, 168,* 918-51. http://dx.doi.org/10.1164/rccm.200304-505SO

Gilleland, L. B., Gilleland, H. E., Gibson, J. A., & Champlin, F. R. (1989). Adaptive resistance to aminoglycoside antibiotics in Pseudomonas aeruginosa. *Journal of medical microbiology, 29*(1), 41-50. http://dx.doi.org/10.1099/00222615-29-1-41

Goranson, J., Hovey, A. K., & Frank, D. W. (1997). Functional analysis of exsC and exsB in regulation of exoenzyme S production by Pseudomonas aeruginosa. *Journal of bacteriology, 179*(5), 1646-1654.

Gotoh, N., Itoh, N., Tsujimoto, H., Yamagishi, J. I., Oyamada, Y., & Nishino, T. (1994). Isolation of OprM - deficient mutants of Pseudomonas aeruginosa by transposon insertion mutagenesis: Evidence of involvement in multiple antibiotic resistance. *FEMS microbiology letters, 122*(3), 267-273. http://dx.doi.org/10.1111/j. 1574-6968.1994.tb07179.x

Govan, J. R., & Deretic, V. (1996). Microbial pathogenesis in cystic fibrosis: mucoid *Pseudomonas aeruginosa* and Burkholderia cepacia. *Microbiol Rev, 60,* 539-74.

Gram. (1997). Chemically reactive intermediates and pulmonary xenobiotic toxicity. *Pharmacol rev, 49*(4), 297-341.

Greene, C. M., Carroll, T. P., Smith, S. G., Taggart, C. C., Devaney, J., Griffin, S., ... & McElvaney, N. G. (2005). TLR-induced inflammation in cystic fibrosis and non-cystic fibrosis airway epithelial cells. *The Journal of Immunology, 174*(3), 1638-1646. http://dx.doi.org/10.4049/jimmunol.174.3.1638

Grimwood, K. (1992). The pathogenesis of *Pseudomonas aeruginosa* lung infections in cystic fibrosis. *J Paediatr Child Health, 28,* 4-11. http://dx.doi.org/10.1111/j.1440-1754.1992.tb02609.x

Hahn, H. P. W. (1997). The type-4 pilus is the major virulence-associated adhesin of *Pseudomonas aeruginosa*--a review. *Gene, 192,* 99-108. http://dx.doi.org/10.1016/S0378-1119(97)00116-9

Hancock, R. E. W. (1997). The bacterial outer membrane as a drug barrier. *Trends Microbiol, 5,* 37-42. http://dx.doi.org/10.1016/S0966-842X(97)81773-8

Hancock, R. E. W. (1998). Resistance mechanisms in Pseudomonas aeruginosa and other nonfermentative gram-negative bacteria. *Clinical Infectious Diseases, 27*(Supplement 1), S93-S99. http://dx.doi.org/10.1086/ 514909

Hancock, R. E., Mutharia, L. M., Chan, L., Darveau, R. P., Speert, D. P., & Pier, G. B. (1983). Pseudomonas aeruginosa isolates from patients with cystic fibrosis: a class of serum-sensitive, nontypable strains deficient in lipopolysaccharide O side chains. *Infection and Immunity, 42*(1), 170-177.

Hanessian, S., Regan, W., Watson, D., & Haskell, T. H. (1971). Isolation and characterization of antigenic components of a new heptavalent Pseudomonas vaccine. *Nat New Biol, 229,* 209-10. http://dx.doi.org/10.1038/newbio229209a0

Hansen, J. K., & Forest, K. T. (2006). Type IV pilin structures: insights on shared architecture, fiber assembly, receptor binding and type II secretion. *J Mol Microbiol Biotechnol, 11,* 192-207. http://dx.doi.org/10.1159/ 000094054

Harmsen, M., Yang, L., Pamp, S. J., & Tolker - Nielsen, T. (2010). An update on Pseudomonas aeruginosa biofilm formation, tolerance, and dispersal. *FEMS Immunology & Medical Microbiology, 59*(3), 253-268.

Hauser, A. R. (2009). The type III secretion system of Pseudomonas aeruginosa: infection by injection. *Nature Reviews Microbiology, 7*(9), 654-665. http://dx.doi.org/10.1038/nrmicro2199

Hauser, R. A., & Ozer, A. E. (2011). Pseudomonas aeruginosa. *Nature Review Microbiology, 9*(3). Poster produced with support from Cubist Pharmaceuticals.

Häußler, S., Ziegler, I., Löttel, A., Götz, F. V., Rohde, M., Wehmhöhner, D., ... & Steinmetz, I. (2003). Highly adherent small-colony variants of Pseudomonas aeruginosa in cystic fibrosis lung infection. *Journal of medical microbiology, 52*(4), 295-301. http://dx.doi.org/10.1099/jmm.0.05069-0

Hayashi, F., Smith, K. D., Ozinsky, A., Hawn, T. R., Yi, E. C., Goodlett, D. R., ... & Aderem, A. (2001). The innate immune response to bacterial flagellin is mediated by Toll-like receptor 5. *Nature, 410*(6832), 1099-1103. http://dx.doi.org/10.1038/35074106

Heine, H., Rietschel, E. T., & Ulmer, A. J. 2001. The biology of endotoxin. *Mol Biotechnol, 19,* 279-96. http://dx.doi.org/10.1385/MB:19:3:279

Henrichfreise, B., Wiegand, I., Pfister, W., & Wiedemann, B. 2007. Resistance mechanisms of multiresistant *Pseudomonas aeruginosa* strains from Germany and correlation with hypermutation. *Antimicrob Agents Chemother, 51,* 4062-70. http://dx.doi.org/10.1128/AAC.00148-07

Heurlier, K., Williams, F., Heeb, S., Dormond, C., Pessi, G., Singer, D., ... & Haas, D. (2004). Positive control of swarming, rhamnolipid synthesis, and lipase production by the posttranscriptional RsmA/RsmZ system in Pseudomonas aeruginosa PAO1. *Journal of bacteriology, 186*(10), 2936-2945. http://dx.doi.org/10.1128/JB.186.10.2936-2945.2004

Hidron, A. I., Edwards, J. R., Patel, J., Horan, T. C., Sievert, D. M., Pollock, D. A., & Fridkin, S. K. (2008). Antimicrobial ‐ resistant pathogens associated with healthcare ‐ associated infections: annual summary of data reported to the National Healthcare Safety Network at the Centers for Disease Control and Prevention, 2006–2007. *infection control and hospital epidemiology, 29*(11), 996-1011. http://dx.doi.org/10.1086/591861

Hirakawa, H., & Tomita, H. (2013). Interference of bacterial cell-to-cell communication: a new concept of antimicrobial chemotherapy breaks antibiotic resistance. *Frontiers in microbiology, 4.* http://dx.doi.org/10.3389/fmicb.2013.00114

Ho, B. T., Dong, T. G., & Mekalanos, J. J. (2014). A view to a kill: the bacterial type VI secretion system. *Cell host & microbe, 15*(1), 9-21. http://dx.doi.org/10.1016/j.chom.2013.11.008

Hodson, M. E. (2000). Treatment of cystic fibrosis in the adult. *Respiration, 67,* 595-607. http://dx.doi.org/10.1159/000056287

Hodson, M., Penketh, A. R. L., & Batten, J. C. (1981). Aerosol carbenicillin and gentamicin treatment of Pseudomonas aeruginosa infection in patients with cystic fibrosis. *The Lancet, 318*(8256), 1137-1139. http://dx.doi.org/10.1016/S0140-6736(81)90588-2

Hoffmann, N., Lee, B., Hentzer, M., Rasmussen, T. B., Song, Z., Johansen, H. K., ... & Høiby, N. (2007). Azithromycin blocks quorum sensing and alginate polymer formation and increases the sensitivity to serum and stationary-growth-phase killing of Pseudomonas aeruginosa and attenuates chronic P. aeruginosa lung infection in Cftr−/− mice. *Antimicrobial agents and chemotherapy, 51*(10), 3677-3687. http://dx.doi.org/10.1128/AAC.01011-06

Hoffmann, N., Rasmussen, T. B., Jensen, P., Stub, C., Hentzer, M., Molin, S., ... & Høiby, N. (2005). Novel mouse model of chronic Pseudomonas aeruginosa lung infection mimicking cystic fibrosis. *Infection and immunity, 73*(4), 2504-2514. http://dx.doi.org/10.1128/IAI.73.8.5290.2005

Hogardt, M., Hoboth, C., Schmoldt, S., Henke, C., Bader, L., & Heesemann, J. (2007). Stage-specific adaptation of hypermutable Pseudomonas aeruginosa isolates during chronic pulmonary infection in patients with cystic fibrosis. *Journal of Infectious Diseases, 195*(1), 70-80. http://dx.doi.org/10.1086/509821

Hollsing, A. E., Granström, M., Vasil, M. L., Wretlind, B., & Strandvik, B. (1987). Prospective study of serum antibodies to Pseudomonas aeruginosa exoproteins in cystic fibrosis. *Journal of clinical microbiology, 25*(10), 1868-1874.

Hoshino, K., Takeuchi, O., Kawai, T., Sanjo, H., Ogawa, T., Takeda, Y., ... & Akira, S. (1999). Cutting edge: Toll-like receptor 4 (TLR4)-deficient mice are hyporesponsive to lipopolysaccharide: evidence for TLR4 as the Lps gene product. *The Journal of Immunology, 162*(7), 3749-3752.

Hoyle, B. D., & Costerton, J. W. (1991). Bacterial resistance to antibiotics: the role of biofilms. *Prog Drug Res, 37*, 91-105.

Hudson, V. L., Wielinski, C. L., & Regelmann, W. E. (1993). Prognostic implications of initial oropharyngeal bacterial flora in patients with cystic fibrosis diagnosed before the age of two years. *J Pediatr, 122*, 854-60. http://dx.doi.org/10.1016/S0022-3476(09)90007-5

Iglewski, B. H., Liu, P. V., & Kabat, D. A. V. I. D. (1977). Mechanism of action of Pseudomonas aeruginosa exotoxin Aiadenosine diphosphate-ribosylation of mammalian elongation factor 2 in vitro and in vivo. *Infection and immunity, 15*(1), 138-144.

Irvin, R. T., Govan, J. W., Fyfe, J. A., & Costerton, J. W. (1981). Heterogeneity of antibiotic resistance in mucoid isolates of Pseudomonas aeruginosa obtained from cystic fibrosis patients: role of outer membrane proteins. *Antimicrobial agents and chemotherapy, 19*(6), 1056-1063. http://dx.doi.org/10.1128/AAC.19.6.1056

Jagger, K. S., Robinson, D. L., Franz, M. N., & Warren, R. L. (1982). Detection by enzyme-linked immunosorbent assays of antibody specific for Pseudomonas proteases and exotoxin A in sera from cystic fibrosis patients. *Journal of clinical microbiology, 15*(6), 1054-1058.

Jensen, P. Ø., Bjarnsholt, T., Phipps, R., Rasmussen, T. B., Calum, H., Christoffersen, L., ... & Høiby, N. (2007). Rapid necrotic killing of polymorphonuclear leukocytes is caused by quorum-sensing-controlled production of rhamnolipid by Pseudomonas aeruginosa. *Microbiology, 153*(5), 1329-1338. http://dx.doi.org/10.1099/mic.0.2006/003863-0

Jiang, F., Waterfield, N. R., Yang, J., Yang, G., & Jin, Q. (2014). A Pseudomonas Aeruginosa Type Vi Secretion Phospholipase D Effector Targets both Prokaryotic and Eukaryotic Cells. *Cell Host & Microbe, 15*(5), 600-610. http://dx.doi.org/10.1016/j.chom.2014.04.010

Karatan, E., & Watnick, P. (2009). Signals, regulatory networks, and materials that build and break bacterial biofilms. *Microbiology and Molecular Biology Reviews, 73*(2), 310-347. http://dx.doi.org/10.1128/MMBR.00041-08

Khoury, A. E., Lam, K., Ellis, B., & Costerton, J. W. (1992). Prevention and control of bacterial infections associated with medical devices. *ASAIO journal, 38*(3), M174-M178.

Kierbel, A., Gassama-Diagne, A., Mostov, K., & Engel, J. N. (2005). The Phosphoinositol-3-Kinase–Protein Kinase B/Akt Pathway is Critical for Pseudomonas Aeruginosa Strain Pak Internalization. *Molecular Biology of the Cell, 16*(5), 2577-2585. http://dx.doi.org/10.1091/mbc.e04-08-0717

Kim, J. J., & Jo, E. K. (2013). NLRP3 inflammasome and host protection against bacterial infection. *Journal of Korean medical science, 28*(10), 1415-1423. http://dx.doi.org/10.3346/jkms.2013.28.10.1415

Kipnis, E., Sawa, T., & Wiener-Kronish, J. (2006). Targeting mechanisms of Pseudomonas aeruginosa pathogenesis. *Medecine et maladies infectieuses, 36*(2), 78-91. http://dx.doi.org/10.1016/j.medmal.2005.10.007

Kirisits, M. J., Prost, L., Starkey, M., & Parsek, M. R. (2005). Characterization of colony morphology variants isolated from Pseudomonas aeruginosa biofilms. *Applied and environmental microbiology, 71*(8), 4809-4821. http://dx.doi.org/10.1128/AEM.71.8.4809-4821.2005

Kirisits, M. J., Prost, L., Starkey, M., & Parsek. M. R. (2005). Characterization Of Colony Morphology Variants Isolated From Pseudomonas Aeruginosa Biofilms. *Appl Environ Microbiol, 71*, 4809-21. http://dx.doi.org/10.1128/aem.71.8.4809-4821.2005

Klausen, M., Aaes - Jørgensen, A., Molin, S., & Tolker - Nielsen, T. (2003). Involvement of bacterial migration in the development of complex multicellular structures in Pseudomonas aeruginosa biofilms. *Molecular microbiology, 50*(1), 61-68. http://dx.doi.org/10.1046/j.1365-2958.2003.03677.x

Klausen, M., Heydorn, A., Ragas, P., Lambertsen, L., Aaes - Jørgensen, A., Molin, S., & Tolker - Nielsen, T. (2003). Biofilm formation by Pseudomonas aeruginosa wild type, flagella and type IV pili mutants. *Molecular microbiology, 48*(6), 1511-1524. http://dx.doi.org/10.1046/j.1365-2958.2003.03525.x

Kobayashi, M., Yoshida, T., Takeuchi, D., Jones, V. C., Shigematsu, K., Herndon, D. N., & Suzuki, F. (2008). Gr-1+ CD11b+ cells as an accelerator of sepsis stemming from Pseudomonas aeruginosa wound infection in thermally injured mice. *Journal of leukocyte biology, 83*(6), 1354-1362. http://dx.doi.org/10.1189/jlb.0 807541

Köhler, T., Curty, L. K., Barja, F., van Delden, C., & Pechère, J. C. (2000). Swarming of Pseudomonas aeruginosa is dependent on cell-to-cell signaling and requires flagella and pili. *Journal of bacteriology, 182*(21), 5990-5996. http:/dx.doi.org/10.1128/JB.182.21.5990-5996.2000

Köhler, T., Michea-Hamzehpour, M., Plesiat, P., Kahr, A. L., & Pechere, J. C. (1997). Differential selection of multidrug efflux systems by quinolones in Pseudomonas aeruginosa. *Antimicrobial agents and chemotherapy*, *41*(11), 2540-2543.

Kollberg, H., Mossberg, B., Afzelius, B. A., Philipson, K., & Camner, P. (1977). Cystic fibrosis compared with the immotile-cilia syndrome. A study of mucociliary clearance, ciliary ultrastructure, clinical picture and ventilatory function. *Scandinavian journal of respiratory diseases*, *59*(6), 297-306. http://dx.doi.org/10.1172/JCI7124

Krachler, A. M., Woolery, A. R., & Orth, K. (2011). Manipulation of Kinase Signaling by Bacterial Pathogens. *The Journal of Cell Biology, 195*(7), 1083-1092. http://dx.doi.org/10.1083/jcb.201107132

Kurahashi, K., Kajikawa, O., Sawa, T., Ohara, M., Gropper, M. A., Frank, D. W., ... & Wiener-Kronish, J. P. (1999). Pathogenesis of septic shock in *Pseudomonas aeruginosa* pneumonia. *The Journal of clinical investigation, 104*(6), 743-750.

Kus, J. V., Tullis, E., Cvitkovitch, D. G., & Burrows, L. L. (2004). Significant differences in type IV pilin allele distribution among Pseudomonas aeruginosa isolates from cystic fibrosis (CF) versus non-CF patients. *Microbiology, 150*(5), 1315-1326. http://dx.doi.org/10.1099/mic.0.26822-0

Lambert, P. A. (2002). Mechanisms of antibiotic resistance in Pseudomonas aeruginosa. *Journal of the Royal Society of Medicine, 95*(Suppl 41), 22.

Lamkanfi, M., & Dixit, V. M. (2009). The inflammasomes. *PLoS pathogens, 5*(12), e1000510. http://dx.doi.org/10.1371/journal.ppat.1000510

Langford, D. T., & Hiller, J. (1984). Prospective, controlled study of a polyvalent Pseudomonas vaccine in cystic fibrosis--three year results. *Arch Dis Child, 59*, 1131-4. http://dx.doi.org/10.1136/adc.59.12.1131

Lartigau, A. J. (1898). A contribution to the study of the pathogenesis of the Bacillus Pyocyaneus, with special reference to its relation to an epidemic of dysentery. *J Exp Med, 3*, 595-609.

Lindestam Arlehamn, C. S., & Evans, T. J. (2011). Pseudomonas aeruginosa pilin activates the inflammasome. *Cellular microbiology, 13*(3), 388-401. http://dx.doi.org/10.1111/j.1462-5822.2010.01541.x

Lister, P. D., Wolter, D. J., & Hanson, N. D. (2009). Antibacterial-resistant Pseudomonas aeruginosa: clinical impact and complex regulation of chromosomally encoded resistance mechanisms. *Clinical microbiology reviews, 22*(4), 582-610. http://dx.doi.org/10.1128/CMR.00040-09

Liu, P. V., & Wang, S. (1990). Three new major somatic antigens of *Pseudomonas aeruginosa*. *J Clin Microbiol, 28*, 922-5.

Lyczak, J. B., Cannon, C. L., & Pier, G. B. (2000). Establishment of *Pseudomonas aeruginosa* infection: lessons from a versatile opportunist. *Microbes Infect, 2*, 1051-60. http://dx.doi.org/10.1016/S1286-4579(00)01259-4

Ma, L., Conover, M., Lu, H., Parsek, M. R., Bayles, K., & Wozniak, D. J. (2009). Assembly and development of the Pseudomonas aeruginosa biofilm matrix. *PLoS pathogens, 5*(3), e1000354. http://dx.doi.org/10.1371/journal.ppat.1000354

Maceachran, D. P., Ye, S., Bomberger, J. M., Hogan, D. A., Swiatecka-Urban, A., Stanton, B. A., & O'Toole, G. A. (2007). The Pseudomonas aeruginosa secreted protein PA2934 decreases apical membrane expression of the cystic fibrosis transmembrane conductance regulator. *Infect Immun, 75*(8), 3902-3912. http://dx.doi.org/10.1128/IAI.00338-07

MacLeod, D. L., Nelson, L. E., Shawar, R. M., Lin, B. B., Lockwood, L. G., Dirks, J. E., ... & Garber, R. L. (2000). Aminoglycoside-resistance mechanisms for cystic fibrosis Pseudomonas aeruginosa isolates are unchanged by long-term, intermittent, inhaled tobramycin treatment. *Journal of Infectious Diseases, 181*(3), 1180-1184. http://dx.doi.org/10.1086/315312

MacMillan, B. G. (1980). Infections following burn injury. *Surg Clin North Am, 60*, 185-96.

Mah, T. F., Pitts, B., Pellock, B., Walker, G. C., Stewart, P. S., & O'Toole, G. A. (2003). A genetic basis for Pseudomonas aeruginosa biofilm antibiotic resistance. *Nature, 426*(6964), 306-310. http://dx.doi.org/10.1038/nature02122

Mahenthiralingam, E., & Speert, D. P. (1995). Nonopsonic phagocytosis of *Pseudomonas aeruginosa* by macrophages and polymorphonuclear leukocytes requires the presence of the bacterial flagellum. *Infect Immun, 63*, 4519-23.

Mahenthiralingam, E., Campbell, M. E., & Speert, D. P. (1994). Nonmotility and phagocytic resistance of Pseudomonas aeruginosa isolates from chronically colonized patients with cystic fibrosis. *Infection and Immunity*, *62*(2), 596-605.

Masuda, N., Sakagawa, E., Ohya, S., Gotoh, N., Tsujimoto, H., & Nishino, T. (2000). Contribution of the MexX-MexY-OprM efflux system to intrinsic resistance in Pseudomonas aeruginosa. *Antimicrobial agents and chemotherapy*, *44*(9), 2242-2246.

Masuda, N., Sakagawa, E., Ohya, S., Gotoh, N., Tsujimoto, H., & Nishino, T. (2000). Substrate specificities of MexAB-OprM, MexCD-OprJ, and MexXY-oprM efflux pumps in Pseudomonas aeruginosa. *Antimicrobial agents and chemotherapy*, *44*(12), 3322-3327. http://dx.doi.org/10.1128/AAC.44.9.2242-2246.2000

Mendelman, P. M., Smith, A. L., Levy, J., Weber, A., Ramsey, B., & Davis, R. L. (1985). Aminoglycoside penetration, inactivation, and efficacy in cystic fibrosis sputum. *The American review of respiratory disease*, *132*(4), 761-765.

Miao, E. A., Ernst, R. K., Dors, M., Mao, D. P., & Aderem, A. (2008). Pseudomonas aeruginosa activates caspase 1 through Ipaf. Proc. Natl. Acad. Sci. USA 105, 2562–2567. http://dx.doi.org/10.1073/pnas.0712183105

Miao, E. A., Ernst, R. K., Dors, M., Mao, D. P., & Aderem, A. (2008). Pseudomonas Aeruginosa activates caspase 1 through Ipaf. *Proc. Natl. Acad. Sci. USA 105*, 2562–2567. http://dx.doi.org/10.1073/pnas.0712183105

Miao, E. A., Mao, D. P., Yudkovsky, N., Bonneau, R., Lorang, C. G., Warren, S. E., ... & Aderem, A. (2010). Innate immune detection of the type III secretion apparatus through the NLRC4 inflammasome. *Proceedings of the National Academy of Sciences*, *107*(7), 3076-3080. http://dx.doi.org/10.1073/pnas.0913087107

Moreau-Marquis, S., Stanton, B. A., & O'Toole, G. A. (2008). *Pseudomonas aeruginosa* biofilm formation in the cystic fibrosis airway. *Pulmonary pharmacology & therapeutics*, *21*(4), 595-599. http://dx.doi.org/10.1016/j.pupt.2007.12.001

Moskowitz, S. M., Ernst, R. K., & Miller, S. I. (2004). PmrAB, a two-component regulatory system of Pseudomonas aeruginosa that modulates resistance to cationic antimicrobial peptides and addition of aminoarabinose to lipid A. *Journal of bacteriology*, *186*(2), 575-579. http://dx.doi.org/10.1128/JB.186.2.575-579.2004

Nixon, G. M., Armstrong, D. S., Carzino, R., Carlin, J. B., Olinsky, A., Robertson, C. F., & Grimwood, K. (2001). Clinical outcome after early *Pseudomonas aeruginosa* infection in cystic fibrosis. *The Journal of pediatrics*, *138*(5), 699-704. http://dx.doi.org/10.1067/mpd.2001.112897

Oliver, A., Cantón, R., Campo, P., Baquero, F., & Blázquez, J. (2000). High frequency of hypermutable Pseudomonas aeruginosa in cystic fibrosis lung infection. *Science*, *288*(5469), 1251-1253. http://dx.doi.org/10.1126/science.288.5469.1251

Otterlei, M., Sundan, A., Skjåk-Bræk, G., Ryan, L., Smidsrød, O., & Espevik, T. (1993). Similar mechanisms of action of defined polysaccharides and lipopolysaccharides: characterization of binding and tumor necrosis factor alpha induction. *Infection and immunity*, *61*(5), 1917-1925.

Pai, V. B., & Nahata, M. C. (2001). Efficacy and safety of aerosolized tobramycin in cystic fibrosis. *Pediatr Pulmonol, 32*, 314-27. http://dx.doi.org/10.1002/ppul.1125

Parad, R. B., Gerard, C. J., Zurakowski, D., Nichols, D. P., & Pier, G. B. (1999). Pulmonary outcome in cystic fibrosis is influenced primarily by mucoid Pseudomonas aeruginosa infection and immune status and only modestly by genotype. *Infection and immunity*, *67*(9), 4744-4750.

Parsek, M. R., & Greenberg, E. P. (2005). Sociomicrobiology: the connections between quorum sensing and biofilms. *Trends in microbiology*, *13*(1), 27-33. http://dx.doi.org/10.1016/j.tim.2004.11.007

Passador, L., & Iglewski, W. (1994). ADP-ribosylating toxins. *Methods Enzymol, 235*, 617-31. http://dx.doi.org/10.1016/0076-6879(94)35175-9

Pearson, J. P., Gray, K. M., Passador, L., Tucker, K. D., Eberhard, A., Iglewski, B. H., & Greenberg, E. P. (1994). Structure of the autoinducer required for expression of Pseudomonas aeruginosa virulence genes. *Proceedings of the National Academy of Sciences*, *91*(1), 197-201. http://dx.doi.org/10.1073/pnas.91.1.197

Pearson, J. P., Passador, L., Iglewski, B. H., & Greenberg, E. P. (1995). A second N-acylhomoserine lactone signal produced by Pseudomonas aeruginosa. *Proceedings of the National Academy of Sciences*, *92*(5), 1490-1494. http://dx.doi.org/10.1073/pnas.92.5.1490

Pennington, J. E. (1979). Lipopolysaccharide Pseudomonas vaccine: efficacy against pulmonary infection with *Pseudomonas aeruginosa. J Infect Dis, 140*, 73-80. http://dx.doi.org/10.1093/infdis/140.1.73

Pennington, J. E., & Miler, J. J. (1979). Evaluation of a new polyvalent Pseudomonas vaccine in respiratory infections. *Infect Immun, 25*, 1029-34.

Pier, G. B. (2003). Promises and pitfalls of *Pseudomonas aeruginosa* lipopolysaccharide as a vaccine antigen. *Carbohydr Res, 338*, 2549-56. http://dx.doi.org/10.1016/S0008-6215(03)00312-4

Pier, G. B., & Ames, P. (1984). Mediation of the killing of rough, mucoid isolates of *Pseudomonas aeruginosa* from patients with cystic fibrosis by the alternative pathway of complement. *J Infect Dis, 150*, 223-8. http://dx.doi.org/10.1093/infdis/150.2.223

Pier, G. B., & Ramphal, R. (2005). *Pseudomonas aeruginosa*. In G. L. Mandell & J. E. Bennett (Ed.), Mandell, Douglas, and Bennett's principles and practice of infectious diseases (vol. 2, pp. 2587-2615). Elsevier/Churchill Livingstone, New York.

Pier, G. B., Coleman, F., Grout, M., Franklin, M., & Ohman, D. E. (2001). Role of alginate O acetylation in resistance of mucoid Pseudomonas aeruginosa to opsonic phagocytosis. *Infection and immunity, 69*(3), 1895-1901. http://dx.doi.org/10.1128/IAI.69.3.1895-1901.2001

Pier, G. B., Grout, M., & Zaidi, T. S. (1997). Cystic fibrosis transmembrane conductance regulator is an epithelial cell receptor for clearance of Pseudomonas aeruginosa from the lung. *Proceedings of the National Academy of Sciences, 94*(22), 12088-12093.

Pier, G. B., Grout, M., Zaidi, T. S., Olsen, J. C., Johnson, L. G., Yankaskas, J. R., & Goldberg, J. B. (1996). Role of mutant CFTR in hypersusceptibility of cystic fibrosis patients to lung infections. *Science, 271*(5245), 64-67. http://dx.doi.org/10.1126/science.271.5245.64

Poltorak, A., He, X., Smirnova, I., Liu, M. Y., Van Huffel, C., Du, X., ... & Beutler, B. (1998). Defective LPS signaling in C3H/HeJ and C57BL/10ScCr mice: mutations in Tlr4 gene. Science, 282(5396), 2085-2088. http://dx.doi.org/10.1126/science.282.5396.2085

Poole, K. (2001). Multidrug efflux pumps and antimicrobial resistance in Pseudomonas aeruginosa and related organisms. *J Mol Microbiol Biotechnol, 3*(2), 255-264.

Preston, M. J., Fleiszig, S. M., Zaidi, T. S., Goldberg, J. B., Shortridge, V. D., Vasil, M. L., & Pier, G. B. (1995). Rapid and sensitive method for evaluating Pseudomonas aeruginosa virulence factors during corneal infections in mice. *Infection and immunity, 63*(9), 3497-3501.

Punsalang, A. P., & Sawyer, W. D. (1973). Role of pili in the virulence of Neisseria gonorrhoeae. *Infection and immunity, 8*(2), 255-263.

Qureshi, S. T., Larivière, L., Leveque, G., Clermont, S., Moore, K. J., Gros, P., & Malo, D. (1999). Endotoxin-tolerant mice have mutations in Toll-like receptor 4 (Tlr4). *The Journal of experimental medicine, 189*(4), 615-625. http://dx.doi.org/10.1084/jem.189.4.615

Rada, B., Gardina, P., Myers, T. G., & Leto, T. L. (2011). Reactive oxygen species mediate inflammatory cytokine release and egfr-dependent mucin secretion in airway epithelial cells exposed to pseudomonas pyocyanin. *Mucosal immunol, 4*(2), 158-171. http://dx.doi.org/10.1038/mi.2010.62

Raetz, C. R., & Whitfield, C. (2002). Lipopolysaccharide endotoxins. *Annu Rev Biochem, 71*, 635-700. http://dx.doi.org/10.1146/annurev.biochem.71.110601.135414

Rahman, I., & Adcock, I. M. (2006). Oxidative stress and redox regulation of lung inflammation in COPD. *European Respiratory Journal, 28*(1), 219-242.

Rahme, L. G., Tan, M. W., Le, L., Wong, S. M., Tompkins, R. G., Calderwood, S. B., & Ausubel, F. M. (1997). Use of model plant hosts to identify Pseudomonas aeruginosa virulence factors. *Proceedings of the National Academy of Sciences, 94*(24), 13245-13250.

Ramsey, B. W., Pepe, M. S., Quan, J. M., Otto, K. L., Montgomery, A. B., Williams-Warren, J., ... & Smith, A. L. (1999). Intermittent administration of inhaled tobramycin in patients with cystic fibrosis. *New England Journal of Medicine, 340*(1), 23-30. http://dx.doi.org/10.1056/NEJM199901073400104

Ratjen, F., & Doring, G. (2003). Cystic fibrosis. *Lancet, 361*, 681-9. http://dx.doi.org/10.1016/S0140-6736 (03)12567-6

Ratjen, F., Döring, G., & Nikolaizik, W. H. (2001). Effect of inhaled tobramycin on early Pseudomonas aeruginosa colonisation in patients with cystic fibrosis. *The Lancet, 358*(9286), 983-984. http://dx.doi.org/10.1016/S0140-6736(01)06124-4

Rello, J., Ricart, M., Mirelis, B., Quintana, E., Gurgui, M., Net, A., & Prats, G. (1994). Nosocomial bacteremia in a medical-surgical intensive care unit: epidemiologic 61 characteristics and factors influencing mortality in 111 episodes. *Intensive Care Med, 20*, 94-8. http://dx.doi.org/10.1007/BF01707661

Richmond, M. H., & Sykes, R. B. (1973). The beta-lactamases of gram-negative bacteria and their possible physiological role. *Adv Microb Physiol, 9*, 31-88. http://dx.doi.org/10.1016/S0065-2911(08)60376-8

Rosenfeld, M., Ramsey, B. W., & Gibson, R. L. (2003). Pseudomonas acquisition in young patients with cystic fibrosis: pathophysiology, diagnosis, and management. *Curr Opin Pulm Med, 9*, 492-7.

Rumbaugh, K. P., Griswold, J. A., & Hamood, A. N. (2000). The role of quorum sensing in the in vivo virulence of *Pseudomonas aeruginosa. Microbes and infection, 2*(14), 1721-1731. http://dx.doi.org/10.1016/S1286-4579(00)01327-7

Ryder, C., Byrd, M., & Wozniak, D. J. (2007). Role of polysaccharides in *Pseudomonas aeruginosa* biofilm development. *Current opinion in microbiology, 10*(6), 644-648. http://dx.doi.org/10.1016/j.mib.2007. 09.010

Salunkhe, P., Smart, C. H., Morgan, J. A. W., Panagea, S., Walshaw, M. J., Hart, C. A., ... & Winstanley, C. (2005). A cystic fibrosis epidemic strain of Pseudomonas aeruginosa displays enhanced virulence and antimicrobial resistance. *Journal of bacteriology, 187*(14), 4908-4920. http://dx.doi.org/10.1128/JB.187.14.4908-4920. 2005

Sana, T. G., Hachani, A., Bucior, I., Soscia, C., Garvis, S., Termine, E., ... & Bleves, S. (2012). The second type VI secretion system of Pseudomonas aeruginosa strain PAO1 is regulated by quorum sensing and Fur and modulates internalization in epithelial cells. *Journal of Biological Chemistry, 287*(32), 27095-27105. http://dx.doi.org/10.1074/jbc.M112.376368

Sanders, C. C. (1992). beta-Lactamases of gram-negative bacteria: new challenges for new drugs. *Clin Infect Dis, 14*, 1089-99.62. http://dx.doi.org/10.1093/clinids/14.5.1089

Sato, H., & Frank, D. W. (2004). ExoU is a potent intracellular phospholipase. *Mol Microbiol, 53*, 1279-90. http://dx.doi.org/10.1111/j.1365-2958.2004.04194.x

Sauer, K., Camper, A. K., Ehrlich, G. D., Costerton, J. W., & Davies, D. G. (2002). Pseudomonas aeruginosa displays multiple phenotypes during development as a biofilm. *Journal of bacteriology, 184*(4), 1140-1154. http://dx.doi.org/10.1128/jb.184.4.1140-1154.2002

Schroeder, T. H., Lee, M. M., Yacono, P. W., Cannon, C. L., Gerçeker, A. A., Golan, D. E., & Pier, G. B. (2002). CFTR is a pattern recognition molecule that extracts Pseudomonas aeruginosa LPS from the outer membrane into epithelial cells and activates NF-κB translocation. *Proceedings of the National Academy of Sciences, 99*(10), 6907-6912. http://dx.doi.org/10.1073/pnas.092160899

Schuster, M., Lostroh, C. P., Ogi, T., & Greenberg, E. P. (2003). Identification, timing, and signal specificity of Pseudomonas aeruginosa quorum-controlled genes: a transcriptome analysis. *Journal of bacteriology, 185*(7), 2066-2079. http://dx.doi.org/10.1128/JB.185.7.2066-2079.2003

Shimazu, R., Akashi, S., Ogata, H., Nagai, Y., Fukudome, K., Miyake, K., & Kimoto, M. (1999). MD-2, a molecule that confers lipopolysaccharide responsiveness on Toll-like receptor 4. The Journal of experimental medicine, 189(11), 1777-1782. http://dx.doi.org/10.1084/jem.189.11.1777

Shime, N., Sawa, T., Fujimoto, J., Faure, K., Allmond, L. R., Karaca, T., ... & Wiener-Kronish, J. P. (2001). Therapeutic administration of anti-PcrV F (ab′) 2 in sepsis associated with Pseudomonas aeruginosa. *The Journal of Immunology, 167*(10), 5880-5886. http://dx.doi.org/10.4049/jimmunol.167.10.5880

Singh, P. K., Schaefer, A. L., Parsek, M. R., Moninger, T. O., Welsh, M. J., & Greenberg, E. P. (2000). Quorum-sensing signals indicate that cystic fibrosis lungs are infected with bacterial biofilms. *Nature, 407*(6805), 762-764. http://dx.doi.org/10.1038/35037627

Skerrett, S. J., Wilson, C. B., Liggitt, H. D., & Hajjar, A. M. (2007). Redundant Toll-like receptor signaling in the pulmonary host response to Pseudomonas aeruginosa. *American Journal of Physiology-Lung Cellular and Molecular Physiology, 292*(1), L312-L322. http://dx.doi.org/10.1152/ajplung.00250.2006

Smith, R. S., Harris, S. G., Phipps, R., & Iglewski, B. (2002). The Pseudomonas aeruginosa quorum-sensing molecule N-(3-oxododecanoyl) homoserine lactone contributes to virulence and induces inflammation in vivo. *Journal of bacteriology, 184*(4), 1132-1139. http://dx.doi.org/10.1128/jb.184.4.1132-1139.2002

Souli, M., Galani, I., & Giamarellou, H. (2008). Emergence of extensively drug-resistant and pandrug-resistant Gram-negative bacilli in Europe. *Euro surveill, 13*(47), 19045.

Speert, D. P., Campbell, M. E., Henry, D. A., Milner, R., Taha, F., Gravelle, A., ... & Mahenthiralingam, E. (2002). Epidemiology of Pseudomonas aeruginosa in cystic fibrosis in British Columbia, Canada. *American journal of respiratory and critical care medicine, 166*(7), 988-993. http://dx.doi.org/10.1164/rccm.2203011

Starner, T. D., & McCray, P. B. (2005). Pathogenesis of early lung disease in cystic fibrosis: a window of opportunity to eradicate bacteria. *Annals of internal medicine, 143*(11), 816-822. http://dx.doi.org/10.7326/0003-4819-143-11-200512060-

Stead, R. J., Hodson, M. E., & Batten, J. C. (1987). Inhaled ceftazidime compared with gentamicin and carbenicillin in older patients with cystic fibrosis infected with *Pseudomonas aeruginosa. British journal of diseases of the chest, 81*, 272-279. http://dx.doi.org/10.1016/0007-0971(87)90161-6

Stover, C. K., Pham, X. Q., Erwin, A. L., Mizoguchi, S. D., Warrener, P., Hickey, M. J., ... & Olson, M. V. (2000). Complete genome sequence of Pseudomonas aeruginosa PAO1, an opportunistic pathogen. *Nature, 406*(6799), 959-964. http://dx.doi.org/10.1038/35023079

Stutz, A., Golenbock, D. T., & Latz, E. (2009). Inflammasomes: too big to miss. *The Journal of clinical investigation, 119*(12), 3502-3511. http://dx.doi.org/10.1172/JCI40599

Sykes, R. B., & Matthew, M. (1976). The beta-lactamases of gram-negative bacteria and their role in resistance to beta-lactam antibiotics. *J Antimicrob Chemother, 2*, 115-57. http://dx.doi.org/10.1093/jac/2.2.115

Tamura, Y., Suzuki, S., & Sawada, T. (1992). Role of elastase as a virulence factor in experimental *Pseudomonas aeruginosa* infection in mice. *Microbial pathogenesis, 12*(3), 237-244. http://dx.doi.org/10.1016/0882-4010(92)90058-V

Tamura, Y., Suzuki, S., Kijima, M., Takahashi, T., & Nakamura, M. (1992). Effect of proteolytic enzyme on experimental infection of mice with *Pseudomonas aeruginosa. J Vet Med Sci, 54*, 597-9. http://dx.doi.org/10.1292/jvms.54.597

Tang, H. B., DiMango, E., Bryan, R., Gambello, M., Iglewski, B. H., Goldberg, J. B., & Prince, A. (1996). Contribution of specific Pseudomonas aeruginosa virulence factors to pathogenesis of pneumonia in a neonatal mouse model of infection. *Infection and immunity, 64*(1), 37-43.

Tart, A. H., Wolfgang, M. C., & Wozniak, D. J. (2005). The alternative sigma factor AlgT represses Pseudomonas aeruginosa flagellum biosynthesis by inhibiting expression of fleQ. *Journal of bacteriology, 187*(23), 7955-7962. http://dx.doi.org/10.1128/JB.187.23.7955-7962.2005

Thuong, M., Arvaniti, K., Ruimy, R., De la Salmoniere, P., Scanvic-Hameg, A., Lucet, J. C., & Regnier, B. (2003). Epidemiology of *Pseudomonas aeruginosa* and risk factors for carriage acquisition in an intensive care unit. *Journal of Hospital Infection, 53*(4), 274-282. http://dx.doi.org/10.1128/JB.187.23.7955-7962. 2005

Trafny, E. A. (1998). Susceptibility of adherent organisms from *Pseudomonas aeruginosa* and Staphylococcus aureus strains isolated from burn wounds to antimicrobial agents. *Int J Antimicrob Agents, 10*, 223-8. http://dx.doi.org/10.1016/S0924-8579(98)00042-9

Vallet, I., Olson, J. W., Lory, S., Lazdunski, A., & Filloux, A. (2001). The chaperone/usher pathways of Pseudomonas aeruginosa: identification of fimbrial gene clusters (cup) and their involvement in biofilm formation. *Proceedings of the National Academy of Sciences, 98*(12), 6911-6916. http://dx.doi.org/10.1073/pnas.111551898

Van Gennip, M., Christensen, L. D., Alhede, M., Phipps, R., Jensen, P. Ø., Christophersen, L., ... & Bjarnsholt, T. (2009). Inactivation of the rhlA gene in Pseudomonas aeruginosa prevents rhamnolipid production, disabling the protection against polymorphonuclear leukocytes. *Apmis, 117*(7), 537-546. http://dx.doi.org/10.1111/j.1600-0463.2009.02466.x

Van Hartingsveldt, J., & Stouthamer, A. H. (1973). Mapping and characterization of mutants of Pseudomonas aeruginosa affected in nitrate respiration in aerobic or anaerobic growth. *Journal of general microbiology, 74*(1), 97-106. http://dx.doi.org/10.1099/00221287-74-1-97

Verma, A., Arora, S. K., Kuravi, S. K., & Ramphal, R. (2005). Roles of specific amino acids in the N terminus of Pseudomonas aeruginosa flagellin and of flagellin glycosylation in the innate immune response. *Infection and immunity, 73*(12), 8237-8246. http://dx.doi.org/10.1128/IAI.73.12.8237-8246.2005

Ward. (2010). Oxidative stress: acute and progressive lung injury. *Ann n y acad sci, 1203*, 53-59.

Wei, Q., & Ma, L. Z. (2013). Biofilm matrix and its regulation in Pseudomonas aeruginosa. *International journal of molecular sciences*, *14*(10), 20983-21005. http://dx.doi.org/10.3390/ijms141020983

Whitchurch, C. B., Tolker-Nielsen, T., Ragas, P. C., & Mattick, J. S. (2002). Extracellular DNA required for bacterial biofilm formation. *Science*, *295*(5559), 1487-1487. http://dx.doi.org/10.1126/science.295. 5559.1487

Whitecar Jr, J. P., Lima, M., & Body, G. P. (1970). Pseudomonas bacteremia in patients with malignant diseases. *The American journal of the medical sciences*, *260*(4), 216-223.

Whitney, J. C., Beck, C. M., Goo, Y. A., Russell, A. B., Harding, B. N., De Leon, J. A., ... & Mougous, J. D. (2014). Genetically Distinct Pathways Guide Effector Export Through the Type Vi Secretion System. Molecular Microbiology, 92(3), 529-542. http://dx.doi.org/10.1111/mmi.12571

Wright, S. D., Ramos, R. A., Tobias, P. S., Ulevitch, R. J., & Mathison, J. C. (1990). CD14, a receptor for complexes of lipopolysaccharide (LPS) and LPS binding protein. *Science, 249*(4975), 1431-1433. http://dx.doi.org/10.1126/science.1698311

Yahr, T. L. (2006). A Critical New Pathway For Toxin Secretion?. *New England Journal Of Medicine, 355*(11), 1171. http://dx.doi.org/10.1056/nejmcibr063931

Yahr, T. L., Mende-Mueller, L. M., Friese, M. B., & Frank, D. W. (1997). Identification of type III secreted products of the Pseudomonas aeruginosa exoenzyme S regulon. *Journal of bacteriology*, *179*(22), 7165-7168.

Yahr, T. L., Vallis, A. J., Hancock, M. K., Barbieri, J. T., & Frank, D. W. (1998). ExoY, an adenylate cyclase secreted by the Pseudomonas aeruginosa type III system. *Proceedings of the National Academy of Sciences*, *95*(23), 13899-13904. http://dx.doi.org/10.1073/pnas.95.23.13899

Yan, F., Li, W., Jono, H., Li, Q., Zhang, S., Li, J. D., & Shen, H. (2008). Reactive oxygen species regulate pseudomonas aeruginosa lipopolysaccharide-induced muc5ac mucin expression via pkc-nadph oxidase-ros-tgf-alpha signaling pathways in human airway epithelial cells. *Biochem biophys res commun, 366*(2), 513-519. http://dx.doi.org/10.1016/j.bbrc.2007.11.172

Zhuo, H., Yang, K., Lynch, S. V., Dotson, R. H., Glidden, D. V., Singh, G., ... & Wiener-Kronish, J. P. (2008). Increased mortality of ventilated patients with endotracheal Pseudomonas aeruginosa without clinical signs of infection. *Critical care medicine, 36*(9), 2495-2503. http://dx.doi.org/10.1097/CCM.0b013e318183f3f8

Permissions

List of Contributors

Marie-Claire Cammaerts
Faculté des Sciences, Université Libre de Bruxelles, Bruxelles, Belgium

Geoffrey Gosset
Faculté des Sciences, Université Libre de Bruxelles, Bruxelles, Belgium

Biao Wu
Key Laboratory of Sustainable Development of Marine Fisheries, Ministry of Agriculture, Yellow Sea Fisheries Research Institute, Chinese Academy of Fishery Sciences, Qingdao 266071, PR China

Aiguo Yang
Key Laboratory of Sustainable Development of Marine Fisheries, Ministry of Agriculture, Yellow Sea Fisheries Research Institute, Chinese Academy of Fishery Sciences, Qingdao 266071, PR China

Jiakun Yan
Key Laboratory of Sustainable Development of Marine Fisheries, Ministry of Agriculture, Yellow Sea Fisheries Research Institute, Chinese Academy of Fishery Sciences, Qingdao 266071, PR China

Wandong Xu
Kenli Prefecture Ocean and Fisheries Bureau, Dongying 257500, PR China

Tao Yu
Changdao Enhancement and Experiment Station, Chinese Academy of Fishery Science, Changdao 265800, PR China

Jiteng Tian
Key Laboratory of Sustainable Development of Marine Fisheries, Ministry of Agriculture, Yellow Sea Fisheries Research Institute, Chinese Academy of Fishery Sciences, Qingdao 266071, PR China

Zhihong Liu
Key Laboratory of Sustainable Development of Marine Fisheries, Ministry of Agriculture, Yellow Sea Fisheries Research Institute, Chinese Academy of Fishery Sciences, Qingdao 266071, PR China

Liqing Zhou
Key Laboratory of Sustainable Development of Marine Fisheries, Ministry of Agriculture, Yellow Sea Fisheries Research Institute, Chinese Academy of Fishery Sciences, Qingdao 266071, PR China

Xiujun Sun
Key Laboratory of Sustainable Development of Marine Fisheries, Ministry of Agriculture, Yellow Sea Fisheries Research Institute, Chinese Academy of Fishery Sciences, Qingdao 266071, PR China

Robert Musundire
Department of Food Science and Postharvest Technology, Chinhoyi University of Technology, Chinhoyi, Zimbabwe

Johnson C. Zvidzai
Department of Food Science and Postharvest Technology, Chinhoyi University of Technology, Chinhoyi, Zimbabwe

Cathrine Chidewe
Department of Biochemistry, University of Zimbabwe, Harare, Zimbabwe

Leilson R. Bezerra
School of Zootecnia, Federal University of Piauí, Bom Jesus, Piauí, Brazil

Cezario B. de Oliveira Neto
School of Zootecnia, Federal University of Piauí, Bom Jesus, Piauí, Brazil

Marcos J. de Araújo
School of Zootecnia, Federal University of Piauí, Bom Jesus, Piauí, Brazil

Ricardo L. Edvan
School of Zootecnia, Federal University of Piauí, Bom Jesus, Piauí, Brazil

Wagner D. C. de Oliveira
School of Zootecnia, Federal University of Piauí, Bom Jesus, Piauí, Brazil

Fabrício B. Pereira
School of Zootecnia, Federal University of Piauí, Bom Jesus, Piauí, Brazil

Zhongqiang Cai
Changdao Enhancement and Experiment Station, Chinese Academy of Fishery Sciences, Changdao, China

Xiujun Sun
Key Laboratory of Sustainable Development of Marine Fisheries, Ministry of Agriculture, Yellow Sea Fisheries Research Institute, Chinese Academy of Fishery Sciences, Qingdao 266071, China

Aiguo Yang
Key Laboratory of Sustainable Development of Marine Fisheries, Ministry of Agriculture, Yellow Sea Fisheries Research Institute, Chinese Academy of Fishery Sciences, Qingdao 266071, China

Hussein H. Abulreesh
Department of Biology, Faculty of Applied Science, Umm Al-Qura University, Saudi Arabia

Beatriz Baselga-Cervera
Genetics. Animal Production. Veterinary Faculty. Complutense University, Madrid, Spain

Camino García-Balboa
Genetics. Animal Production. Veterinary Faculty. Complutense University, Madrid, Spain

Eduardo Costas
Genetics. Animal Production. Veterinary Faculty. Complutense University, Madrid, Spain

Victoria López-Rodas
Genetics. Animal Production. Veterinary Faculty. Complutense University, Madrid, Spain

Purwanta
Agriculture Extension College, Gowa, South Sulawesi, Indonesia

Mihrani
Agriculture Extension College, Gowa, South Sulawesi, Indonesia

Sartika Juwita
Agriculture Extension College, Gowa, South Sulawesi, Indonesia

Ahmad Nadif
Agriculture Quarantine of Pare-Pare, South Sulawesi, Indonesia

Ali Ma'shum
Agriculture Extension College, Gowa, South Sulawesi, Indonesia

Muh Arby Hamire
Agriculture Extension College, Gowa, South Sulawesi, Indonesia

Courtney S. Cave
Annis Water Resources Institute, Grand Valley State University, Muskegon, Michigan, United States of America

Kevin B. Strychar
Annis Water Resources Institute, Grand Valley State University, Muskegon, Michigan, United States of America

Michelle Zammit
Department of Pharmacy, Faculty of Medicine and Surgery, University of Malta, Msida, Malta

Claire Shoemake
Department of Pharmacy, Faculty of Medicine and Surgery, University of Malta, Msida, Malta

Everaldo Attard
Division of Rural Sciences and Food Systems, Institute of Earth Systems, University of Malta, Msida, Malta

Lilian M. Azzopardi
Department of Pharmacy, Faculty of Medicine and Surgery, University of Malta, Msida, Malta

E. A. Artemieva
Ulyanovsk State Pedagogical University of I. N. Ulyanov, Russia

I. V. Muraviev
Ulyanovsk State Pedagogical University of I. N. Ulyanov, Russia

Eguono Esther Anomohanran
Department of Microbiology, Delta State University, Abraka, Nigeria

Alla Vereschagina
Russian Academy of Sciences, All Russian Institute for Plant Protection (VIZR), Saint-Petersburg, Pushkin, Russia

Elena Gandrabur
Russian Academy of Sciences, All Russian Institute for Plant Protection (VIZR), Saint-Petersburg, Pushkin, Russia

Angélica Massaroli
Department of Biological Sciences, University of Mato Grosso State, Brazil

Alessandra Regina Butnariu
Department of Biological Sciences, University of Mato Grosso State, Brazil

Augusta Karkow Doetzer
Department of Biological Sciences, University of Mato Grosso State, Brazil

Mahipal Singh
Animal Biotechnology Laboratory, Agricultural Research Station, Fort Valley State University, Fort Valley, GA, USA

Xiaoling Ma
Animal Biotechnology Laboratory, Agricultural Research Station, Fort Valley State University, Fort Valley, GA, USA

Hamed Soleyman Dehkordi
Under Graduated Student of Veterinary Medicine, College of Veterinary Medicine, Islamic Azad University, Shahrekord Branch, Shahrekord, Iran
Young Researchers and Elite Club, Islamic Azad University, Shahrekord Branch, Shahrekord, Iran

Hamid Iranpour Mobarakeh
Under Graduated Student of Veterinary Medicine, College of Veterinary Medicine, Islamic Azad University, Shahrekord Branch, Shahrekord, Iran

Mohsen Jafarian Dehkordi
Department of Pathology, Faculty of Veterinary Medicine, Islamic Azad University, Shahrekord Branch, Shahrekord, Iran

Faham Khamesipour
Under Graduated Student of Veterinary Medicine, College of Veterinary Medicine, Islamic Azad University, Shahrekord Branch, Shahrekord, Iran
Young Researchers and Elite Club, Islamic Azad University, Shahrekord Branch, Shahrekord, Iran

Lee-Hsueh Lee
Department of Landscape Architecture, Chung Hua University, Hsinchu, Taiwan

Jun-Cheng Lin
Department of Landscape Architecture, Chung Hua University, Hsinchu, Taiwan

Tanjina Rahman
Department of Food Technology and Nutrition Science, Noakhali Science and Technology University, Noakhali 3814, Bangladesh

Mohammed Mehadi Hassan Chowdhury
Department of Microbiology, Noakhali Science and Technology University, Noakhali 3814, Bangladesh

Md. Tazul Islam
Department of Food Technology and Nutrition Science, Noakhali Science and Technology University, Noakhali 3814, Bangladesh

M. Akhtaruzzaman
Institute of Nutrition and Food Science, University of Dhaka, Dhaka-1000, Bangladesh

Tidaporn Chaweepack
Chanthaburi Coastal Fisheries Research and Development Center, Department of Fisheries, Muang, Chanthaburi, 22000, Thailand

Boonyee Muenthaisong
Chanthaburi Coastal Fisheries Research and Development Center, Department of Fisheries, Muang, Chanthaburi, 22000, Thailand

Surachart Chaweepack
Chanthaburi Coastal Fisheries Research and Development Center, Department of Fisheries, Muang, Chanthaburi, 22000, Thailand

Kaeko Kamei
Department of Biomolecular Engineering, Kyoto Institute of Technology, Matsugasaki, Sakyo-ku, Kyoto 606-8585, Japan

Alaa Alhazmi
Department of Biology, Lakehead University, Thunder Bay, Ontario, Canada
Department of Medical Laboratory Technology, Jazan University, Jazan, Kingdom of Saudi Arabia

www.ingramcontent.com/pod-product-compliance
Lightning Source LLC
Chambersburg PA
CBHW050450200326
41458CB00014B/5126